Solid Mechanics with Python

Cartesian Coordinates

Series in Computational Methods

Series Editor: Gui-Rong Liu *(University of Cincinnati, USA)*

Published

Vol. 5 *Solid Mechanics with Python: Cartesian Coordinates*
by G R Liu

Vol. 4 *Vector Calculus: Formulations, Applications and Python Codes*
by G R Liu

Vol. 3 *Mechanics of Materials: Formulations and Solutions with Python*
by G R Liu

Vol. 2 *Calculus: Formulations and Solutions with Python*
by G R Liu

Vol. 1 *Numbers and Functions: Theory, Formulation, and Python Codes*
by G R Liu

Series in Computational Methods
Volume 5

Solid Mechanics with Python

Cartesian Coordinates

G. R. Liu
University of Cincinnati, USA

World Scientific

NEW JERSEY • LONDON • SINGAPORE • BEIJING • SHANGHAI • HONG KONG • TAIPEI • CHENNAI • TOKYO

Published by

World Scientific Publishing Co. Pte. Ltd.
5 Toh Tuck Link, Singapore 596224
USA office: 27 Warren Street, Suite 401-402, Hackensack, NJ 07601
UK office: 57 Shelton Street, Covent Garden, London WC2H 9HE

Library of Congress Control Number: 2025035020

British Library Cataloguing-in-Publication Data
A catalogue record for this book is available from the British Library.

Series in Computational Methods — Vol. 5
SOLID MECHANICS WITH PYTHON
Cartesian Coordinates

ISBN 978-981-98-1642-2 (hardcover)
ISBN 978-981-98-1643-9 (ebook for institutions)
ISBN 978-981-98-1644-6 (ebook for individuals)

For any available supplementary material, please visit
https://www.worldscientific.com/worldscibooks/10.1142/14400#t=suppl

Desk Editors: Nambirajan Karuppiah/Steven Patt

Typeset by Stallion Press
Email: enquiries@stallionpress.com

About the Author

Gui-Rong Liu received his Ph.D. from Tohoku University, Japan, in 1991. He was a Postdoctoral Fellow at Northwestern University, USA, from 1991 to 1993. He was a Professor at the National University of Singapore until 2010. He is currently a Professor at the Department of Aerospace Engineering and Engineering Mechanics, University of Cincinnati, USA. He was the Founder of the Association for Computational Mechanics (Singapore) (SACM) and served as the President of SACM until 2010. He served as the President of the Asia-Pacific Association for Computational Mechanics (APACM) (2010–2013) and an Executive Council Member of the International Association for Computational Mechanics (IACM) (2005–2010; 2020–2026). He authored a large number of journal papers and books, including two bestsellers, *Mesh Free Method: Moving Beyond the Finite Element Method* and *Smoothed Particle Hydrodynamics: A Meshfree Particle Methods.* He is the Editor-in-Chief of the *International Journal of Computational Methods* and served as an Associate Editor for *IPSE* and *MANO.* He is the recipient of numerous awards, including the Singapore Defence Technology Prize, NUS Outstanding University Researcher Award, NUS Best Teacher Award, APACM Computational Mechanics Award, JSME Computational Mechanics Award, ASME Ted Belytschko Applied Mechanics Award, Zienkiewicz Medal from APACM, AJCM Computational Mechanics Award, Humboldt Research Award, SACM Medal from the Association of Computational Mechanics (Singapore), and the Master Engineering Educator Award (UC, CEAS). He has been listed as one among the world's top 1% most influential scientists (Highly Cited Researchers) by Thomson Reuters for a number of years.

Contents

Chapter 1

Introduction

1.1 Computational methods

This book is part of a series that explores the exciting field of computational methods. The goal is to provide a solid foundation in this discipline, combining theoretical principles with practical applications.

The series encompasses general methods and techniques used across science, technology, engineering, and mathematics (STEM) education, as well as in research within various scientific and engineering disciplines. It serves as your comprehensive resource for computational methods, covering the theory, formulation, and coding techniques needed to get started. The series is designed to be accessible to readers who have completed primary and lower secondary education (roughly nine years of schooling), beginning with foundational volumes on mathematics. As you progress through the mechanics volumes, you will be equipped to tackle research and design projects that leverage computational methods.

Taking advantage of rapid advancements in computer technology, code examples will be integrated throughout the series, enabling concepts, theories, and formulations to be vividly demonstrated through practical examples with clear visualizations. Initially, Python will be the primary language used in the foundational volumes, with the potential for introducing other languages for more advanced topics. By effectively utilizing code, more time can be devoted to understanding core principles, as the computer will handle complex calculations, tedious derivations, time-consuming operations, and extensive data processing, allowing focus to remain on the broader concepts.

This approach empowers exploration of the fascinating world of computational methods with both a strong theoretical grounding and practical skills.

1.2 Why start and contribute to this book series

The editor-in-chief of this book series, Dr. G.R. Liu, has been working in areas related to Computational Methods for over 45 years. He developed his first FEM code for nonlinear problems in 1980 and has published more than 600 journal papers and more than a dozen monographs in this area since then. After all these years of studying, using, and developing computational methods, he started to think about a means to help other interested individuals learn computational methods in a more effective, systematic, and smooth way. He has concluded that developing this book series is the best way to achieve this objective.

1.3 Solid mechanics

Volume 1 of this book series covers two of the most important building blocks in computational methods, *Numbers and Functions: Theory, Formulation, and Python Codes* [1]. Volume 2 covers *Calculus: Formulations and Solutions with Python* [2], which provides powerful tools for differentiation and integration. Volume 3 covers the *Mechanics of Materials: Formulations and Solutions with Python* [3]. Volume 4 covers *Vector Calculus: Formulations, Applications and Python Codes*, which offers powerful tools to examine the critical effects of a scalar or vector function and reveals a mathematical description of fundamental physical laws [4]. This volume focuses on solid mechanics, analyzing stresses and displacements in loaded linear elastic solids.

All physical devices and structures used in our everyday lives are made of solids. This book introduces the subject of **Solid Mechanics** for linear elastic materials, which is essential for safe and effective analysis and design in engineering and scientific fields. When a solid material is loaded with forces, it experiences stresses, strains, and deformation, all of which are expressed as distributed functions. If a device is not designed properly, it may either fail to carry the desired load or be too heavy, with some material in the solid not being fully utilized. Solid mechanics studies the relationships between forces, stresses, strains, and deformation inside solid materials.

This book covers the fundamental principles of solid mechanics, focusing on the mechanical behavior of solids in two and three dimensions under various types of loads. The solids are defined in a Cartesian coordinate system and may have specific shapes and constraints.

The methodology and techniques for solving solid mechanics problems are explained in detail, formulated, and demonstrated with numerous examples and illustrations. Two major types of methods, the stress method and the displacement method, will be introduced. Python is used both for deriving formulations, finding solutions symbolically or numerically, and plotting results graphically.

The materials in this book can be digested through different learning processes, including classroom teaching, online courses, and self-study. Since Python codes are provided, readers can easily see how the theory is formulated and how solutions are obtained in terms of formulas, numerical values, and graphs. Readers may also deepen their understanding by experimenting with the code and further developing their own code for solving related problems.

The book is written in Jupyter Notebook format, so that theory, formulation, and coding are all presented in a unified document. This provides an environment for easy reading, exercises, practice, and further exploration.

This book is written with reference to [5, 6], which was the textbook used by the author as a university student, [7,8], which were frequently used reference books during his PhD and research work, and [9, 10], which were used as textbooks and references when teaching at universities. In addition, Wikipedia has been a valuable reference in writing this book. Both NumPy and SymPy were utilized in developing code for the demonstration examples. Discussions with ChatGPT, Gemini, and DeepSeek were also helpful in coding and in preparing this book.

Numerous examples are presented in this book; however, the applications of vector calculus are vast, and the examples and case studies provided are not exhaustive.

1.4 Who may read this book

This book is written for beginners interested in learning computational methods for solving problems in science, engineering, and nature. The target audience includes university students, graduate students, researchers, and professionals in any STEM discipline. Engineers and practitioners may also find the book useful for establishing a strong foundation in the core concepts of computational methods.

Readers are recommended to first read Volume 3 of this book series, *Mechanics of Materials* [3], or study an introductory course in Mechanics of Materials.

This book is written for human readers. It may not be used for training or testing any automated systems without the author's permission.

1.5 Codes used in this book

Readers who purchase the book may contact the author directly at liugr100@gmail.com to request a free soft copy of the book in Jupyter Notebook format with codes (which may be updated) for academic use after registration.

Conditions of use: The following conditions apply to the use of the book and the codes developed by the author in hard copy, soft copy, and any other media or format:

- **User responsibility:** Users are solely responsible for any risks associated with utilizing any part of the codes and techniques.
- **Code purpose:** The codes are primarily designed to demonstrate concepts and may not be optimized for efficiency or robustness. Many of the codes have not been tested by a third party.
- **Limited use:** The book and codes are for personal use only. Redistribution without the author's permission is not allowed.
- **No user support:** No support is provided for the codes.
- **Citation:** Proper citation and acknowledgment are required when using the book, codes, ideas, and techniques.

These codes often rely on various external packages and modules. Their behavior can be influenced by the versions of Python and these packages. A version mismatch could be the reason why a code doesn't run as expected. If you encounter issues running the code, check the versions of the packages and modules used. You can verify the versions within a code cell in the Jupyter Notebook. For example, to check the current Python environment version, you can use the following:

```
!python -V                  #! is used to execute an external command
```

```
Python 3.9.16
```

```
!jupyter notebook --version
```

```
6.1.5
```

1.6 Troubleshooting version issues

If you encounter a version mismatch error, you can either adapt the code to match the specific version or install the required version on your system. It is often helpful to search the web using the error message for potential solutions or leads. Additionally, large language models can be useful in this regard. This is the author's preferred method when dealing with code execution issues.

Since our descriptions are interspersed with code cells, a code cell can often depend on previous code cells. If an error is encountered when running a cell, try running the cells before that one to ensure all dependencies are met.

1.7 Learning Python

This book does not cover Python basics, as many resources are available online. Interested readers can refer to Chapters 2 and 3 in Ref. [11] for a concise overview of using Python for scientific computations. Since Python is relatively easy to learn, you can use the codes and examples in this book to get a head start on learning Python while studying the technical subjects.

1.8 Use of external modules or dependencies

Module imports in Jupyter Notebooks:

Frequently used modules in a chapter are imported in the first code cell, which should be executed at the beginning. If you encounter a `name error` (such as `NameError: name 'sp' is not defined`), return to the beginning of the chapter and execute the cell, which typically reads as follows:

```
#Often used external modules are in commonImports placed in folder grbin
#Place cursor in this cell, and press Ctrl+Enter to import dependencies.

import sys                          # for accessing the computer system
sys.path.append('../grbin/')     # Change to the directory in your system

from commonImports import *         # Import dependencies from '../grbin/'
import grcodes as gr                       # Import the module of the author
importlib.reload(gr)              # For in case, when grcodes is modified

from continuum_mechanics import vector
from continuum_mechanics.solids import sym_grad, strain_stress
init_printing(use_unicode=True)            # For latex-like quality printing

# Digits in print-outs
np.set_printoptions(precision=4,suppress=True,
                    formatter={'float_kind': '{:.4e}'.format})
```

The above cell includes a common import file, `commonImports.py` (given below). This file contains all the frequently used modules in this book, helping to avoid repeatedly importing lengthy external modules and reducing redundancy. By executing this cell at the beginning of each chapter, you'll ensure that all necessary modules are available throughout the notebook:

```
# commonImports.py:
from __future__ import print_function
import numpy as np                  # for numerical computation
import sympy as sp                  # sympy module for computation
import numpy.linalg as lg           # numpy linear algebra module
import scipy.linalg as sg           # scipy linear algebra module
import scipy.integrate as si
from scipy.stats import ortho_group
import importlib
import itertools
import inspect
import csv
import pandas as pd
import random
from IPython.display import display, Math

import autograd.numpy as anp                    # Thinly-wrapped numpy
from autograd import grad

from grcodes import drawArrow, plotfig, printM, printx # frequently used

import math as ma
from sympy import sin, cos, exp, symbols, lambdify, init_printing, latex
from sympy import pi, Matrix, sqrt, oo, integrate, diff, Derivative
from sympy import MatrixSymbol, simplify, nsimplify, Function
from sympy import factor, expand, nsimplify, Matrix, ordered, hessian
from sympy.plotting import plot as splt
init_printing(use_unicode=True) # for latex-like quality printing format

from matplotlib.ticker import MultipleLocator
import matplotlib.pyplot as plt     # for plotting figures
import matplotlib as mpl
```

Module reloading: While importing the same module multiple times generally doesn't cause issues, Python uses a cache and ignores subsequent imports if the module is already loaded. However, if you modify an imported module and want those changes to be reflected in your code, you need to reload the module. This can be done using the `importlib.reload()` function:

```
# grcodes was imported as gr, reload it when grcodes is modified
importlib.reload(gr)      # This is also included in the first code cell
```

Inspection of imported modules:

To view the code in the imported module, you can simply use the following by uncommenting it (it may produce a long output):

```
import inspect
#source_code = inspect.getsource(gr)   # to view everything in gr module
source_code = inspect.getsource(gr.displacement2stress)   # view a func.
print(source_code)
```

```
def displacement2stress(X, G, ν, U):
    '''Compute the stresses for given displacements
       U: array like with u, v, w.
       G: shear modulus, ν: Poisson's ratio.
       It returns 6 stress components in a list.
    '''
    x, y, z = X
    u, v, w = U
    #G = E/(2*(1+ν))                    # Use G for more concise formulas
    e = u.diff(x) + v.diff(y) + w.diff(z)   # volume strain (disp. div.)
    σxx = (2*G*(ν/(1-2*ν)*e+u.diff(x)))
    σyy = (2*G*(ν/(1-2*ν)*e+v.diff(y)))
    σzz = (2*G*(ν/(1-2*ν)*e+w.diff(z)))
    σyz = (G*(w.diff(y)+v.diff(z)))
    σxz = (G*(w.diff(x)+u.diff(z)))
    σxy = (G*(v.diff(x)+u.diff(y)))

    return [σxx, σyy, σzz, σyz, σxz, σxy]
```

Editing codes:

- You can use any text editor to view and modify the code.
- Alternatively, copy the code into a cell within a Jupyter Notebook and edit it directly.

Using functions from imported modules: To use a function from an imported module, you typically combine the module name with the function name using a dot (.). For example, if you import a module named `gr` containing a function called `printx`, you would use it like this:

```
x = 8
gr.printx('x')      # when gr. is used, printx() is from grcodes module
```

```
x = 8
```

1.9 Use of help()

Exploring modules and objects: To access detailed information about a module or an object within a module after importing it, you can use the `help()` function. This is a useful way to explore the functionalities and attributes available in different modules. For example:

```
help(gr.strain2stress)
```

```
Help on function strain2stress in module grcodes:

strain2stress(E, ν, ε)
    3D constitutive equation, from strains to stresses
```

Code comments: The code cells throughout the book include comments for additional explanations. These comments, marked with a hash symbol (#), are placed on the right side of the cell to minimize disruption while providing quick reference and guidance.

References

[1] G.R. Liu, *Numbers and Functions: Theory, Formulation, and Python Codes*, World Scientific, New Jersey, 2024.

[2] G.R. Liu, *Calculus: Formulations and Solutions with Python*, World Scientific, New Jersey, 2025.

[3] G.R. Liu, *Mechanics of Materials: Formulations and Solutions with Python*, World Scientific, New Jersey, 2024.

[4] G.R. Liu, *Vector Calculus: Formulation, Applications, and Python Codes*, World Scientific, New Jersey, 2025.

[5] S. Timoshenko and J.N. Goodier, *Theory of Elasticity*, McGraw-Hill, New York, 1970. Available online http://books.google.com/books?id=yFISAAAAIAAJ&dq=theory+of+elasticity&ei=ICiKSsr3G4jwkQSbxMyPCg.

[6] Z.L. Xu, *Elasticity*, Vol. 1 & 2, People's Publisher, China, 1979.

[7] J.D. Achenbach, *Wave Propagation in Elastic Solids*, North-Holland, Amsterdam, 1984.

[8] Y.C. Fung, *Foundations of Solid Mechanics*, Prentice-Hall of India, New Delhi, 1968.

[9] A.P. Boresi, K. Chong and J.D. Lee, *Elasticity in Engineering Mechanics*, John Wiley & Sons, New York, 2010.

[10] A.P. Boresi, R.J. Schmidt and O.M. Sidebottom, *Advanced Mechanics of Materials*, John Wiley & Sons, New York, 1985.

[11] G.R. Liu, *Machine Learning with Python: Theory and Applications*, World Scientific, New Jersey, 2023.

Chapter 2

Stress Methods for 2D Solids

We import the necessary external modules and dependence for later use in this chapter.

```
# Place cursor in this cell, and press Ctrl+Enter to import dependencies.
import sys                          # For accessing the computer system
sys.path.append('../grbin/')  # Change to the directory in your system

from commonImports import *          # Import dependencies from '../grbin/'
import grcodes as gr                    # Import the module of the author
importlib.reload(gr)                 # When grcodes is modified, reload it

init_printing(use_unicode=True)        # For latex-like quality printing
np.set_printoptions(precision=4,suppress=True,
                    formatter={'float_kind': '{:.4e}'.format})
```

2.1 Introduction

In engineering applications, most problems in the mechanics of solids are formulated and solved in Cartesian coordinates. These problems include one-dimensional (1D), two-dimensional (2D), and three-dimensional (3D) cases. Finding solutions to these problems involves determining the displacements and stresses that must satisfy the governing equilibrium equations and compatibility equations, along with the stress boundary conditions (SBCs) and displacement boundary conditions (DBCs). This chapter deals with various techniques used to solve these problems, focusing on 2D problems and the stress methods defined in the Cartesian coordinate system. The problem domain is confined by three boundary segments in a wedge shape or four segments in a rectangular shape.

This chapter is written with reference to [1,2], which were the textbooks when the author was a university student; [3, 4], frequently used reference books during the author's PhD and research work; and [5,6], which were used as additional textbooks and reference books when teaching at universities.

For 1D problems, we have systematic methods and codes to find solutions, as discussed in detail in Ref. [7]. This chapter focuses on 2D problems. The procedure becomes significantly more complicated, but the basic strategy is to convert the 2D problems into 1D-like ones, which can then be solved. With this approach, we can navigate the complexities of the stress method more effectively.

A typical stress method uses a function called the **stress function**, which allows the governing equations to be reduced to simpler forms. Here, we present the stress method for solving a few problems with rectangular domains, including the well-known classic 2D cantilever problem [1].

2.2 Basic equations

For an analytical approach to be effective, we often need to manipulate equations individually. Therefore, we will use explicit notations for these equations rather than the concise indicial notation. This section lists the necessary equations for this chapter. Detailed derivations of these equations can be found in Chapter 6 of Ref. [7].

2.2.1 *Equilibrium equations*

Consider a stress analysis problem defined in a 2D domain Ω in Cartesian coordinates x and y. The **equilibrium equations** are written explicitly in terms of stresses as

$$\frac{\partial}{\partial x}\sigma_{xx}(x,y) + \frac{\partial}{\partial y}\sigma_{yx}(x,y) + b_x = 0 \tag{2.1a}$$

$$\frac{\partial}{\partial x}\sigma_{xy}(x,y) + \frac{\partial}{\partial y}\sigma_{yy}(x,y) + b_y = 0 \tag{2.1b}$$

where $\sigma_{ij}(i,j = x,y)$ is the stress component on plane i with direction j and $b_i(i = x,y)$ is the body force component in direction i that is distributed over the 2D domain Ω. Equation (2.1) is a set of two partial differential equations (PDEs) of first order in terms of stresses.

2.2.2 *Boundary conditions*

In general, a solid mechanics problem typically involves boundary conditions and is considered a boundary value problem. On the boundary Γ that encloses the domain Ω of the 2D solid, tractions may be applied to some parts, while displacements are specified on others. These conditions must be enforced when solving the problem. Therefore, the boundary Γ is usually divided into two parts: Γ_t, where traction is applied, and Γ_u, where the displacements are specified.

2.2.2.1 *Stress boundary conditions*

Stress boundary conditions (SBCs) for 2D problems can be written as

$$\begin{aligned} t_x &= n_1\sigma_{xx} + n_2\sigma_{yx} \\ t_y &= n_1\sigma_{xy} + n_2\sigma_{yy} \end{aligned} \tag{2.2}$$

where t_i $(i = x, y)$ is the traction on the boundary Γ_t, and n_1 and n_2 are the direction cosines of the outward normal to the boundary curve at the point of interest.

To find displacement solutions, displacement boundary conditions must also be specified.

2.2.2.2 *Displacement boundary conditions*

The **displacement boundary conditions** (DBCs) can be expressed as follows:

$$u_i = u_{\Gamma i} \quad (i = x, y) \quad \text{on } \Gamma_u \subset \Gamma \tag{2.3}$$

where $u_{\Gamma i}$ are the specified displacements u_i on the boundary Γ_u, which may or may not be zero.

In some cases, we may specify the derivatives of the displacement components on parts of the boundary:

$$\frac{\partial u_i}{\partial x_j} = \frac{\partial u_{\Gamma i}}{\partial x_j} \quad (i, j = x, y) \quad \text{on } \Gamma_u \subset \Gamma \tag{2.4}$$

where $\frac{\partial u_{\Gamma i}}{\partial x_j}$ are the specified derivatives of displacements u_i with respect to x_j on the boundary Γ_u, which may be zero but not necessarily so.

For a 2D solid, there are generally three degrees of freedom related to rigid body movements. Therefore, to obtain a well-determined displacement for a static problem, at least three DBCs are needed. For our study in this chapter, we usually fix both u and v and one of their partial derivatives at a particular point. Alternatively, one may fix both u and v at one point and then fix u or v at another point. The key point is that the solid cannot undergo rigid body movement after the DBCs are imposed, ensuring that any displacement in the solid results in strains and, consequently, stresses.

2.2.3 *Constitutive equations*

Solid mechanics problems in 2D are largely classified into two types: **plane stress** and **plane strain** problems. In general, a solution found for one type can be converted to the other. Detailed discussions on their differences and methods for conversion can be found in Ref. [7]. Here, we list the **constitutive equations** for both.

Consider 2D plane stress problems, where the nonzero stresses are all within the x–y plane. Typically, the 2D solid is a thin plate loaded with forces only in the x–y plane. The constitutive equations, or strain–stress relations, can be written as

$$\begin{aligned} \varepsilon_{xx} &= \frac{1}{E}(\sigma_{xx} - \nu\sigma_{yy}) \\ \varepsilon_{yy} &= \frac{1}{E}(\sigma_{yy} - \nu\sigma_{xx}) \\ \varepsilon_{xy} &= \frac{1}{2G}\sigma_{xy} = \frac{1+\nu}{E}\sigma_{xy} \end{aligned} \tag{2.5}$$

where E is Young's modulus, G is the shear modulus, ν is the Poisson's ratio of the material, and σ_{ij} $(i, j = x, y)$ is the stress component acting in the i-direction due to deformation in the j-direction [7].

For plane strain problems, where the nonzero strains are all within the x–y plane, this situation typically occurs in long, uniform 3D solids subjected to forces that remain constant along the z-axis, such as in the case of a water dam. The constitutive equations can be written as

$$\begin{aligned} \varepsilon_{xx} &= \frac{1}{E}[(1-\nu^2)\sigma_{xx} - \nu(1+\nu)\sigma_{yy}] \\ \varepsilon_{yy} &= \frac{1}{E}[(1-\nu^2)\sigma_{yy} - \nu(1+\nu)\sigma_{xx}] \\ \varepsilon_{xy} &= \frac{1+\nu}{E}\sigma_{xy} \end{aligned} \tag{2.6}$$

As seen in Eqs. (2.5) and (2.6), the differences lie in the material coefficients associated with these normal stresses.

For stable elastic solid materials (nonzero E and G), these **strain–stress** relations can be simply inverted to obtain the **stress–strain** relations, and the detailed expressions can be found in Chapter 5 of Ref. [7].

2.2.4 *Strain–displacement relations*

Once the strains are determined from stresses using the constitutive equations in a stress method, the displacements can be obtained. This is done by solving the **strain–displacement relations**, which represent the symmetric part of the displacement gradient (see Chapter 4 of Ref. [7]):

$$\begin{aligned} \varepsilon_{xx} &= \frac{\partial u}{\partial x}, \quad \varepsilon_{yy} = \frac{\partial v}{\partial y} \\ \varepsilon_{xy} &= \frac{1}{2}\left(\frac{\partial u}{\partial y} + \frac{\partial v}{\partial x}\right) \end{aligned} \tag{2.7}$$

This is, in fact, a set of first-order PDEs. To solve for the displacement components, assuming the strains are given, we need to perform integration, which can be easily carried out using SymPy [8]. Such integration introduces **unknown constants** or **unknown functions**.

2.2.5 *Compatibility equations in strains and stresses*

The **compatibility equation in strains** is written explicitly in terms of strains as

$$\frac{\partial^2 \varepsilon_{xx}}{\partial y^2} + \frac{\partial^2 \varepsilon_{yy}}{\partial x^2} = 2\frac{\partial^2 \varepsilon_{xy}}{\partial x \partial y} \tag{2.8}$$

which can be derived using Eq. (2.7) by simple differentiations as detailed in Ref. [7]. The following process expresses it in terms of stresses.

Substituting Eq. (2.5) into Eq. (2.8) gives

$$\frac{\partial^2}{\partial y^2}(\sigma_{xx} - \nu\sigma_{yy}) + \frac{\partial^2}{\partial x^2}(\sigma_{yy} - \nu\sigma_{xx}) = 2(1+\nu)\frac{\partial^2}{\partial x \partial y}\sigma_{xy} \tag{2.9}$$

Differentiate equilibrium Eq. (2.1a) with respect to x and equilibrium Eq. (2.1b) with respect to y, then sum the resulting equations to obtain

$$2\frac{\partial^2}{\partial x \partial y}\sigma_{xy} = -\frac{\partial^2}{\partial x^2}\sigma_{xx} - \frac{\partial^2}{\partial y^2}\sigma_{yy} - \left(\frac{\partial b_x}{\partial x} + \frac{\partial b_y}{\partial y}\right) \tag{2.10}$$

Substituting Eq. (2.10) into Eq. (2.9) leads to

$$\frac{\partial^2(\sigma_{xx} - \nu\sigma_{yy})}{\partial y^2} + \frac{\partial^2(\sigma_{yy} - \nu\sigma_{xx})}{\partial x^2} = (-1 - \nu)\left(\frac{\partial^2\sigma_{xx}}{\partial x^2} + \frac{\partial^2\sigma_{yy}}{\partial y^2}\right) - (1+\nu)\left(\frac{\partial b_x}{\partial x} + \frac{\partial b_y}{\partial y}\right) \tag{2.11}$$

All the terms with red ν cancel each other out, which leads to

$$\frac{\partial^2\sigma_{xx}}{\partial y^2} + \frac{\partial^2\sigma_{yy}}{\partial x^2} = -\left(\frac{\partial^2\sigma_{xx}}{\partial x^2} + \frac{\partial^2\sigma_{yy}}{\partial y^2}\right) - (1+\nu)\left(\frac{\partial b_x}{\partial x} + \frac{\partial b_y}{\partial y}\right) \tag{2.12}$$

Simplification gives

$$\underbrace{\left(\frac{\partial^2}{\partial x^2} + \frac{\partial^2}{\partial y^2}\right)}_{\nabla^2}(\sigma_{xx} + \sigma_{yy}) = -(1+\nu)\left(\frac{\partial b_x}{\partial x} + \frac{\partial b_y}{\partial y}\right) \tag{2.13}$$

Or

$$\nabla^2(\sigma_{xx} + \sigma_{yy}) = -\underbrace{(1+\nu)}_{\kappa_\nu}\left(\frac{\partial b_x}{\partial x} + \frac{\partial b_y}{\partial y}\right) \tag{2.14}$$

where ∇^2 is the **Laplace operator**. Equation (2.14) is the **compatibility equation in stresses** for **plane stress** problems, which must be satisfied when using a stress method. If the body forces (b_i) are zero or constants, we have

$$\nabla^2(\sigma_{xx} + \sigma_{yy}) = 0 \tag{2.15}$$

This is the well-known Laplace equation for $(\sigma_{xx} + \sigma_{yy})$, which represents the trace of the 2D stress tensor.

A similar procedure can be applied to derive the **compatibility equation in stresses** for **plane strain** problems. In this case, we use the strain–stress relation given in Eq. (2.6). The result is given as follows:

$$\nabla^2(\sigma_{xx} + \sigma_{yy}) = -\underbrace{\frac{1}{1-\nu}}_{\kappa_\nu}\left(\frac{\partial b_x}{\partial x} + \frac{\partial b_y}{\partial y}\right) \tag{2.16}$$

This result can also be obtained by replacing ν in Eq. (2.14) with $\frac{\nu}{(1-\nu)}$.

Hereafter, we use the parameter κ_ν, which depends only on Poisson's ratio of the material, to ensure that the derived expressions apply to both plane stress and plane strain problems.

$$\begin{aligned} \kappa_\nu &= 1+\nu \qquad \text{plane stress} \\ \kappa_\nu &= \frac{1}{1-\nu} \qquad \text{plane strain} \end{aligned} \tag{2.17}$$

The unified compatibility equation in terms of stresses is given as follows:

$$\nabla^2(\sigma_{xx}+\sigma_{yy}) = -\kappa_\nu\left(\frac{\partial b_x}{\partial x}+\frac{\partial b_y}{\partial y}\right) \tag{2.18}$$

Therefore, if Eq. (2.18) is solved, the solution applies to both plane stress and plane strain problems.

Note that the right-hand side of Eqs. (2.14) and (2.16) is the divergence of the body force vector (scaled by the constant $-\kappa_\nu$). This means that the Laplacian of the sum of the normal stresses is proportional to the divergence of the body force. Since the divergence of a vector is an intrinsic physical property at a point in the solid — indicating whether the vector is converging inward or diverging outward — it remains unaffected by the choice of coordinates (see Chapter 2 in Ref. [9]). This feature can be utilized when performing coordinate transformations to obtain equations in a polar coordinate system [12].

2.3 Stress methods for solid mechanics problems

The equations to be solved in a stress method for a given solid mechanics problem include:

- Equilibrium equations in differential form, Eq. (2.1), which relate the body forces to the internal stresses.
- SBCs, Eq. (2.2), which relate the boundary tractions to the stresses on the boundary.
- Compatibility Eq. (2.8) or (2.18), which imposes constraints on the strain or stress components.
- Integral equilibrium equations, which relate the forces on the boundary to the internal forces inside the solid. These are also called force boundary conditions (FBCs).
- DBCs, Eqs. (2.3) and (2.4), which impose constraints on displacements when a displacement solution is required.

As a general rule, SBCs are imposed on longer boundaries and must hold at every point on the boundary, whereas FBCs are enforced along shorter boundaries in an integral form (not at every point).

Since this system of equations is written in terms of stresses, there are three unknown stress components. This approach, which treats stresses as the primary unknowns, is known as the **stress method**. It typically introduces a stress function, $\phi(x, y)$, known as the **Airy stress function**. All three unknown stress components are expressed in terms of this single stress function through differential equations, transforming the problem into one of solving for the stress function.

2.3.1 *Potential body forces*

A force applied to solids is a vector quantity that forms a force field. A special but frequently encountered type of force field is called a gradient field. Since such a field has a scalar potential function, it is also referred to as a potential force field. Studies of gradient fields and the types of vector functions associated with these fields can be found in Ref. [9].

Consider cases where the body force has a potential; they can be written as

$$b_x = -\frac{\partial \varphi}{\partial x}, \quad b_y = -\frac{\partial \varphi}{\partial y} \tag{2.19}$$

where $\varphi(x, y)$ is the potential function for the body force. It is a scalar function and can be determined for any given potential force.

Taking partial derivatives of Eq. (2.19), we obtain

$$\frac{\partial b_x}{\partial x} + \frac{\partial b_y}{\partial y} = -\nabla^2 \varphi \tag{2.20}$$

This means that the divergence of a potential body force vector is the negative Laplacian of its potential function.

2.3.2 *Stresses written in stress functions*

Using Eq. (2.19), the equilibrium Eq. (2.1) becomes

$$\begin{aligned} &\frac{\partial}{\partial x}(\sigma_{xx} - \varphi) + \frac{\partial}{\partial y}\sigma_{yx} = 0 \qquad \text{(a)} \\ &\frac{\partial}{\partial y}(\sigma_{yy} - \varphi) + \frac{\partial}{\partial y}\sigma_{xy} = 0 \qquad \text{(b)} \end{aligned} \tag{2.21}$$

It is clear that if we treat $(\sigma_{xx} - \varphi)$, $(\sigma_{yy} - \varphi)$, and σ_{xy} as a new set of stresses, we obtain a set of equilibrium equations with zero body force.

Note that the left-hand sides of the foregoing equations are all partial derivatives, and the right-hand sides are zero, hence it is called a homogeneous equation. We can thus write a possible stress solution with an arbitrary function $\phi(x, y)$ as follows:

$$\sigma_{xx} - \varphi = \frac{\partial^2 \phi}{\partial y^2}, \quad \sigma_{yy} - \varphi = \frac{\partial^2 \phi}{\partial x^2}, \quad \sigma_{xy} = -\frac{\partial^2 \phi}{\partial x \partial y} \tag{2.22}$$

This can be easily confirmed by substituting Eq. (2.22) back into Eq. (2.21). The above equations show that the stresses can be computed from the second derivatives of the stress function.

Note that:

Any linear function $ax + by + c$ contained in a stress function will contribute nothing to the stresses due to the second partial derivatives.

The sum of the first two equations in Eq. (2.22) gives

$$\sigma_{xx} + \sigma_{yy} = \frac{\partial^2 \phi}{\partial x^2} + \frac{\partial^2 \phi}{\partial y^2} + 2\varphi = \nabla^2 \phi + 2\varphi \tag{2.23}$$

The trace of the stress is now written in terms of the stress function (φ is usually given).

2.3.3 *Governing equation of the stress functions*

It is clear that the stresses in Eq. (2.22) satisfy the equilibrium Eq. (2.21). However, the solution must also satisfy the compatibility equation. Therefore, we can write the compatibility equation in terms of the stress function as well.

Substituting Eq. (2.23) into Eq. (2.18), we obtain

$$\begin{aligned} \nabla^2(\sigma_{xx} + \sigma_{yy}) &= \nabla^2(\nabla^2 \phi + 2\varphi) \\ &= \nabla^4 \phi + 2\nabla^2 \varphi = -\kappa_\nu \left(\frac{\partial b_x}{\partial x} + \frac{\partial b_y}{\partial y} \right) \end{aligned} \tag{2.24}$$

which gives (with the help of Eqs. (2.19) and (2.20))

$$\nabla^4 \phi = -2\nabla^2 \varphi - \kappa_\nu \left(\frac{\partial b_x}{\partial x} + \frac{\partial b_y}{\partial y} \right) = -\kappa'_\nu \left(\frac{\partial b_x}{\partial x} + \frac{\partial b_y}{\partial y} \right) \tag{2.25}$$

where $\kappa'_\nu = \kappa_\nu - 2$, which depends only on Poisson's ratio of the material. Using Eq. (2.17), we obtain

$$\begin{aligned} \kappa'_\nu &= \nu - 1 \quad \text{plane stress} \\ \kappa'_\nu &= \frac{2\nu - 1}{1 - \nu} \quad \text{plane strain} \end{aligned} \tag{2.26}$$

Equation (2.25) is call the **biharmonic equation**. Its right side is known for given body forces.

The biharmonic operator ∇^4 is given as follows:

$$\begin{aligned} \nabla^4 &= \underbrace{\left(\frac{\partial^2}{\partial x^2} + \frac{\partial^2}{\partial y^2}\right)}_{\nabla^2} \underbrace{\left(\frac{\partial^2}{\partial x^2} + \frac{\partial^2}{\partial y^2}\right)}_{\nabla^2} \\ &= \frac{\partial^4}{\partial x^4} + 2\frac{\partial^4}{\partial x^2 \partial y^2} + \frac{\partial^4}{\partial y^4} \end{aligned} \tag{2.27}$$

Equation (2.25) indicates that the biharmonic of the stress function is proportional to the divergence of the force vector field. This can also be written as

$$\nabla^4 \phi = \kappa'_\nu \nabla^2 \varphi \tag{2.28}$$

Equation (2.25) or (2.28) governs the behavior of the stress function ϕ. It represents the compatibility equation expressed in terms of ϕ. When the body forces are constants, Eq. (2.28) reduces to

$$\nabla^4 \phi = 0 \tag{2.29}$$

This is a homogeneous equation, where $\nabla^4 \phi$ is the biharmonic of ϕ. This means the biharmonic of ϕ must vanish for a constant (including zero) body force vector field.

2.3.4 *Some stress functions*

Now, our problem becomes finding the stress function ϕ that satisfies Eq. (2.29). As discussed in Ref. [10], we can use basis functions to form the stress function. Therefore, we need to find the possible basis functions for ϕ. We term such basis functions as biharmonic basis functions (BBFs). Each BBF satisfies the biharmonic equation, and hence any linear combination of a set of BBFs forms a candidate stress function that satisfies the biharmonic equation. Possible BBFs include polynomial BBFs and biharmonic function BBFs.

2.3.5 *Polynomial basis functions*

Some known polynomial BBFs that satisfy Eq. (2.29) are listed below. Any terms of these BBFs can be used as a candidate to form a stress function $\phi(x, y)$ through linear combinations:

$$\begin{aligned}
&1,\ x,\ y,\ xy,\ x^2,\ y^2,\ x^3,\ yx^2,\ xy^2,\ y^3 && \leq \text{third-order}\\
&(3x^2y^2 - x^4),\ x^3y,\ xy^3,\ (3x^2y^2 - y^4),\ (x^4 - y^4) && \text{fourth-order}\\
&(5x^2y^3 - y^5),\ (x^3y^2 - xy^4),\ (5x^4y - y^5),\ (x^5 - 5xy^4) && \text{fifth-order}\\
&(5x^3y^3 - 3x^5y),\ (x^6 - 10x^4y^2 + 5x^2y^4),\ (xy^5 - x^5y),\\
&\quad (2x^6 - 15x^4y^2 + y^6) && \text{sixth-order}\\
&(35x^4y^3 - 42x^2y^5 + 3y^7),\ (3x^5y^2 - 10x^3y^4 + 3xy^6),\\
&\quad (7x^6y - 21x^2y^5 + 2y^7),\ (x^7 - 35x^3y^4 + 14xy^6) && \text{seventh-order}\\
&(7x^5y^3 - 14x^3y^5 + 3xy^7), (7x^6y^2 - 35x^4y^4 + 21x^2y^6 - y^8),\\
&\quad (x^7y - 7x^3y^5 + 2xy^7), (x^8 - 70x^4y^4 + 56x^2y^6 - 3y^8) && \text{eighth-order}\\
&\ldots\ldots
\end{aligned} \tag{2.30}$$

The BBFs are generated from a set of complete polynomials of a given order, say order n. A complete set of such polynomials will contain $n + 1$ binomial terms (see Pascal's triangle or Yanghui's triangle). For example, to generate the fifth-order BBFs, we use the set

$$(y^5, xy^4, x^2y^3, x^3y^2, x^4y, x^5) \tag{2.31}$$

where the order sum for x and y in a binomial is always 5, and the set includes all possible such terms. We have found that there are always 4 basis polynomial functions for orders greater than or equal to 5 that satisfy $\nabla^4\phi = 0$. This is because the biharmonic operator is a complete differential of the fourth order. Therefore, the order of the remaining binomial set will always be $(n+1) - 4$. The condition $\nabla^4\phi = 0$ results in $n - 3$ equations. As a result, the biharmonic basis polynomial functions will always be $(n+1) - (n-3) = 4$ polynomials.

The process for generating and testing the BBFs is provided in the Python code in the next section. One can use the `create_nbf()` function to generate BBFs of any order. We also note the following:

1. A stress function is not unique. Any of the terms up to third-order monomials (in green and blue fonts in Eq. (2.30)) and their arbitrary linear combinations can be added to a stress function.
2. Any of the first three bases: 1, x, y (in green fonts in Eq. (2.30)) and their linear combinations will not contribute anything to the stresses because the stresses relate to the second-order partial derivatives of ϕ, as shown in Eq. (2.22). Therefore, when using a stress function for solid mechanics problems, these terms can be omitted. In other words, we use only the terms in blue and black.

Biharmonic functions that consist of trigonometric and hyperbolic functions with arbitrary constant λ can also be used as basis functions to form the stress function. For example:

$$\begin{array}{llll} \cos(x\lambda)\cosh(y\lambda), & \cos(y\lambda)\cosh(x\lambda), & \sin(x\lambda)\cosh(y\lambda), & \cos(y\lambda)\sinh(x\lambda), \\ x\cos(x\lambda)\cosh(y\lambda), & x\cos(y\lambda)\cosh(x\lambda), & y\cos(x\lambda)\cosh(y\lambda), & y\cos(y\lambda)\cosh(x\lambda), \\ x\sin(x\lambda)\cosh(y\lambda), & x\cos(y\lambda)\sinh(x\lambda), & \sin(x\lambda)\cosh(y\lambda), & y\cos(y\lambda)\sinh(x\lambda), \\ \ldots\ldots & & & \end{array} \tag{2.32}$$

These BBFs satisfy the biharmonic equation, which can be confirmed using the following snippet. Note that the first four are also harmonic basis functions. The utility snippet **nabla_f()** for testing harmonic and biharmonic functions is given as follows and is self-explanatory.

```
def nabla_f(f, X, dfn = 2, smplfy=1):
    '''Compute the nabla operator ∇^2 (dfn = 2) or ∇^4 (dfn = 4) on
       (sympy) function f defined in 2D (dim = 2) or 3D (dim = 3).
       smplfy=1: use sp.simplify(); =2 use sp.expand() for polynomials.
       return: nf - the nabla of f.
    '''
    #x, y, z = sp.symbols("x, y, z")
    dim = len(X)
    smpfy = sp.simplify if smplfy == 1 else sp.expand
    if dim == 2:                                                # 2D
        x, y = X
        if dfn == 2:
            n2f=smpfy((f.diff(x,2)+f.diff(y,2)))                # ∇^2
            nf = n2f
        elif dfn == 4:                                          # ∇^4
            n2f=smpfy((f.diff(x,2)+f.diff(y,2)))                # ∇^2
            n4f=smpfy((n2f.diff(x,2)+n2f.diff(y,2)))
            nf = n4f
        else: print(f'dfn must be set either 2 or 4')
    elif dim == 3:                                              # 3D
```

```
        x, y, z = X
        if dfn == 2:
            n2f=smpfy((f.diff(x,2)+f.diff(y,2)+f.diff(z,2)))      # ∇^2
            nf = n2f
        elif dfn == 4:                                            # ∇^4
            n2f=smpfy((f.diff(x,2)+f.diff(y,2)+f.diff(z,2)))      # ∇^2
            n4f=smpfy((n2f.diff(x,2)+n2f.diff(y,2)+n2f.diff(z,2)))
            nf = n4f
        else: print(f'dfn must be set either 2 or 4')
    else: print(f'dim must be set either 2 or 3')
    return nf
```

```
# Perform some tests:

x, y, z = sp.symbols("x, y, z")
λ = sp.symbols("λ", real=True)
X = sp.Matrix([x, y])
BBFs = sp.Matrix([ cos(λ*x)*cosh(λ*y),  cos(λ*y)*cosh(λ*x),    #∇^4 BBFs.
                   sin(λ*x)*cosh(λ*y),  cos(λ*y)*sinh(λ*x),
                 x*cos(λ*x)*cosh(λ*y), x*cos(λ*y)*cosh(λ*x),
                 y*cos(λ*x)*cosh(λ*y), y*cos(λ*y)*cosh(λ*x),
                 x*sin(λ*x)*cosh(λ*y), x*cos(λ*y)*sinh(λ*x),
                 y*sin(λ*x)*cosh(λ*y), y*cos(λ*y)*sinh(λ*x)])
#display(Math(f" \\text{{Biharmonic BBFs}}={latex(BBFs.T)}"))

HBFs = sp.Matrix(BBFs[:4])   # first 4 are also harmonic basis functions

# Check whether any f satisfies  ∇^4 f = 0. Example:
d4BBFs = gr.nabla_f(BBFs, X, dfn=4, smplfy=2)
display(Math(f" \\text{{$∇^4$ (BBFs) = 0? }}{latex(d4BBFs.T)}"))
d2HBFs = nabla_f(HBFs, X, dfn=2, smplfy=2)
display(Math(f" \\text{{$∇^2$ (HBFs) = 0? }}{latex(d2HBFs.T)}"))
```

$$\nabla^4 \text{ (BBFs)} = 0? \; \begin{bmatrix} 0 & 0 & 0 & 0 & 0 & 0 & 0 & 0 & 0 & 0 & 0 & 0 \end{bmatrix}$$

$$\nabla^2 \text{ (HBFs)} = 0? \; \begin{bmatrix} 0 & 0 & 0 & 0 \end{bmatrix}$$

2.3.6 *Utilities for biharmonic basis functions*

To develop the function `create_nbf()` efficiently, we have written a few utility codes to assist. The `nabla_f()` (given above) computes the nabla operator ∇^2 or ∇^4 on any given SymPy function defined in 2D or 3D. For generating BBFs, as mentioned above, we need only ∇^4 in 2D. In later chapters, these Python functions will be used for 3D problems.

Another utility code is for creating monomial bases of a given order for 2D or 3D functions, **polyBases()**, which is also self-explanatory.

```
def polyBases(order, X):
    '''Create polynomial bases of a given order for 2D (dim=2)
       or 3D (dim=3)
    '''
    dim = len(X)
    if dim == 2:
        x, y = X
        basis = Matrix([x**i * y**(order-i) for i in range(order+1)])
    elif dim == 3:
        x, y, z = X
        basis = Matrix([x**i*y**j*z**k for i in range(order+1) for j in
                  range(order+1) for k in range(order+1) if i+j+k==order])
    else: print(f'dim must be set either 2 or 3')
    #print(f"Polynomials bases, complete order {order}:\n {basis.T}")
    return basis
```

2.3.7 *Code for creating polynomial basis functions*

The following function **create_nbf()** creates polynomial basis functions that satisfy $\nabla^2 f_b = 0$ or $\nabla^4 f_b = 0$ defined in 2D or 3D domains. It will be used for creating BBFs for stress functions and, in Chapter 4, for displacement potential and displacement functions.

```
def create_nbf(order, X, dfn=4, dim=3):
    '''Creates polynomial basis functions that satisfy
       ∇^2=0  (dfn=2, order>=2) or ∇^4=0 (dfn=4, order>=4) defined in
       2D (dim=2) or 3D (dim=3).
       return: polynomial basis functions, nbf (sympy Matrix).
       '''
    dim = len(X)
    print(f'Dimension={dim}, order={order}, ∇^={dfn}')
    if dim == 3:
        x, y, z = X
        ψ = sp.Function("ψ")(x, y, z)          # a displacement-function
    if dim == 2:
        x, y = X
        ψ = sp.Function("ψ")(x, y)             # a displacement-function

    c = sp.symbols("c0:99")     # coeffs., define more than what we need

    # create polynomial bases of the same given order
    basis = polyBases(order, X)
    #print(f"Polynomials bases, complete order {order}:\n {basis.T}")
```

```
cs = Matrix([c[i] for i in range(len(basis))])      # coeffs. needed

# compute total number of constants:
tn_cs = (order+2)*(order+1)/2 if dim==3 else order+1
print(f'No. of constants={len(cs)},  should be {tn_cs}')

ψ = (cs.T@basis)[0]      # create the function by linear combination
#print(f"Initial ψ with complete order {order} for ψ:\n {ψ}")

# form the order-reduced polynomial bases after applying ∇^2:
basisR = polyBases(order-dfn, X)
#print(f"Polynomials bases after ∇^2:\n {basisR.T} {len(basisR)}")

nabψ = nabla_f(ψ, X, dfn=dfn, smplfy=2)   # compute the nabla of ψ

# collect terms for order-reduced bases, to group the coeffs.
for i in range(len(basisR)):
    nabψ = nabψ.collect(basisR[i]) #print(nabψ)

# form equations of relations of the coeffs.
eqs = [nabψ.coeff(basisR[i]) for i in range(len(basisR))]
if len(eqs)==1: eqs = [nabψ]

sln_cs=sp.solve(eqs, cs)  #print(sln_cs)
keys_s = list(sln_cs.keys())                             # solution keys

# get the remaining coefficients in the expression
keys_r  = list([x for x in cs if x not in keys_s])

ns = order-dfn+1                         # number of solved coefficients
nr = dfn                              # number of remaining coefficients
if dim == 3:
    ns = (order-dfn+2)*(order-dfn+1)/2
    nr = tn_cs-ns
#print(len(keys_s)," should be",ns," ",len(keys_r)," should be",nr)

ψ_ci = ψ.subs(sln_cs).expand()              # get the ψ with keys_r

# collect terms based on keys_r, to group polynomials
for i in range(len(keys_r)):
    ψ_ci = ψ_ci.collect(keys_r[i])

# group polynomials, based on keys_r
nbf=Matrix([(ψ_ci.coeff(keys_r[i])).expand()
               for i in range(len(keys_r))])

nabψ = nabla_f(nbf, X, dfn=dfn, smplfy=2) # compute the nabla of ψ

print(f'Is ∇^{dfn} ψ=0? {nabψ.T@nabψ}! Number of bases={len(nbf)}')
return nbf
```

2.3.8 *Example: Generating biharmonic basis functions (BBFs)*

The following snippet generates BBFs using the `create_nbf()` function:

```
x, y, z = sp.symbols("x, y, z")
X = sp.Matrix([x, y])
order = 9                                    # order of biharmonic basis
BBFs = create_nbf(order, X, dfn=4)            # ∇^4 biharmonic basis fun.
display(Math(f" \\text{{9th-order BBFs, 1, 2}}={latex(BBFs.T[:2])}"))
display(Math(f" \\text{{9th-order BBFs, 3, 4}}={latex(BBFs.T[2:])}"))
```

```
Dimension=2, order=9, ∇^=4
No. of constants=10,  should be 10
Is ∇^4 ψ=0? Matrix([[0]])! Number of bases=4
```

Ninth-order BBFs, 1, 2

$$= \left[x^6y^3 - 3x^4y^5 + \frac{9x^2y^7}{7} - \frac{y^9}{21},\ x^7y^2 - 7x^5y^4 + 7x^3y^6 - xy^8\right]$$

Ninth-order BBFs, 3, 4

$$= \left[x^8y - 14x^4y^5 + 8x^2y^7 - \frac{y^9}{3},\ x^9 - 126x^5y^4 + 168x^3y^6 - 27xy^8\right]$$

One may multiply 21 by the first term and 3 by the third term to remove the fractions. Since these terms are used as basis functions, any constant factor does not affect their use. It is good practice to check whether a function in question is biharmonic. The example snippet is as follows:

```
# Check whether an f satisfies  ∇^4 f = 0. Example:
d4BBFs = nabla_f(BBFs, X, dfn=4, smplfy=2)
display(Math(f" \\text{{$∇^4$ (BBFs) = 0? }}{latex(d4BBFs.T)}"))
```

∇^4 (BBFs) $= 0?$ $\left[0\ 0\ 0\ 0\right]$

Note that the lower-order monomials given in Eq. (2.30) cannot be generated using `create_nbf()`. These are written out directly. We can check whether these terms satisfy Eq. (2.29) using:

```
bfs = polyBases(3, X)                          # generate basis functions
display(Math(f" \\text{{ The monomials = }}{latex(bfs.T)}"))
d4bfs = nabla_f(bfs, X, dfn=4, smplfy=2)             # Check if ∇^4 f = 0.
display(Math(f" \\text{{$∇^4$ (monomials) = 0? }}{latex(d4bfs.T)}"))
```

The monomials = $\left[y^3\ xy^2\ x^2y\ x^3\right]$
∇^4 (monomials) = 0? $\left[0\ 0\ 0\ 0\right]$

2.4 Solution procedure using stress functions

The general solution procedure using stress functions is as follows:

1. Find a stress function via a linear combination of some BBFs given in Eq. (2.30) or generated using `create_nbf()` with unknown constants. Better choices of BBFs can be made based on the problem setup and from experience.
2. Use Eq. (2.22) to find the stresses, which involve only differentiations. The stresses are expressed with constants.
3. Determine these constants using the stress boundary and equilibrium conditions.
4. Find the strains using the constitutive equations and the stresses obtained, which involve only algebraic operations.
5. Find the displacements through integration, using the strain–displacement relations, which introduce integral functions and constants.
6. Determine these constants using the DBCs.

In the following studies, we assume the problems are 2D plane stress, and the dimension in the depth (perpendicular to the 2D plane) is unity: $b = 1$.

We first use the following example problem to demonstrate the solution process by writing a Python code. This code can also be used (with modifications) for other similar problems.

We present solutions for a number of problems solved in Cartesian coordinates. These problems are classical and can be found in other texts. This section references a widely used classic text [11]. The following sections present the general method using SymPy as a systematic solution procedure for these and other problems.

2.5 A simplest case: Stress state due to gravity

Consider solids occupying a 2D half-space in a given gravity field g, with positive y downward, as shown in Fig. 2.1.

In this case, the body force is simply the weight and is constant:

$$b_x = 0, \quad b_y = \rho g \tag{2.33}$$

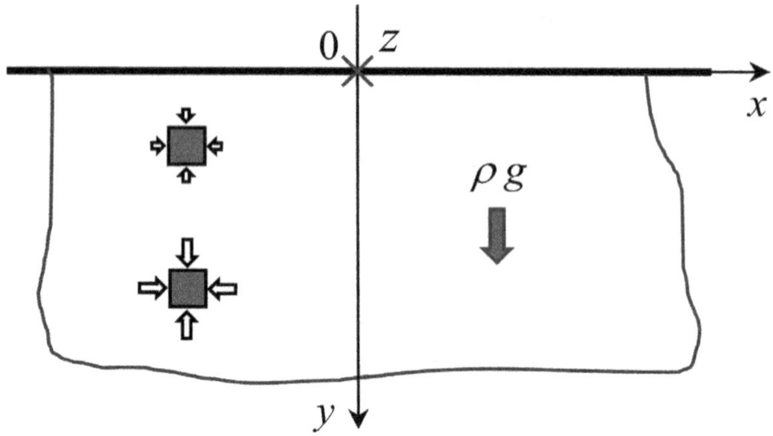

Figure 2.1. A 2D half-space solid of density ρ subjected to a given gravity field.

where ρ is the density of the solid. Clearly, the body forces are constants. The potential function φ of the body force vector can be found using Eq. (2.19) via integrations:

$$\varphi = \int b_y \, dy + C = -\rho g y + C \tag{2.34}$$

where C is the constant of integration. Setting φ to zero at $y = 0$, where the body force begins to take effect, gives $C = 0$. We thus have

$$\varphi = -\rho g y \tag{2.35}$$

For this simple problem, we use the zero stress function, $\phi = 0$, which clearly satisfies Eq. (2.29), and the stress solutions are found using Eq. (2.22) as

$$\sigma_{xx} = -\rho g y; \quad \sigma_{yy} = -\rho g y; \quad \sigma_{xy} = 0 \tag{2.36}$$

This is the stress state in half-space solids under all-direction compression (or tension when g is negative). The shear stress is zero. All the normal stresses are zero on the surface $y = 0$ and increase linearly downward, as shown in Fig. 2.1. This represents the idealized stress state in the soil/rock underneath and near the Earth's surface if the standard gravity constant on the Earth's surface, say 9.807 m/s^2, is used.

The stresses given in Eq. (2.36) also represent the stress state of a finite solid under hydrostatic pressure of $\rho g y$ with zero stresses at $y = 0$.

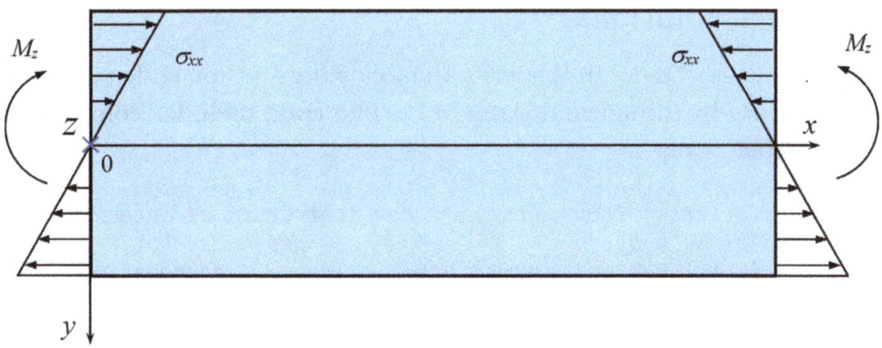

Figure 2.2. A rectangular solid subjected to pure bending.

2.6 Rectangle 2D solids subject to pure bending

2.6.1 *Setting of the problem*

Consider a solid occupying a rectangular domain that is subjected to pure bending: an external moment M with respect to only the z-axis, as shown in Fig. 2.2.

2.6.2 *Stresses with unknown constants*

In this case, the body forces are zero. We choose the following two monomial BBFs from Eq. (2.30):

$$1,\ x,\ y,\ xy,\ x^2,\ y^2,\ x^3,\ yx^2,\ xy^2,\ y^3 \quad \leq \text{third-order} \tag{2.37}$$

The stress function is the linear combination of these two terms:

$$\phi = \frac{a_3}{6}x^3 + \frac{d_3}{6}y^3 \tag{2.38}$$

where a_3 and d_3 are constants. The stress expressions are found using Eq. (2.22) as

$$\sigma_{xx} = \frac{\partial^2 \phi}{\partial y^2} = d_3 y; \quad \sigma_{yy} = \frac{\partial^2 \phi}{\partial x^2} = a_3 x; \quad \sigma_{xy} = 0 \tag{2.39}$$

We find pure bending stress states with $\sigma_{xy} = 0$:

- When $a_3 = 0$ and $d_3 \neq 0$, it is a pure bending state with normal stress σ_{xx}, as shown in Fig. 2.2.
- When $d_3 = 0$ and $a_3 \neq 0$, it is a pure bending state with normal stress σ_{yy} on the upper and lower surfaces.

2.6.3 *Stress solutions*

Consider the case of $a_3 = 0$. We write the following Python code to determine the constant d_3 in the normal stress σ_{xx}. The code includes comments and is largely explanatory.

```
# The possible stress components with constant d3 to be determined.
x, y = symbols('x, y')                            # define symbolic variables
E, G, I, h, l, nu, M = sp.symbols('E,G,I,h,l,nu,M', nonnegative=True)
c0, c1, d3, cF = symbols('c0, c1, d3, c_F')           # integral constants

# define stress components as symbolic variable
σxx, σxy, σyy = symbols('σxx, σxy, σyy')
σxx, σyy, σxy = d3*y, 0, 0                        # the possible normal stress
σ = Matrix([σxx, σyy, σxy])
σ.T
```

$[d_3 y\ 0\ 0]$

We obtained the expressions for all the stress components. The only nonzero stress is the normal stress σ_{xx}. Clearly, this set of stresses satisfies the **equilibrium equations** (2.1) and **SBCs** (2.2) on both the top and bottom surfaces (where $n_1 = 0, n_2 = \pm 1$), regardless of the value of d_3.

Let us now compute the resultant internal moment due to the normal stress. This involves an integration of the normal stress over any cross-section of the rectangle. The formula is as follows:

$$M_z = \int_A y\sigma_{xx}\mathrm{d}A = \int_{-h/2}^{h/2} y\sigma_{xx}b\mathrm{d}y \tag{2.40}$$

Here, we assumed that the depth (in the z-direction) of the 2D solid is $b = 1$.

```
Mz = sp.integrate(σxx*y,(y, -h/2, h/2))        # find the internal moment

cs = sp.solve([Mz-M], [d3])              # M: assumed given external moment
cs                                    # the constants in the assumed solution
display(Math(f" \\text{{ Coefficient $d_3$}}: {latex(cs)}"))
#display(Math(f" \\text{{ Normal stress σ$_{{xx}}$}}=\
#                                          {latex(σxx.subs(cs))}"))
```

Coefficient d_3 : $\left\{d_3 : \dfrac{12M}{h^3}\right\}$

Since M_z does not change with x, Eq. (2.40) is used directly to solve for d_3. Substituting d_3 back into the first equation in Eq. (2.39), we find the normal stress as

$$\sigma_{xx} = \frac{12M}{h^3} y \tag{2.41}$$

Its distribution is linear, as shown in Fig. 2.2.

2.6.4 *Strain solutions*

With the stresses found, we can now determine the strains using Eq. (2.5), which contains only second-order derivatives, considering a plane stress problem. Given that $\sigma_{yy} = \sigma_{xy} = 0$, we have

```
# define strain components as symbolic variable
εxx, εyy, γxy = symbols('ε_xx, ε_yy, γ_xy')
εxx, εyy, γxy = σxx/E, -nu*σxx/E, 0          # Plane stress and σyy=σyy=0
ε = Matrix([εxx, εyy, γxy])                   # strains in the x-y plane
display(Math(f" \\text{{ Strains }}={latex(ε.T)}"))
```

$$\text{Strains} = \begin{bmatrix} \frac{d_3 y}{E} & -\frac{d_3 \nu y}{E} & 0 \end{bmatrix}$$

These strains are linear and hence satisfy the **compatibility conditions** Eq. (2.8). Using the second moment of area for a beam with unit depth ($b = 1$), given by $I = h^3/12$, the normal stress and strain can be written as follows:

```
σxx = σxx.subs(d3, M/I)    # substitute the constant back to the stress
εxx = εxx.subs(d3, M/I)                                      # I=h^3/12
εyy = εyy.subs(d3, M/I)
display(Math(f" \\text{{ Normal stress σ$_{{xx}}$}}={latex(σxx)}"))
display(Math(f" \\text{{ Normal strain ε$_{{xx}}$}}={latex(εxx)}"))
display(Math(f" \\text{{ Normal strain ε$_{{yy}}$}}={latex(εyy)}"))
```

$$\text{Normal stress, } \sigma_{xx} = \frac{My}{I}$$

$$\text{Normal strain, } \varepsilon_{xx} = \frac{My}{EI}$$

$$\text{Normal strain, } \varepsilon_{yy} = -\frac{M\nu y}{EI}$$

The result is the same as that obtained using the Euler–Bernoulli thin beam theory [7]. Here, we obtain these results by solving the 2D plane stress problem using stress functions.

2.6.5 *Displacement solution via integration*

With the strains found, we can now determine the displacements using the strain–displacement relation in Eq. (2.7). Since the strains are compatible, the displacement must exist.

Let us first find the displacement component $u(x, y)$ using the first equation in Eq. (2.7). The displacement can be determined using indefinite integration with respect to x, which produces an arbitrary unknown function of y. The process is as follows:

```
f = sp.Function('f')(y)                 # define an arbitrary function of y

dudx = εxx
u_with_f = sp.integrate(dudx,(x)) + f    # f: integral constant w.r.t x
u_with_f = u_with_f.expand().doit()
display(Math(f" \\text{{Displacement $u$ with unknown function}}=\
                                                    {latex(u_with_f)}"))
```

$$\text{Displacement } u \text{ with unknown function} = f(y) + \frac{Mxy}{EI}$$

We found the displacement component u, but with $f(y)$, which is an unknown function of y only.

Now, let's determine the displacement component v. This time, we use the second equation in Eq. (2.7). The displacement v is obtained by integrating indefinitely with respect to y, which produces another arbitrary function of x only:

```
g = sp.Function('g')(x)                 # define an arbitrary function of x

dvdy = εyy
v_with_g = sp.integrate(dvdy,(y)) + g
v_with_g = v_with_g.expand().doit()
display(Math(f" \\text{{Displacement $v$ with unknown function}}=\
                                                    {latex(v_with_g)}"))
```

$$\text{Displacement } v \text{ with unknown function} = g(x) - \frac{M\nu y^2}{2EI}$$

We found the displacement component v, but with an unknown function $g(x)$.

Next, we determine the unknown functions $g(x)$ and $f(y)$. This can be done using the strain–displacement relation for γ_{xy}, which is the third equation in Eq. (2.7) and involves differentiation. The corresponding Python snippet is as follows:

```
expr = (diff(u_with_f,y)+diff(v_with_g,x) - γxy).expand() # the 3rd eq.
display(Math(f" \\text{{ Expression from the shear strain}}=\
                                                    {latex(expr)}"))
```

Expression from the shear strain $= \frac{d}{dy}f(y) + \frac{d}{dx}g(x) + \frac{Mx}{EI}$

This is a 2D PDE with two unknown functions. To solve it, we convert it into 1D differential equations (DEs). This process is as follows.

Rearrange the terms in the given expression as

$$\underbrace{\frac{d}{dy}f(y)}_{F(y)} + \underbrace{\frac{d}{dx}g(x) + \frac{M}{EI}x}_{G(x)} = 0 \tag{2.42}$$

We observe that $F(y)$ depends only on y, and $G(x)$ depends only on x. Since their sum is zero, $F(y)$ and $G(x)$ must be equal to a common constant with opposite signs. Let us denote $F(y) = c_F$ and $G(x) = -c_F$. This gives the following two separate equations:

$$\frac{d}{dx}g(x) = -c_F - \frac{Mx}{EI} \tag{2.43}$$

$$\frac{d}{dy}f(y) = c_F \tag{2.44}$$

Both equations are now in 1D, and each contains only one unknown function. We have successfully converted the original 2D PDE into two 1D DEs. We can now solve them one by one using SymPy.

First, integrate Eq. (2.43), which introduces a new constant:

```
g = sp.integrate(-cF-expr.args[0],(x))+c0     # add an integral constant
g = g.expand().doit()
display(Math(f" \\text{{ Function $g(x)$}}={latex(g)}"))
```

Function $g(x) = c_0 - c_F x - \frac{Mx^2}{2EI}$

Next, integrate (2.44), which introduces another constant:

```
f = sp.integrate(cF, (y)) + c1            # add another integral constant
f = f.expand().doit()
display(Math(f" \\text{{ Function $f(x)$}}={latex(f)}"))
```

Function $f(x) = c_1 + c_F y$

Now that both unknown functions are found, the displacement components u and v can be written as follows:

```
u = sp.integrate(dudx, (x)) + f                    # re-form displacement u
display(Math(f" \\text{{ Displacement $u(x,y)$}}={latex(u)}"))

v = sp.integrate(dvdy, (y)) + g                    # re-form displacement v
display(Math(f" \\text{{ Displacement $v(x,y)$}}={latex(v)}"))
```

Displacement $u(x, y) = c_1 + c_F y + \frac{Mxy}{EI}$

Displacement $v(x, y) = c_0 - c_F x - \frac{M\nu y^2}{2EI} - \frac{Mx^2}{2EI}$

2.6.6 *Determine the constants for displacements*

We have now obtained the general solution for displacements u and v, with the integral constants c_0, c_1, and c_F. There are three unknown constants. We need the DBCs to determine them.

Assuming at point ($x = 0$, $y = 0$), we have (see Fig. 2.2)

$$u = 0; \quad v = 0; \quad \frac{\partial v}{\partial x} = 0 \tag{2.45}$$

This gives us three conditions, providing the right number of equations to determine the three constants in the displacement expressions. Let us write a SymPy code to determine these constants:

```
dic = {x:0, y:0}
expressions = [u.subs(dic),v.subs(dic), diff(v,x).subs(dic)]
unknown_cs = [c0,c1,cF]                              # unknown_constants
cs=sp.solve(expressions, unknown_constants)
cs
```

$\{c_0 : 0, \ c_1 : 0, \ c_F : 0\}$

It is found that, for this set of boundary conditions, all three constants are zero. We finally arrive at the solutions for the displacements by substituting the constants found back into the displacement expressions:

```
# final solution for the displacement v:
v_final = v.subs(cs).expand()
display(Math(f" \\text{{ Displacement $v(x,y)$}}={latex(v_final)}"))

# final solution for the displacement u:
u_final = u.subs(cs).expand()
display(Math(f" \\text{{ Displacement $u(x,y)$}}={latex(u_final)}"))
```

Displacement $v(x, y) = -\frac{M\nu y^2}{2EI} - \frac{Mx^2}{2EI}$

Displacement $u(x, y) = \frac{Mxy}{EI}$

It is good practice to check the results at various stages. Here, we check the strain–displacement relations:

```
print(f'Is εxx=∂u/∂x? {εxx == diff(u_final,x)}')        # check results
print(f'Is εyy=∂v/∂y? {εyy == diff(v_final,y)}')        # check results
print(f'γ_xy=∂u/∂y+∂v/∂x? {γxy ==diff(u_final,y)+diff(v_final,x)}')
```

```
Is εxx=∂u/∂x? True
Is εyy=∂v/∂y? True
γ_xy=∂u/∂y+∂v/∂x? True
```

We obtained all the solutions for stress, strain, and displacements as functions of x and y. These satisfy all the necessary conditions.

Let us check the solution at some critical locations. Along the neutral axis, $y = 0$, the deflection (displacement v) is as follows:

```
v_final.subs(y,0)
display(Math(f" \\text{{ Displacement $v(x,0)$ along $y=0$}}: \
                                        {latex(v_final.subs(y,0))}"))
# At $x=0, y=0$, the deflection is:
display(Math(f" \\text{{ Displacement $v(0,0)$ at the beam root}}:\
                                        {latex(v_final.subs({x:0,y:0}))}"))
# at x = l, the deflection is:
display(Math(f" \\text{{ Displacement $v(l,0)$ at tip of the beam}}:\
                                        {latex(v_final.subs({x:l,y:0}))}"))
```

Displacement $v(x, 0)$ along $y = 0$: $-\frac{Mx^2}{2EI}$

Displacement $v(0, 0)$ at the beam root: 0

Displacement $v(l, 0)$ at tip of the beam: $-\frac{Ml^2}{2EI}$

The displacements along the right edge ($x = l$) of the rectangle can be found as follows:

```
# at x = l, the displacement u is:
display(Math(f" \\text{{ Displacement $u(l,y)$ along beam's free end}}:\
                                        {latex(u_final.subs({x:l}))}"))
# at x = l, the displacement v is:
display(Math(f" \\text{{ Displacement $v(l,y)$ along beam's free end}}:\
                                        {latex(v_final.subs({x:l}))}"))
```

Displacement $u(l, y)$ along beam's free end: $\frac{Mly}{EI}$

Displacement $v(l, y)$ along beam's free end: $-\frac{Ml^2}{2EI} - \frac{M\nu y^2}{2EI}$

We found the full set of solutions, with their correctness checked, for a beam under pure bending. Readers may compare this set of solutions with that obtained in the Euler–Bernoulli thin beam theory.

2.7 Cantilever rectangle 2D solid

Consider a 2D thin cantilever solid occupying a rectangular domain with unit thickness (or depth) $b = 1$, which is small compared to its height h and length l. It is constrained at the right end and subjected to a downward shear force P at the left end, as shown in Fig. 2.3. The solid is free from body forces, and the upper and lower surfaces are free from any external forces.

Figure 2.3. A 2D rectangular thin cantilever with dimensions of $l \times h$ and unit thickness $b = 1$, subjected to a downward force P at the left end.

Our task is to find the solutions for stress, strain, and displacements as functions of x and y, assuming that Young's modulus E and Poisson's ratio ν of the material are given.

2.7.1 *Stress function construction*

Since the body forces are zero, the stress function needs to satisfy Eq. (2.29). We shall choose a candidate stress function through the following considerations.

Consider a cantilever thin beam (based on the Euler–Bernoulli theory) subjected to a point load at its tip, and we make the following observations [7]:

1. The shear stress σ_{xy} on the cross-section of a beam changes quadratically in y and is constant in x. From Eq. (2.22), we know that σ_{xy} is the first derivative of ϕ with respect to y and x. Therefore, ϕ must be at least cubic in y and linear in x.
2. The normal stress σ_{xx} changes linearly in both y and x. From Eq. (2.22), we know that σ_{xx} is the second derivative of ϕ with respect to y. Thus, ϕ must also be at least cubic in y and linear in x.

Based on this simple analysis, we consider using up to second order for x, and up to fourth order for y from Eq. (2.30). This should give us a candidate ϕ sufficient for our cantilever 2D solid. The selected BBFs are shown in red in the following:

$$\begin{aligned} &1,\ x,\ y,\ xy,\ x^2,\ y^2,\ x^3,\ yx^2,\ xy^2,\ y^3 && \leq \text{third-order} \\ &(3x^2y^2 - x^4),\ x^3y,\ xy^3,\ (3x^2y^2 - y^4),\ (x^4 - y^4) && \text{fourth-order} \end{aligned} \tag{2.46}$$

Note that we excluded the linear bases because stresses are computed via second-order differentiations of the stress function. These linear terms do not contribute to the stress. Our stress function can thus be expressed as a linear combination of these eight selected BBFs:

$$\begin{aligned} \varphi = a_0xy + a_1x^2 + a_2y^2 + a_3yx^2 + a_4xy^2 + a_5y^3 \\ + a_6xy^3 + a_7(3x^2y^2 - y^4) \end{aligned} \tag{2.47}$$

where a_i $(i = 0, 1, \ldots, 7)$ are arbitrary unknown constants.

Since we are using only BBFs, ϕ given in Eq. (2.47) automatically satisfies Eq. (2.29). The stress function is defined in Python using the following snippet:

```
E, G, I, h, l, nu, P = sp.symbols('E,G,I,h,l,nu,P', nonnegative=True)
x,  y  = symbols('x, y')                          # define symbolic variables

# unknown constants for BBFs:
a0, a1, a2, a3, a4, a5, a6, a7 = symbols('a0,a1,a2,a3,a4,a5,a6,a7')
σxx, σxy, σyy = symbols('σ_xx, σ_xy, σ_yy')

ϕ = sp.Function('ϕ')(x, y)                        # define symbolic functions
ϕ = a0*x*y + a1*x**2 + a2*y**2 + a3*y*x**2 + a4*x*y**2 + a5*y**3\
    + a6*x*y**3 + a7*(3*x**2*y**2-y**4)
display(Math(f" \\text{{ Candidate ϕ}}={latex(ϕ)}"))
```

Candidate $\phi = a_0xy + a_1x^2 + a_2y^2 + a_3x^2y + a_4xy^2 + a_5y^3 + a_6xy^3 + a_7 \cdot (3x^2y^2 - y^4)$

2.7.2 *Stress solution*

Using the stress function and Eq. (2.22), the general expressions for stresses are given as follows:

```
σxx = ϕ.diff(y,2)
σyy = ϕ.diff(x,2)
σxy = -(ϕ.diff(x,1)).diff(y,1)

σ_ai = Matrix([σxx, σyy, σxy])                    # Stresses with constants a_i
display(Math(f" \\text{{ Vector of stresses}}={latex(σ_ai)}"))
```

$$\text{Vector of stresses} = \begin{bmatrix} 2\left(a_2 + a_4x + 3a_5y + 3a_6xy + 3a_7\left(x^2 - 2y^2\right)\right) \\ 2\left(a_1 + a_3y + 3a_7y^2\right) \\ -a_0 - 2a_3x - 2a_4y - 3a_6y^2 - 12a_7xy \end{bmatrix}$$

Boundary conditions for σ_{xx} on the left edge

For our cantilever solid, the left edge ($x = 0$) is free of normal stress σ_{xx}:

$$\sigma_{xx}\Big|_{x=0} \quad \text{for all } y \tag{2.48}$$

Let us compute σ_{xx} along $x = 0$:

```
σ_ai[0].subs(x,0).expand()                                    # σ_{xx} at x=0
```

$2a_2 + 6a_5y - 12a_7y^2$

As demanded by Eq. (2.48), the foregoing expression must be zero for all y. Hence, the coefficients of y^2, y^1, and y^0 must be zero. This leads to $a_2 = a_5 = a_7 = 0$:

```
σ_ai = σ_ai.subs({a2:0, a5:0, a7:0})
display(Math(f" \\text{{ Simplified stresses}}={latex(σ_ai)}"))
ϕ_reduced = ϕ.subs({a2:0, a5:0, a7:0})
display(Math(f" \\text{{ Reduced ϕ}}={latex(ϕ_reduced)}"))
```

$$\text{Simplified stresses} = \begin{bmatrix} 2a_4x + 6a_6xy \\ 2a_1 + 2a_3y \\ -a_0 - 2a_3x - 2a_4y - 3a_6y^2 \end{bmatrix}$$

$$\text{Reduced } \phi = a_0xy + a_1x^2 + a_3x^2y + a_4xy^2 + a_6xy^3$$

We obtained a much simpler expression for these stresses, but it still has five unknown constants.

Boundary conditions on the top and bottom edges

We shall now apply the boundary conditions along the longer sides of the problem domain, which are the top and bottom edges (or surfaces):

$$\begin{aligned} \sigma_{yy}\big|_{y=-h/2} &= -p; \quad \sigma_{xy}\big|_{y=-h/2} = 0 \quad \text{on top edge} \\ \sigma_{yy}\big|_{y=h/2} &= 0; \quad \sigma_{xy}\big|_{y=h/2} = 0 \quad \text{on bottom edge} \end{aligned} \tag{2.49}$$

This provides four conditions, which can be used to determine the four constants. Let us examine the expressions for these conditions:

```
bc1 = σ_ai[1].subs(y, h/2)
bc2 = σ_ai[1].subs(y,-h/2)
bc3 = σ_ai[2].subs(y, h/2)
bc4 = σ_ai[2].subs(y,-h/2)
bc1, bc2, bc3, bc4
```

$$\left(2a_1 + a_3h,\ 2a_1 - a_3h,\ -a_0 - 2a_3x - a_4h - \frac{3a_6h^2}{4},\right.$$
$$\left. -a_0 - 2a_3x + a_4h - \frac{3a_6h^2}{4}\right)$$

Since Eq. (2.49) needs to be satisfied at any point on the surfaces of the beam, all these foregoing expressions must be zero for any x. Hence, the coefficients of x^1 and x^0 must be zero. Clearly, the coefficient of x in the third (and fourth) condition must be zero. The Python snippet to get all these done is as follows:

```
bc_a3 = bc3.coeff(x,1)
SBCs = [bc1, bc2, bc3, bc4, bc_a3]
unknown_ai = [a0, a1, a3, a4, a6]
a_sln = sp.solve(SBCs, unknown_ai)
display(Math(f" \\text{{ Solutions for $a_i$}} ={latex(a_sln)}"))

σ_ai = σ_ai.subs(a_sln)
display(Math(f" \\text{{ Constants reduced stresses}}={latex(σ_ai.T)}"))
```

$$\text{Solutions for } a_i = \left\{a_0 : -\frac{3a_6h^2}{4},\ a_1 : 0,\ a_3 : 0,\ a_4 : 0\right\}$$

$$\text{Constants reduced stresses} = \begin{bmatrix} 6a_6xy & 0 & \frac{3a_6h^2}{4} - 3a_6y^2 \end{bmatrix}$$

The stress expressions still contain one constant, a_6. We shall now use the integral equilibrium condition for the boundary at the left edge of the problem domain (in general, at any cross-section on the beam). Since the shear stress along this edge must balance the externally applied force P, the formula is as follows:

$$\int_{-h/2}^{h/2} \left(\sigma_{xy}\big|_{x=0}\right) dy + P = 0 \tag{2.50}$$

This gives one condition, which can be used to determine the constant a_6. The snippet is as follows:

```
bc5 = sp.integrate(σ_ai[2].subs(x, 0),    (y, -h/2, h/2)) + P

a6_sln = sp.solve([bc5], [a6])
display(Math(f" \\text{{ Solutions for $a_6$}} ={latex(a6_sln)}"))

σ = sp.simplify(σ_ai.subs(a6_sln))
display(Math(f" \\text{{ Final solution for stresses}}={latex(σ.T)}"))
```

$$\text{Solutions for } a_6 = \left\{a_6 : -\frac{2P}{h^3}\right\}$$

$$\text{Final solution for stresses} = \begin{bmatrix} -\frac{12Pxy}{h^3} & 0 & \frac{3P\left(-h^2+4y^2\right)}{2h^3} \end{bmatrix}$$

Finally, we obtain all the stress components, which are rewritten as follows:

$$\sigma_{xx} = -\frac{Pxy}{I}; \quad \sigma_{yy} = 0; \quad \sigma_{xy} = -\frac{Ph^2}{8I}\left(1 - \frac{4y^2}{h^2}\right) \tag{2.51}$$

where I is the second moment of the cross-section of the cantilever (with unit thickness): $I = \frac{h^3}{12}$. We make the following observations:

1. σ_{xx} is linear in both x and y directions (bilinear).
2. σ_{yy} is zero at any point in the rectangle.
3. σ_{xy} is independent of x and parabolic along y with zero at both the upper and lower surfaces and maximum at the middle line.

This stress solution agrees well with that from the thin beam theory. It is exact if the actual shear stress distribution at both ends of the cantilever follows σ_{xy} as given in Eq. (2.51). If the stresses have other distributions, the solution is still applicable in areas away from the ends, based on Saint-Venant's principle. This is because the overall equilibrium is satisfied at either end in an integral sense by the stress solution obtained and by any other distribution for the same cantilever in equilibrium.

Let us plot the distributions of the stresses:

```
# Re-define the stresses using the solution obtained:
σ = sp.Matrix([[-12*P*x*y/h**3], [0], [3*P*(-h**2 + 4*y**2)/(2*h**3)]])

# Generate the plots for these stresses:
fij = [1, 1, 1, "xl", "yl"]                          # define figure grids

plt.ioff()
#plt.ion()                        # uncomment this to display the plots
#plt.rcParams.update({'font.size': 5})            # settings for plots
fig_s = plt.figure()#figsize=(4,2.5))

P_val = 1; l_val = 1; h_val = 1
print(f" P = {P_val}, l = {l_val}, h = {h_val}")
dic_ = {P:P_val, l:l_val, h:h_val, x:l_val/2}
np_σxx = sp.lambdify((y), σ[0].subs(dic_),'numpy')
np_σxy = sp.lambdify((y), σ[2].subs(dic_),'numpy')

labels=[r'$σ_{xx}$ at $x=l/2$', r'$σ_{xy}$ at $x=l/2$']

y_data = np.linspace(-h_val/2, h_val/2, 400)           # y-axis in plot
f_data = [np_σxx(y_data), np_σxy(y_data)]

gr.plot_fij(fig_s, y_data, f_data, labels, '$y$', 'Stresses', fij)
plt.savefig('imagesSM/Canti_stresses_y.png',dpi=500,bbox_inches='tight')
#plt.show()                       # uncomment this to display the plots
```

```
P=1,l=1,h=1
```

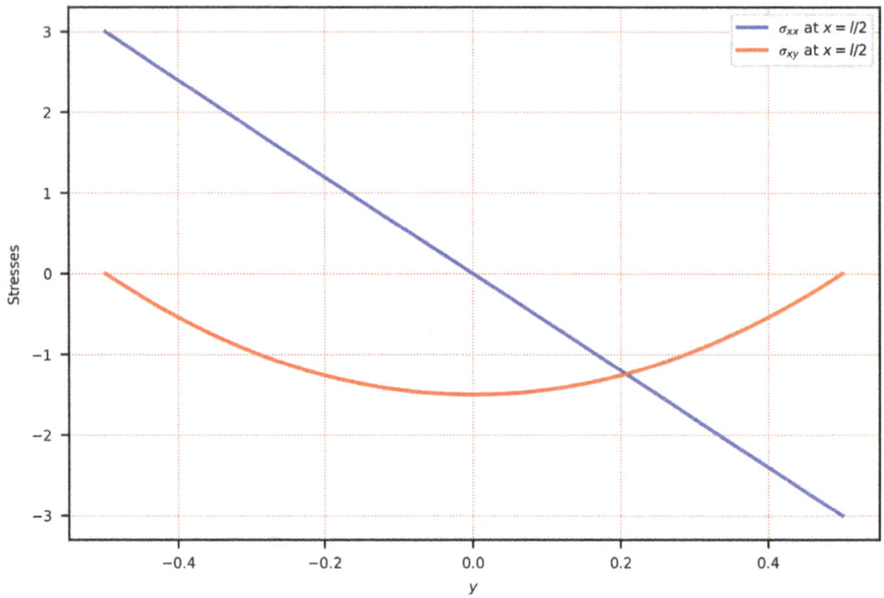

Figure 2.4. Stress distribution along the cross-section at $x = l/2$ in a 2D rectangular solid with dimensions $l \times h$. It is subjected to a vertical downward force P on the left edge on the top surface ($x = 0$). Parameters used are $P = 1, l = 1, h = 1$.

Distributions of these stresses along the cross-section at $x = l/2$ are plotted in Fig. 2.4. It can be seen that the largest stress is the normal stress σ_{xx} for this set of parameters. It is linearly distributed along the cross-section, which is the same as that from the thin beam theory. The shear stress σ_{xy} has a parabolic distribution and is zero on both surfaces, which is also the same as the result from the thin beam theory.

Our next step is to obtain the solution for displacements.

2.7.3 *Displacement solution*

Since the displacements relate to the strains, we first obtain the strains using the already obtained stresses. Using the constitutive Eq. (2.5) and the strain–displacement relation Eq. (2.7), we have

$$\begin{aligned}
&\frac{\partial u}{\partial x} = \varepsilon_{xx} = \frac{1}{E}(\sigma_{xx} - \nu\sigma_{yy}) = -\frac{P}{EI}xy && (a)\\
&\frac{\partial v}{\partial y} = \varepsilon_{yy} = \frac{1}{E}(\sigma_{yy} - \nu\sigma_{xx}) = \frac{\nu P}{EI}xy && (b)\\
&\frac{\partial u}{\partial y} + \frac{\partial v}{\partial x} = \gamma_{xy} = \frac{1}{G}\sigma_{xy} = -\frac{Ph^2}{8GI}\left(1 - \frac{4y^2}{h^2}\right) && (c)
\end{aligned} \tag{2.52}$$

First, let us compute the strains using the stresses obtained:

```
# define stress components as symbolic variables
σxx, σyy, σxy = -P/I*x*y, 0, -P*h**2/(8*I)*(1-4*y**2/h**2)   # stresses
σ = Matrix([σxx, σyy, σxy])
display(Math(f" \\text{{ Vector formed by stresses}}={latex(σ.T)}"))

# define strain components as symbolic variables
εxx, εyy, γxy = symbols('ε_xx, ε_yy, γ_xy')

# strains via the constitutive equations
εxx, εyy, γxy = σxx/E, -nu*σxx/E, -P*h**2/(8*G*I)*(1-4*y**2/h**2)
ε = Matrix([εxx, εyy, γxy])
display(Math(f" \\text{{ Vector formed by strains}}={latex(ε.T )}"))
```

$$\text{Vector formed by stresses} = \begin{bmatrix} -\frac{Pxy}{I} & 0 & -\frac{Ph^2\cdot\left(1-\frac{4y^2}{h^2}\right)}{8I} \end{bmatrix}$$

$$\text{Vector formed by strains} = \begin{bmatrix} -\frac{Pxy}{EI} & \frac{P\nu xy}{EI} & -\frac{Ph^2\cdot\left(1-\frac{4y^2}{h^2}\right)}{8GI} \end{bmatrix}$$

Before computing the displacements, it is good practice to check whether the given strains satisfy the compatibility condition Eq. (2.8), as this is necessary for the existence of a displacement field corresponding to the strain field.

```
comptblty = (εxx.diff(y,2)+εyy.diff(x,2)-γxy.diff(x).diff(y)).simplify()
print(f"Is compatibility condition satisfied? {comptblty==0}")
```

```
Is compatibility condition satisfied? True
```

Since the compatibility condition is satisfied, we can now proceed to compute the displacements by integrating the first two equations in Eq. (2.52):

```
# find the displacements via integrations
f = sp.Function('f')(y)                          # prepare a integral function
dudx = εxx
u_with_f = sp.integrate(dudx,(x)) + f  # integral constant w.r.t x: f(y)
u_with_f = u_with_f.expand().doit()
display(Math(f" \\text{{ Displacement $u$ with a unknown function}}=\
                                       \\small {latex(u_with_f)}"))
g = sp.Function('g')(x)
dvdy = εyy
v_with_g = sp.integrate(dvdy,(y))+ g   # integral constant w.r.t y: g(x)
v_with_g = v_with_g.expand().doit()
display(Math(f" \\text{{ Displacement $v$ with a unknown function}}=\
                                       \\small {latex(v_with_g)}"))
```

Displacement u with an unknown function $= f(y) - \dfrac{Px^2y}{2EI}$

Displacement v with an unknown function $= g(x) + \dfrac{P\nu xy^2}{2EI}$

Substituting u and v into the third equation in Eq. (2.52), we obtain

```
expr = (diff(u_with_f,y)+diff(v_with_g,x) - γxy).expand()
display(Math(f" \\text{{ Expression from the strain relation}}:\
                                          {latex(expr)}"))
```

Expression from the strain relation:
$\frac{d}{dy}f(y) + \frac{d}{dx}g(x) + \frac{Ph^2}{8GI} - \frac{Py^2}{2GI} + \frac{P\nu y^2}{2EI} - \frac{Px^2}{2EI}$

The above equation must be zero. We regroup the terms as follows:

$$\underbrace{\frac{d}{dy}f(y) - \frac{Py^2}{2GI} + \frac{P\nu y^2}{2EI}}_{F(y)} + \underbrace{\frac{d}{dx}g(x) - \frac{Px^2}{2EI}}_{G(x)} = \underbrace{-\frac{Ph^2}{8GI}}_{\text{constant}} \tag{2.53}$$

It can be seen that $F(y)$ is a function of only y, and $G(x)$ is a function of only x. However, the right side of the equation is a constant, which is not possible unless both $F(y)$ and $G(x)$ are constants. Therefore, we denote $F(y) = c_F$ and $G(x) = c_G$. This gives us the following three separate equations:

$$\frac{d}{dx}g(x) - \frac{Px^2}{2EI} = c_G \tag{2.54}$$

$$\frac{d}{dy}f(y) - \frac{Py^2}{2GI} + \frac{P\nu y^2}{2EI} = c_F \tag{2.55}$$

$$c_F + c_G = -\frac{Ph^2}{8GI} \tag{2.56}$$

We observe that Eqs. (2.54) and (2.55) are 1D differential equations, each containing only one unknown function, while Eq. (2.56) is a simple algebraic equation. We begin by integrating these two differential equations. Let us do this using SymPy, starting by displaying the terms of the expression:

```
# Use sp.args() to locate the terms for G(x):
expr.args
```

$$\left(-\frac{Px^2}{2EI},\ -\frac{Py^2}{2GI},\ \frac{Ph^2}{8GI},\ \frac{P\nu y^2}{2EI},\ \frac{d}{dy}f(y),\ \frac{d}{dx}g(x)\right)$$

The first element, **expr.args[0]**, is the term we need for Eq. (2.54). Let's integrate Eq. (2.54), which introduces a new unknown constant:

```
c0, c1, cF, cG = symbols('c0, c1, c_F, c_G')          #integral constants

# find g(x) via integration:
g = sp.integrate(cG-expr.args[0],(x))+c0       # add an integral constant
g = g.expand().doit()
display(Math(f" \\text{{ Function $g(x)$}}={latex(g)}"))
```

Function $g(x) = c_0 + c_G x + \frac{Px^3}{6EI}$

Let us integrate Eq. (2.55), which involves **expr.args[1]** and **expr.args[3]**, resulting in another unknown constant:

```
# find f(y) via integration:
f = sp.integrate(cF - expr.args[1] - expr.args[3],(y)) + c1
f = f.expand().doit()
display(Math(f" \\text{{ Function $f(x)$}}={latex(f)}"))
```

Function $f(x) = c_1 + c_F y + \frac{Py^3}{6GI} - \frac{P\nu y^3}{6EI}$

We can now form the expressions for displacements u and v:

```
# Form the expression for displacement $u$:
u = sp.integrate(dudx, (x)) + f
display(Math(f" \\text{{ Displacement $u(x,y)$}}={latex(u)}"))

#Form the expression for displacement $v$:
v = sp.integrate(dvdy,(y)) + g
display(Math(f" \\text{{ Displacement $v(x,y)$}}={latex(v)}"))
```

Displacement $u(x, y) = c_1 + c_F y + \frac{Py^3}{6GI} - \frac{P\nu y^3}{6EI} - \frac{Px^2 y}{2EI}$

Displacement $v(x, y) = c_0 + c_G x + \frac{P\nu x y^2}{2EI} + \frac{Px^3}{6EI}$

2.7.3.1 *Determine displacements using BC, case 1*

We have now obtained the general solution for displacements u and v, with integral constants c_0 and c_1, and unknown constants c_F and c_G. There are a total of four constants. We now need DBCs to determine these constants.

For boundary condition case 1, we assume at point A ($x = l, y = 0$) (see Fig. 2.3):

$$u = 0; \quad v = 0; \quad \frac{\partial v}{\partial x} = 0 \tag{2.57}$$

This gives us three conditions. Together with Eq. (2.56), we have the right number of equations to determine the four constants. Let us write a SymPy code to calculate these constants and then substitute them back into the displacement expression:

```
# DBC case 1:
cs=sp.solve([u.subs({x:l,y:0}),         v.subs({x:l,y:0}),
             diff(v,x).subs({x:l,y:0}),cF+cG+P*h**2/(8*G*I)],
            [c0, c1, cF, cG])
display(Math(f" \\text{{ Constants found}}: {latex(cs)}"))
```

$$\text{Constants found}: \left\{c_0 : \frac{Pl^3}{3EI},\ c_1 : 0,\ c_F : \frac{-EPh^2 + 4GPl^2}{8EGI},\ c_G : -\frac{Pl^2}{2EI}\right\}$$

```
# Substitute the constants found back into the displacement expressions:
u_final = u.subs(cs).expand()
display(Math(f" \\text{{ Displacement solution $u(x,y)$}}=\
                                                      {latex(u_final)}"))
print(f'Is εxx=∂u/∂x? {εxx == diff(u_final,x)}')          # check results

v_final = v.subs(cs).expand()
display(Math(f" \\text{{ Displacement solution $v(x,y)$}}=\
                                                      {latex(v_final)}"))
print(f'Is εyy=∂v/∂y? {εyy == diff(v_final,y)}')          # check results

print(f'γ_xy=∂u/∂y+∂v/∂x? \
 {γxy.expand()==diff(u_final,y)+diff(v_final,x)}')
```

$$\text{Displacement solution } u(x, y) = -\frac{Ph^2y}{8GI} + \frac{Py^3}{6GI} + \frac{Pl^2y}{2EI} - \frac{P\nu y^3}{6EI} - \frac{Px^2y}{2EI}$$

```
Is εxx=∂u/∂x? True
```

$$\text{Displacement solution } v(x, y) = \frac{Pl^3}{3EI} - \frac{Pl^2x}{2EI} + \frac{P\nu xy^2}{2EI} + \frac{Px^3}{6EI}$$

```
Is εyy=∂v/∂y? True
γ_xy=∂u/∂y+∂v/∂x?  True
```

We have obtained the complete solutions for stress, strain, and displacements as functions of x and y. Now, let's check the solution at some critical locations:

```
# Along neutral axis, $y=0$, the deflection (displacement v):
display(Math(f" \\text{{ Displacement $v(x,0)$ along the cantilever \
                              axis $y=0$ is}}: {latex(v_final.subs(y,0))}"))
# At $x=0, y=0$, the deflection:
display(Math(f" \\text{{ Displacement $v(0,0)$ at tip of the beam is}}:\
                                        {latex(v_final.subs({x:0,y:0}))}"))
# at x = l, the deflection v:
display(Math(f" \\text{{ Displacement $v(l,0)$ at the beam root is}}:\
                                        {latex(v_final.subs({x:l,y:0}))}"))
```

Displacement $v(x, 0)$ along the cantilever axis $y = 0$ is: $\frac{Pl^3}{3EI} - \frac{Pl^2x}{2EI} + \frac{Px^3}{6EI}$

Displacement $v(0, 0)$ at tip of the beam is: $\frac{Pl^3}{3EI}$

Displacement $v(l, 0)$ at the beam root is: 0

It is observed that the displacement $v(x, 0)$ along the beam axis $y = 0$ (including the displacement $v(0, 0)$ at the tip of the beam) matches the result obtained from the thin beam theory based on the Euler–Bernoulli assumption. Additionally, the displacement at the beam root, $v(l, 0)$, satisfies the boundary condition. This indicates that the Euler–Bernoulli assumption is valid for capturing these results in this boundary condition case. However, it fails to capture the detailed variation of displacement across the entire rectangular domain. For instance:

```
# at x = l, the displacement u:
display(Math(f" \\text{{ Displacement $u(l,y)$ along beam's free end}}:\
                                                {latex(u_final.subs({x:l}))}"))
# at x = l, the displacement v:
display(Math(f" \\text{{ Displacement $v(l,y)$ along beam's free end}}:\
                                                {latex(v_final.subs({x:l}))}"))
```

Displacement $u(l, y)$ along beam's free end: $-\frac{Ph^2y}{8GI} + \frac{Py^3}{6GI} - \frac{P\nu y^3}{6EI}$

Displacement $v(l, y)$ along beam's free end: $\frac{Pl\nu y^2}{2EI}$

These results cannot be obtained using the Euler–Bernoulli assumption, which is valid only for 1D beam theory.

2.7.3.2 *Code to plot the results*

We now define a Python function to plot the distribution of the solutions and identify the maximum values of the solution along with their respective locations:

```
def plot2curveS(u, xL, xR, title="f_title"):
    dx = 0.01
    xi = np.arange(xL, xR+dx, dx)
    uf = lambdify((x), u[0], 'numpy')      # convert SymPy f to numpy f
    yi = uf(xi)
    if type(yi) == int: # or len(yi) == 1:
        yi = float(yi)*np.ones_like(xi)

    fig, ax1 = plt.subplots(figsize=(5.,1.), dpi=300)
    fs = 8     # fontsize
    color = 'black'
    ax1.set_xlabel('location', fontsize=fs)
    ax1.set_ylabel(title[0], color=color, fontsize=fs)
    ax1.plot(xi, yi, color=color)
    ax1.grid(color='r',ls=':',lw=.3, which='both') # Use both tick
    ax1.tick_params(axis='x', labelcolor=color, labelsize=fs)
    ax1.tick_params(axis='y', labelcolor=color, labelsize=fs)

    uf = lambdify((x), u[1], 'numpy')      # convert SymPy f to numpy f
    yi2 = uf(xi)
    if type(yi2) == int: # or len(yi2) == 1:
        yi2 = float(yi2)*np.ones_like(xi)

    m1, m2, m3 = np.partition(abs(yi2), 2)[0:3]
    msl=[np.where(abs(yi2)==m1)[0][0],np.where(abs(yi2)==m2)[0][0],
             np.where(abs(yi2)==m3)[0][0]]

    vmax = yi[yi.argmax()]
    max_l = np.argwhere(yi == vmax)
    ax1.plot(xi[max_l],  yi[max_l], 'r*', markersize=4)
    print(f'Maximum {title[0]} value={vmax:.3e}, at x={xi[max_l][0][0]}')
    vmax = yi2[yi2.argmax()]
    max_l = np.argwhere(yi2 == vmax)
    print(f'Maximum {title[1]} value={vmax:.3e}, at x={xi[max_l][0][0]}')

    if abs(xi[msl[2]]-xi[msl[1]])<2*dx:
        if abs(yi2[msl[2]]-0.)<abs(yi2[msl[1]]-0.): msl.pop(1)
        else: msl.pop(2)
    if len(msl) > 2:
        if abs(xi[msl[2]]-xi[msl[0]])<2*dx:
            if abs(yi2[msl[2]]-0.)<abs(yi2[msl[0]]-0.): msl.pop(0)
            else: msl.pop(2)
    if len(msl) > 1:
        if abs(xi[msl[1]]-xi[msl[0]])<2*dx:
```

```
if abs(yi2[msl[1]]-0.)<abs(yi2[msl[0]]-0.): msl.pop(0)
else: msl.pop(1)

ax2 = ax1.twinx() # instantiate second axes sharing the same x-axis

color = 'blue'
ax2.set_ylabel(title[1], color=color, fontsize=fs)
ax2.plot(xi, yi2, color=color)
ax2.plot(xi[max_l], yi2[max_l], 'r*', markersize=4)
ax2.plot(xi[msl], yi2[msl], 'r*', markersize=4)
ax1.plot(xi[msl],  yi[msl], 'r*', markersize=4)
ax2.plot(xi[0], yi2[0], 'ro', markersize=2)
ax1.plot(xi[0],  yi[0], 'ro', markersize=2)
ax2.plot(xi[-1], yi2[-1], 'ro', markersize=2)
ax1.plot(xi[-1],  yi[-1], 'ro', markersize=2)
ax2.grid(color='r',ls=':',lw=.5, which='both') # Use both tick
ax2.tick_params(axis='x', labelcolor=color, labelsize=fs)
ax2.tick_params(axis='y', labelcolor=color, labelsize=fs)
np.set_printoptions(formatter={'float': '{: 0.3e}'.format})
print(f'Extreme {title[0]} values={yi[msl]},\n    at x={xi[msl]}')
print(f'Critical {title[1]} values={yi2[msl]},\n    at x={xi[msl]}')
print(f'{title[0]} values at boundary ={yi[0], yi[-1]}')
print(f'{title[1]} values at boundary ={yi2[0], yi2[-1]}\n')
```

The following snippet plots the displacement distribution and prints out the maximum values along with their respective locations:

```
# compute displacement u(y) at the right end of the cantilever
ul = u_final.subs({x:l})
dict_ = {E:1,I:1,l:1,P:1,nu:1/3,h:1}
uy = nsimplify((ul.subs({G:E/(1+nu)/2})).subs(dict_))
display(Math(f" \\text{{Distribution of $u(l,y)$}}=\\small{latex(uy)}"))

vl = v_final.subs({x:l})
vy = nsimplify(vl.subs(dict_))
display(Math(f" \\text{{Distribution of $v(l,y)$}}=\\small{latex(vy)}"))
```

Distribution of $u(l, y) = \dfrac{7y^3}{18} - \dfrac{y}{3}$

Distribution of $v(l, y) = \dfrac{y^2}{6}$

```
# plot these two curves:
title = ["u(l,y)", "v(l,y)"]
plt.ion()
plot2curveS((uy.subs(y,x),vy.subs(y,x)), -.1, .1, title=title)
```

```
Maximum u(l,y) value=3.294e-02, at x=-0.1
Maximum v(l,y) value=1.667e-03, at x=-0.1
Extreme u(l,y) values=[ 1.850e-17],
    at x=[-5.551e-17]
Critical v(l,y) values=[ 5.136e-34],
    at x=[-5.551e-17]
u(l,y) values at boundary =(0.03294444444444444, -0.03294444444444441)
v(l,y) values at boundary =(0.001666666666666667, 0.001666666666666663)
```

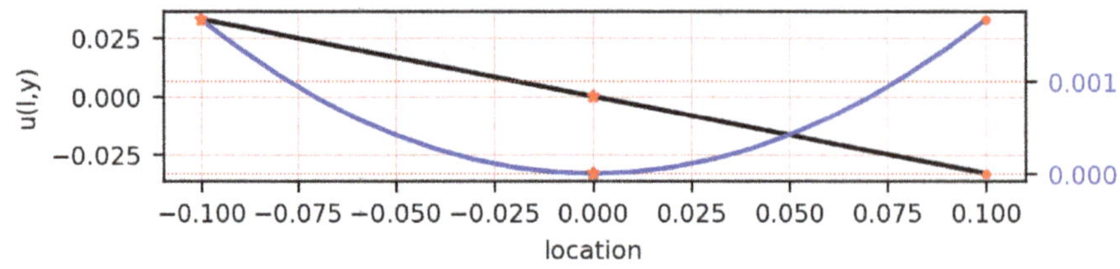

This figure shows the distributions of $u(l, y)$ (black) and $v(l, y)$ (blue) at $x = l$ along different values of y.

2.7.3.3 *Determine displacements using BC, case 2*

Consider boundary condition case 2: at point A ($x = l, y = 0$, see Fig. 2.3),

$$u = 0; \quad v = 0; \quad \frac{\partial u}{\partial y} = 0 \tag{2.58}$$

Using these three conditions and the same general solution for the displacements, the particular solution for this BC case can be found using

```
# Determine these constants using DBCs case-2:
cs=sp.solve([u.subs({x:l,y:0}),           v.subs({x:l,y:0}),
             diff(u,y).subs({x:l,y:0}), cF+cG+P*h**2/(8*G*I)],
            [c0,c1,cF,cG])
cs
```

$$\left\{c_0 : \frac{3EPh^2l + 8GPl^3}{24EGI}, \; c_1 : 0, \; c_F : \frac{Pl^2}{2EI}, \; c_G : \frac{-EPh^2 - 4GPl^2}{8EGI}\right\}$$

```
# Compute the displacements for DBC case-2:
u2_final = u.subs(cs).expand()
display(Math(f" \\text{{Displacement for BC case-2, $u(x,y)$}}=\
                                            {latex(u2_final)}"))
print(f'Is εxx=∂u/∂x? {εxx == diff(u2_final,x)}')         # check results

v2_final = v.subs(cs).expand()
display(Math(f" \\text{{Displacement for BC case-2, $v(x,y)$}}=\
                                            {latex(v2_final)}"))
print(f'Is εyy=∂v/∂y? {εyy == diff(v2_final,y)}')         # check results
print(f'γ_xy=∂u/∂y+∂v/∂x? '\
      f'{γxy.expand()==diff(u2_final,y)+diff(v2_final,x)}')
```

Displacement for BC case-2, $u(x,y) = \frac{Py^3}{6GI} + \frac{Pl^2y}{2EI} - \frac{P\nu y^3}{6EI} - \frac{Px^2y}{2EI}$

```
Is εxx=∂u/∂x? True
```

Displacement for BC case-2, $v(x,y) = \frac{Ph^2l}{8GI} - \frac{Ph^2x}{8GI} + \frac{Pl^3}{3EI} - \frac{Pl^2x}{2EI} + \frac{P\nu xy^2}{2EI} + \frac{Px^3}{6EI}$

```
Is εyy=∂v/∂y? True
γ_xy=∂u/∂y+∂v/∂x? True
```

Let us check the solution at some critical locations:

```
display(Math(f" \\text{{ Displacement $v(x,0)$ along $y=0$ is}}: \
                                  {latex(v2_final.subs(y,0))}"))
# At $x=0, y=0$, the deflection is
display(Math(f" \\text{{ Displacement $v(0,0)$ at the beam free tip}}:\
                                  {latex(v2_final.subs({x:0,y:0}))}"))
# at x = l, the deflection v is:
display(Math(f" \\text{{ Displacement $v(l,0)$ at root of the beam}}:\
                                  {latex(v2_final.subs({x:l,y:0}))}"))
```

Displacement $v(x,0)$ along $y=0$ is: $\frac{Ph^2l}{8GI} - \frac{Ph^2x}{8GI} + \frac{Pl^3}{3EI} - \frac{Pl^2x}{2EI} + \frac{Px^3}{6EI}$

Displacement $v(0,0)$ at the beam free tip: $\frac{Ph^2l}{8GI} + \frac{Pl^3}{3EI}$

Displacement $v(l,0)$ at root of the beam: 0

This time, the displacement $v(x, 0)$ along the beam axis $y = 0$ differs from that predicted by the thin beam theory based on the Euler–Bernoulli assumption. This indicates that the Euler–Bernoulli assumption is not valid for this set of boundary conditions. Furthermore, it cannot capture the detailed variation of displacement over the entire rectangle. For example:

```
# at x = l, the displacement u is:
display(Math(f" \\text{{ Displacement $u(l,y)$ along beam root edge}}:\
                                        {latex(u2_final.subs({x:l}))}"))
# at x = l, the displacement v is:
display(Math(f" \\text{{ Displacement $v(l,y)$ along beam root edge}}:\
                                        {latex(v2_final.subs({x:l}))}"))
```

Displacement $u(l, y)$ along beam root edge: $\frac{Py^3}{6GI} - \frac{P\nu y^3}{6EI}$

Displacement $v(l, y)$ along beam root edge: $\frac{Pl\nu y^2}{2EI}$

The following snippet plots the distribution of displacements, along with the maximum values and their corresponding locations:

```
ul2 = u2_final.subs({x:l})
uy2 = nsimplify((ul2.subs({G:E/(1+nu)/2})).subs(dict_))
display(Math(f" \\text{{ Displacement $v(l,y)$ along beam root edge}}:\
                                                       {latex(uy2)}"))
vl2 = v2_final.subs({x:l})
vy2 = nsimplify(vl2.subs(dict_))
display(Math(f" \\text{{ Displacement $v(l,y)$ along beam root edge}}:\
                                                       {latex(vy2)}"))
```

Displacement $v(l, y)$ along beam root edge: $\frac{7y^3}{18}$

Displacement $v(l, y)$ along beam root edge: $\frac{y^2}{6}$

In comparison with $u(x = l, y)$ obtained using $\frac{\partial v}{\partial x} = 0$, the linear dependence on y disappears, ensuring that $\frac{\partial u}{\partial y} = 0$.

```
# Generate the plot for u and v:
title = ["u(l,y)", "v(l,y)"]
plot2curveS((uy2.subs(y,x),vy2.subs(y,x)), -.1, .1, title=title)
```

```
Maximum u(l,y) value=3.889e-04, at x=0.0999999999999999
Maximum v(l,y) value=1.667e-03, at x=-0.1
Extreme u(l,y) values=[-6.652e-50],at x=[-5.551e-17]
Critical v(l,y) values=[ 5.136e-34],at x=[-5.551e-17]
u(l,y) values at boundary =(-0.000388888888888889, 0.000388888888888876)
v(l,y) values at boundary =(0.00166666666666667, 0.00166666666666663)
```

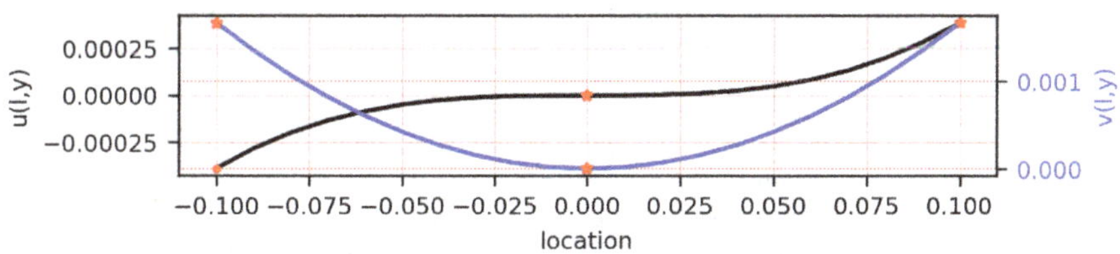

The figure above shows the distributions of $u(l, y)$ (black) and $v(l, y)$ (blue) at $x = l$ along various y-locations for case 2 boundary conditions (DBCs).

2.7.4 *Comparison of different boundary conditions*

Next, we plot the displacement distributions at the root of the cantilever for two different boundary condition cases, all in the same figure for easy comparison:

```
title = ["Case 1 u(l,y)", "Case 2 u(l,y)"]
plot2curveS((uy.subs(y,x),uy2.subs(y,x)), -.1, .1, title=title)
```

```
Maximum Case 1 u(l,y) value=3.294e-02, at x=-0.1
Maximum Case 2 u(l,y) value=3.889e-04, at x=0.0999999999999999
Extreme Case 1 u(l,y) values=[ 1.850e-17],at x=[-5.551e-17]
Critical Case 2 u(l,y) values=[-6.652e-50],at x=[-5.551e-17]
Case 1 u(l,y) values at boundary =(0.0329444444444444,-0.0329444444444441)
Case 2 u(l,y) values at boundary =(-0.000388888888888889,0.000388888888888876)
```

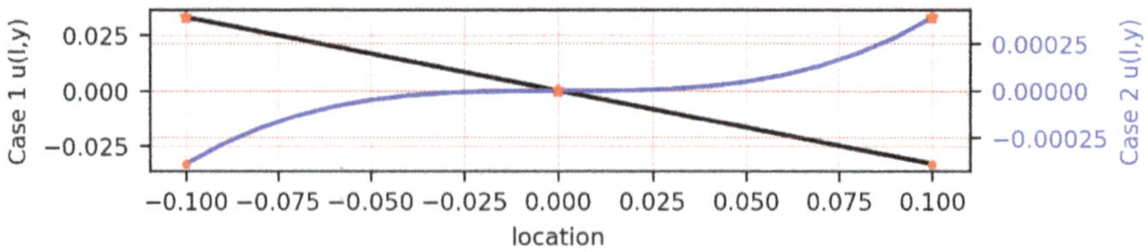

The figure above compares the distributions of $u(l, y)$ for Case 1 DBCs (black) with those for Case 2 DBCs (blue) at $x = l$ along different y-locations.

```
title = ["Case 1 v(l,y)", "Case 2 v(l,y)"]   # these two curves overlap
plot2curveS((vy.subs(y,x),vy2.subs(y,x)), -.1, .1, title=title)
```

```
Maximum Case 1 v(l,y) value=1.667e-03, at x=-0.1
Maximum Case 2 v(l,y) value=1.667e-03, at x=-0.1
Extreme Case 1 v(l,y) values=[ 5.136e-34],at x=[-5.551e-17]
Critical Case 2 v(l,y) values=[ 5.136e-34], at x=[-5.551e-17]
Case 1 v(l,y) values at boundary =(0.001666666666666667, 0.0016666666666666663)
Case 2 v(l,y) values at boundary =(0.001666666666666667, 0.0016666666666666663)
```

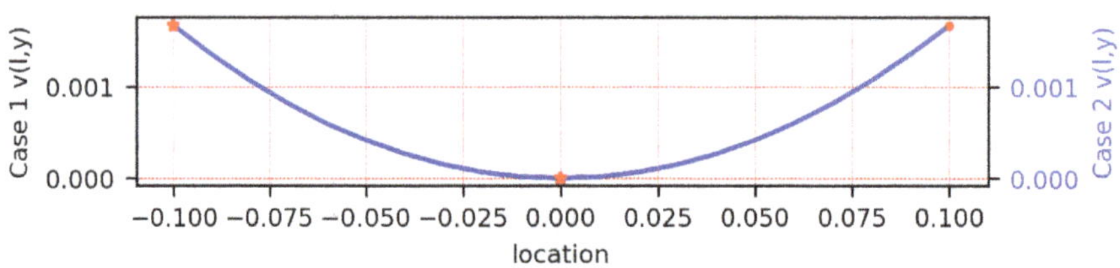

The figure above compares the distributions of $v(l, y)$ for Case 1 DBCs (black) with those for Case 2 DBCs (blue) at $x = l$ along different y-locations. The two curves overlap.

2.7.5 *Contour plots of displacement fields*

Next, we plot the displacement fields in the form of contour plots using the following snippets:

```
def field_plot(sp_f, X, x_mesh, y_mesh, title='', flip='No'):
    """Plot a given field defined by a sympy function `sp_f`
    in the x-y plane defined by X=Matrix([x, y]), and meshed to
    x_mesh, y_mesh.
    """
    np_f = lambdify(X, sp_f, "numpy")                    # convert to numpy
    f_mesh = np_f(x_mesh, y_mesh)

    vmin = np.min(f_mesh); vmax = np.max(f_mesh)          # find extrema
    print(f"Values in the domain: Maximum={vmax:g}; Minimum={vmin:g}")

    fig = plt.gcf()                                        # create plot
    cntur = plt.contourf(x_mesh, y_mesh,f_mesh,cmap="seismic",levels=50)
    cbar = fig.colorbar(cntur)

    # flip the plot, if needed:
    if flip =='y':
        plt.gca().invert_yaxis()
    elif flip == 'x':
        plt.gca().invert_xaxis()
    elif flip == 'xy':
        plt.gca().invert_xaxis()
        plt.gca().invert_yaxis()

    plt.axis("equal")
    plt.title(title)
    return cntur
```

```
# Plot the contour for the displacements:

from collections import OrderedDict

# Re-define using the solution obtained earlier:
u_final =-P*h**2*y/(8*G*I) + P*y**3/(6*G*I) + P*l**2*y/(2*E*I) \
        -P*nu*y**3/(6*E*I) - P*x**2*y/(2*E*I)
v_final = P*l**3/(3*E*I) - P*l**2*x/(2*E*I) + P*nu*x*y**2/(2*E*I) \
        + P*x**3/(6*E*I)

x, y  = symbols('x, y')
X = sp.Matrix([x, y])
l_val = 2; h_val = 1
# Create mesh grid:
x_mesh, y_mesh = np.mgrid[0.:l_val:100j, -h_val/2:h_val/2:100j]

dict_= {P:1, E:1,I:1, nu:1/3, G:3/8, l:l_val, h:h_val}
di = sp.nsimplify(u_final.subs(dict_))
#di = sp.nsimplify(v_final.subs(dict_))          # To plot displacement v
display(Math(f" \\text{{ Displacement $u(x,y)$}}={latex(di)}"))
```

```
plt.figure()
plt.ioff()
#plt.ion()                          # uncomment this to display the plots
field_plot(di, X, x_mesh, y_mesh, title=' Displacement $u$', flip='y')
plt.savefig('imagesSM/CantiLever2D_u.png',dpi=500,bbox_inches='tight')
#plt.show();                        # uncomment this to display the plots
```

Displacement $u(x, y) = -\frac{x^2 y}{2} + \frac{7y^3}{18} + \frac{5y}{3}$

```
Values in the domain: Maximum=0.881944; Minimum=-0.881944
```

The contour plot of the horizontal displacement u for the 2D solid cantilever is shown in Fig. 2.5. As observed, at $x = 0$, the horizontal displacement u on the upper surface is negative, indicating a movement opposite to the x-direction, which is expected. Conversely, on the bottom surface, the displacement is in the positive x-direction. Readers may plot the vertical displacement v using the same Python function:

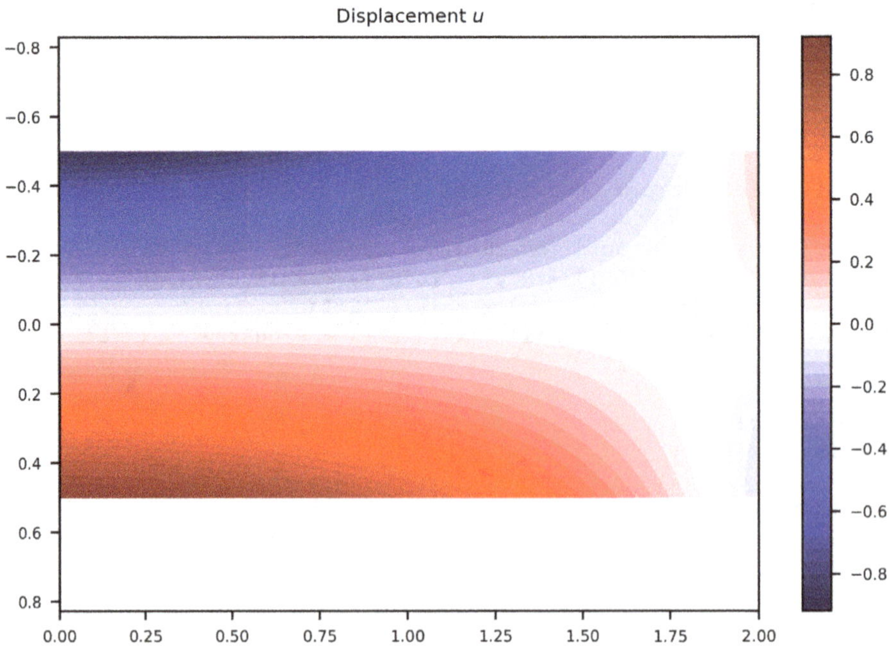

Figure 2.5. Contour plot of the horizontal displacement solution in the 2D rectangular cantilever solid.

Next, let us plot the normal stress σ_{xx} using the **field_plot()** function.

```
# Re-define the stresses using the solution obtained:
σ = sp.Matrix([[-12*P*x*y/h**3], [0], [3*P*(-h**2 + 4*y**2)/(2*h**3)]])

σxxi = sp.nsimplify(σ[0].subs(dict_))
display(Math(f" \\text{{$σ_{{xx}}(x,y)$ to be plotted}}={latex(σxxi)}"))

plt.figure()
plt.ioff()
#plt.ion()                          # uncomment this to display the plots
field_plot(σxxi, X, x_mesh, y_mesh, title=' Stress $σ_{xx}$')
plt.savefig('imagesSM/CantiLever_Sxx.png',dpi=500,bbox_inches='tight')
#plt.show();                        # uncomment this to display the plots
```

$\sigma_{xx}(x, y)$ to be plotted $= -12xy$

```
Values in the domain: Maximum=12; Minimum=-12
```

The contour plot of the normal stress σ_{xx} in the 2D rectangular cantilever solid is shown in Fig. 2.6. It is observed that the normal stress σ_{xx} is zero at $x = 0$ and increases gradually (linearly) with x, as expected. The highest levels of normal stress are found on both the upper and lower surfaces.

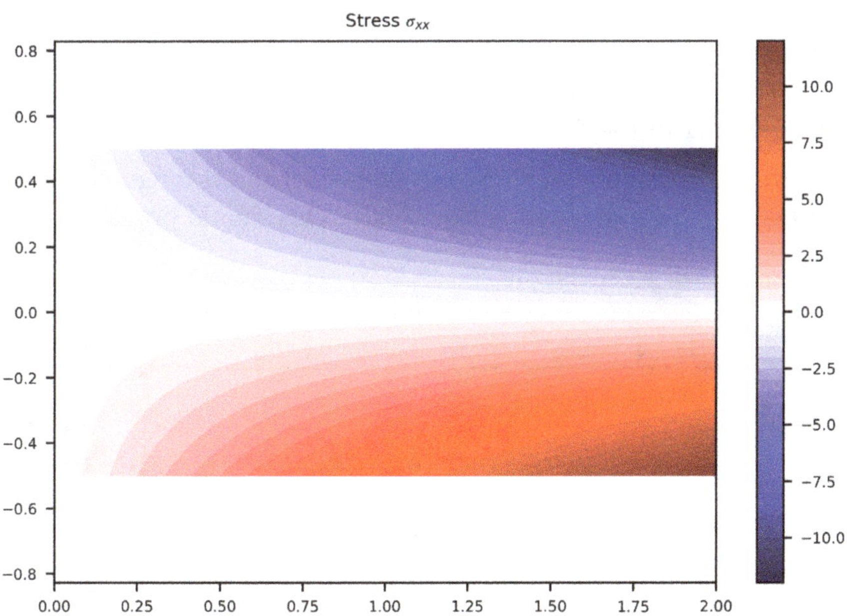

Figure 2.6. Contour plot of the normal stress solution in the 2D rectangular cantilever solid.

Readers can plot the contour of the shear stress σ_{xy} by slightly modifying the code.

2.8 Simply supported rectangle under a uniform load

Consider a thin rectangular 2D solid subjected to a uniformly distributed load p on the top surface, as shown in Fig. 2.7. The solid is free from body forces and is simply supported, implying that the vertical displacements at points A and B are fixed to zero, and the sum of the horizontal displacements at points A and B is also zero.

Our task is to find the solutions for stresses, strains, and displacements as functions of x and y, assuming that Young's modulus E and Poisson's ratio ν of the material are known.

2.8.1 *Stress function construction*

Since there are no body forces, the stress function must satisfy Eq. (2.29). We will select a candidate stress function based on the following considerations.

Given the problem setup shown in Fig. 2.7 and the cantilever rectangle problem from the previous section, we know that the order of the stress function ϕ in the shorter dimension x is two orders lower than that in the longer dimension y. For the cantilever, the order of x in ϕ is 1, and the order of y is 3. Considering the uniformly distributed constant force p, we will increase the order of the function by at least one.

Thus, we can consider using up to second-order terms for x and up to fifth-order terms for y, as indicated by Eq. (2.30). This should provide a sufficient

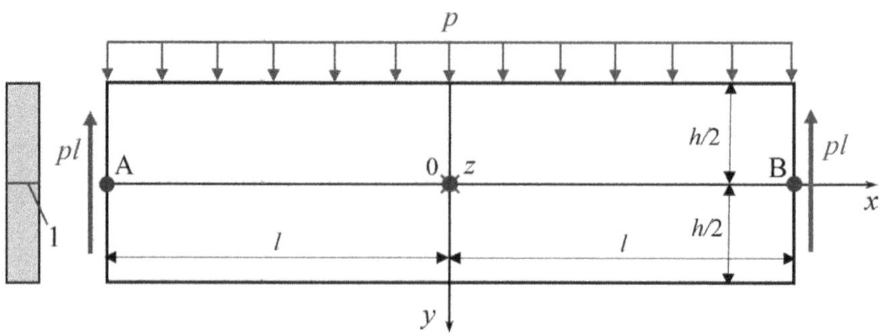

Figure 2.7. A 2D rectangular thin solid with dimensions $2l \times h$ and unit thickness $b = 1$, subjected to a downward uniformly distributed load p on the top surface.

number of basis functions for ϕ. The selected BBFs (basis functions) are shown in red in the following:

$$
\begin{aligned}
&1,\ x,\ y,\ xy,\ x^2,\ y^2,\ x^3,\ yx^2,\ xy^2,\ y^3 && \leq \text{third-order}\\
&(3x^2y^2 - x^4),\ x^3y,\ xy^3,\ (3x^2y^2 - y^4),\ (x^4 - y^4) && \text{fourth-order}\\
&(5x^2y^3 - y^5),\ (x^3y^2 - xy^4),\ (5x^4y - y^5),\ (x^5 - 5xy^4) && \text{fifth-order}
\end{aligned}
\tag{2.59}
$$

Note again that we excluded the linear bases because these linear terms do not contribute to the stresses. Therefore, our stress function can be expressed as a linear combination of the following nine selected BBFs:

$$
\begin{aligned}
\phi = {} & a_0xy + a_1x^2 + a_2y^2 + a_3yx^2 + a_4xy^2 + a_5y^3\\
& + a_6xy^3 + a_7(3x^2y^2 - y^4) + a_8(5x^2y^3 - y^5)
\end{aligned}
\tag{2.60}
$$

where a_i $(i = 0, 1, \ldots, 8)$ are unknown constants.

Since we are using only BBFs, ϕ given in Eq. (2.60) automatically satisfies Eq. (2.29). The stress function is defined using the following snippet:

```
E, G, I, h, l, nu, p = sp.symbols('E,G,I,h,l,nu,p', nonnegative=True)
x, y  = symbols('x, y')                         # define symbolic variables
X = sp.Matrix([x, y])

# unknown constants for BBFs:
a0, a1, a2, a3, a4, a5, a6, a7, a8=symbols('a0,a1,a2,a3,a4,a5,a6,a7,a8')
σxx, σxy, σyy = symbols('σ_xx, σ_xy, σ_yy')

ϕ = sp.Function('ϕ')(x, y)               # define symbolic stress function
ϕ = a0*x*y + a1*x**2 + a2*y**2+a3*y*x**2+a4*x*y**2+a5*y**3\
    +a6*x*y**3+a7*(3*x**2*y**2-y**4)+a8*(5*x**2*y**3-y**5)
ϕ                                              # constructed stress function
```

$$
a_0xy + a_1x^2 + a_2y^2 + a_3x^2y + a_4xy^2 + a_5y^3 + a_6xy^3 + a_7\cdot\left(3x^2y^2 - y^4\right) + a_8\cdot\left(5x^2y^3 - y^5\right)
$$

2.8.2 *Stress solution*

Using the stress function and Eq. (2.22), the general expressions for stresses are given as follows:

```
σxx = ϕ.diff(y,2)
σyy = ϕ.diff(x,2)
σxy = -(ϕ.diff(x,1)).diff(y,1)

σ_ai = Matrix([σxx, σyy, σxy])              # Stresses with constants a_i
display(Math(f" \\text{{ Vector of stresses}}={latex(σ_ai)}"))
```

Vector of stresses

$$= \begin{bmatrix} 2\left(a_2 + a_4 x + 3a_5 y + 3a_6 xy + 3a_7\left(x^2 - 2y^2\right) + 5a_8 y\left(3x^2 - 2y^2\right)\right) \\ 2\left(a_1 + a_3 y + 3a_7 y^2 + 5a_8 y^3\right) \\ -a_0 - 2a_3 x - 2a_4 y - 3a_6 y^2 - 12a_7 xy - 30a_8 xy^2 \end{bmatrix}$$

Symmetric conditions

To determine these constants in an easier manner, we first use the symmetric conditions. Based on the problem setting given in Fig. 2.7, we know that the normal stresses σ_{xx} and σ_{yy} should be symmetric, and the shear stress σ_{xy} should be antisymmetric with respect to x. Therefore, a_0, a_4, and a_6 must be zero.

```
σ_ai[0] = σ_ai[0].subs({a4:0, a6:0})
σ_ai[2] = σ_ai[2].subs({a0:0, a4:0, a6:0})
display(Math(f" \\text{{ Simplified stresses}}={latex(σ_ai)}"))
```

$$\text{Simplified stresses} = \begin{bmatrix} 2a_2 + 6a_5 y + 6a_7\left(x^2 - 2y^2\right) + 10a_8 y\left(3x^2 - 2y^2\right) \\ 2\left(a_1 + a_3 y + 3a_7 y^2 + 5a_8 y^3\right) \\ -2a_3 x - 12a_7 xy - 30a_8 xy^2 \end{bmatrix}$$

Boundary conditions on the top and bottom edges

We obtained the simplified stresses, which still contain six unknown constants. We can now use the SBCs for boundaries along the longer side of the problem domain, which is the top and bottom edges (or surfaces):

$$\begin{aligned} \sigma_{yy}\big|_{y=-h/2} &= -p; \quad \sigma_{xy}\big|_{y=-h/2} = 0 \quad \text{on top edge} \\ \sigma_{yy}\big|_{y=h/2} &= 0; \quad \sigma_{xy}\big|_{y=h/2} = 0 \quad \text{on bottom edge} \end{aligned} \tag{2.61}$$

This gives four SBC conditions, which can be used to determine four constants. The snippet is as follows:

```
SBCs = []
SBCs.append(σ_ai[1].subs(y, h/2))
SBCs.append(σ_ai[1].subs(y,-h/2) + p)
SBCs.append(σ_ai[2].subs(y, h/2))
SBCs.append(σ_ai[2].subs(y,-h/2))
display(SBCs[:2])
display(SBCs[2:])
```

$$\left[2a_1 + a_3h + \frac{3a_7h^2}{2} + \frac{5a_8h^3}{4},\ 2a_1 - a_3h + \frac{3a_7h^2}{2} - \frac{5a_8h^3}{4} + p\right]$$

$$\left[-2a_3x - 6a_7hx - \frac{15a_8h^2x}{2},\ -2a_3x + 6a_7hx - \frac{15a_8h^2x}{2}\right]$$

Since Eq. (2.61) needs to be satisfied at any point on the surfaces of the beam, all these foregoing expressions must be zero for any x. Hence, the coefficients of x^1 and x^0 must be zero. Clearly, the coefficients of x in the third and fourth conditions have to be zero. The Python snippet to get all these done is as follows:

```
SBCs.append(SBCs[ 2].coeff(x,1))                    # add in two conditions
SBCs.append(SBCs[-1].coeff(x,1))

sln_SBC = sp.solve(SBCs, [a1, a2, a3, a5, a7, a8])
display(Math(f" \\text{{ Solutions for $a_i$}} ={latex(sln_SBC)}"))

σ_ai = σ_ai.subs(sln_SBC)
display(Math(f" \\text{{ Constants reduced stresses}}={latex(σ_ai)}"))
```

$$\text{Solutions for } a_i = \left\{a_1 : -\frac{p}{4},\ a_3 : \frac{3p}{4h},\ a_7 : 0,\ a_8 : -\frac{p}{5h^3}\right\}$$

$$\text{Constants reduced stresses} = \begin{bmatrix} 2a_2 + 6a_5y - \frac{2py\left(3x^2-2y^2\right)}{h^3} \\ -\frac{p}{2} + \frac{3py}{2h} - \frac{2py^3}{h^3} \\ -\frac{3px}{2h} + \frac{6pxy^2}{h^3} \end{bmatrix}$$

The stress expressions still contain two constants. This time, we shall use the force boundary conditions (FBCs) for boundaries along the shorter sides of the problem domain, which is the left and right edges of the rectangle. Since the stresses along these edges are functions of x, it is not possible to have the SBCs satisfied at every point along the edges. We thus use the integral form conditions, implying the equilibrium is in terms of forces. This is why these conditions are called force boundary conditions. Also, since we used the symmetric conditions earlier, we now need only to impose the integral equilibrium conditions on one edge, say the right edge. The formulas are as follows:

$$\begin{aligned} &\int_{-h/2}^{h/2} \sigma_{xx}\Big|_{x=l} dy = 0 \quad \text{total normal stress vanishes} \\ &\int_{-h/2}^{h/2} y\,\sigma_{xx}\Big|_{x=l} dy = 0 \quad \text{total moment w.r.t } z \text{ vanishes} \end{aligned} \tag{2.62}$$

This gives two FBC conditions, which can be used to determine the remaining two constants. The snippet is as follows:

```
FBCs = []
FBCs.append(sp.integrate(  σ_ai[0].subs(x, l), (y, -h/2, h/2)))
FBCs.append(sp.integrate(y*σ_ai[0].subs(x, l), (y, -h/2, h/2)))

sln_FBC = sp.solve(FBCs, [a2, a5])
display(Math(f" \\text{{ Solutions for $a_i$}} ={latex(sln_FBC)}"))

σ = sp.simplify(σ_ai.subs(sln_FBC))
display(Math(f" \\text{{ Final solution for stresses}}={latex(σ)}"))
```

$$\text{Solutions for } a_i = \left\{a_2 : 0,\ a_5 : \frac{-h^2p + 10l^2p}{10h^3}\right\}$$

$$\text{Final solution for stresses} = \begin{bmatrix} \frac{py\left(-3h^2+30l^2-30x^2+20y^2\right)}{5h^3} \\ \frac{p\left(-h^3+3h^2y-4y^3\right)}{2h^3} \\ \frac{3px\left(-h^2+4y^2\right)}{2h^3} \end{bmatrix}$$

We found the solution for all the stresses. These are all in polynomial form. The normal stress σ_{xx} changes quadratically in x and cubically in y. The normal stress σ_{yy} does not change with x, but changes cubically with y. The shear stress σ_{xy} changes linearly in x and quadratically in y. The stress solution can be rearranged manually to a more readable form:

$$\begin{aligned} \sigma_{xx} &= -\frac{6py}{h^3}(l^2 - x^2) + \frac{py}{h}\left(\frac{4y^2}{h^2} - \frac{3}{5}\right) \\ \sigma_{yy} &= -\frac{p}{2}\left(1 + \frac{y}{h}\right)\left(1 - \frac{2y}{h}\right)^2 \\ \sigma_{xy} &= -\frac{6px}{h^3}\left(\frac{h^2}{4} - y^2\right) \end{aligned} \tag{2.63}$$

These should satisfy all the necessary conditions. It is a good practice to confirm this after stress solutions are obtained, and it is easy to do with Python code.

2.8.3 *Verification of the solutions*

First, let us check the equilibrium equations:

```
equ_x = (σ[0].diff(x)+σ[2].diff(y)).simplify()
display(Math(f"\\text{{Equilibrium in the $x$-direction?}}:\
                                        {latex(equ_x==0)}"))
equ_y = (σ[1].diff(y)+σ[2].diff(x)).simplify()
display(Math(f"\\text{{Equilibrium in the $y$-direction?}}:\
                                        {latex(equ_y==0)}"))
```

Equilibrium in the x-direction?: True
Equilibrium in the y-direction?: True

```
Next, check the compatibility condition:
```

```
Laplacian_σ = (σ[0]+σ[1]).diff(x,2) + (σ[0]+σ[1]).diff(y,2)     # ∇^2σ

display(Math(f"\\text{{ Is compatibility holds?}}:\
                                     {latex(Laplacian_σ==0)}"))
```

Is compatibility holds?: True

The following snippet confirms that the boundary conditions on the top and bottom edges are all satisfied:

```
σ_top = sp.simplify(σ.subs(y,-h/2))
σ_bottom = sp.simplify(σ.subs(y,h/2))
display(Math(f" \\text{{ Stresses on top edge}}={latex(σ_top)}; \\;\\;\
                \\text{{ Stresses on bottom edge}}={latex(σ_bottom)}"))
```

$$\text{Stresses on top edge} = \begin{bmatrix} \frac{p\left(-h^2-15l^2+15x^2\right)}{5h^2} \\ -p \\ 0 \end{bmatrix};$$

$$\text{Stresses on bottom edge} = \begin{bmatrix} \frac{p\left(h^2+15l^2-15x^2\right)}{5h^2} \\ 0 \\ 0 \end{bmatrix}$$

On the left and right edges, the boundary conditions for the normal stresses are satisfied in an integral sense. The same applies to the shear stresses there. The integral of the shear stress will give the total vertical shear forces (often referred to as support forces), which can be computed using the following expression:

```
F_L = sp.integrate(σ[2].subs(x, -1), (y, -h/2, h/2))     # force on left
F_R = sp.integrate(σ[2].subs(x,  1), (y, -h/2, h/2))    # force on right
display(Math(f"\\text{{Support force at left  edge}}={latex(F_L)}"))
display(Math(f"\\text{{Support force at right edge}}={latex(F_R)}"))
```

Support force at left edge $= lp$

Support force at right edge $= -lp$

As shown in Fig. 2.7, the equilibrium requires

$$p(2l) - F_L + F_R = 0 \tag{2.64}$$

The following snippet confirms this:

```
equ_all = (p*2*l - F_L + F_R).simplify()
display(Math(f"\\text{{Equilibrium of all forces in $y$-direction?}}:\
                                        {latex(equ_all==0)}"))
```

Equilibrium of all forces in y-direction?: True

The following snippet displays the stress components at the left edge of the solid:

```
σxx_L = sp.simplify(σ[0].subs(x,-1))
σxy_L = sp.simplify(σ[2].subs(x,-1))
display(Math(f"\\text{{$σ_{{xx}}$ at left edge}}={latex(σxx_L)};\\;\\;"
             f"\\text{{$σ_{{xy}}$ at left edge}}={latex(σxy_L)}"))
```

σ_{xx} at left edge $= \dfrac{py\left(-3h^2+20y^2\right)}{5h^3}$; σ_{xy} at left edge $= \dfrac{3lp\left(h^2-4y^2\right)}{2h^3}$

As seen from the output of the previous code cell, the normal stress σ_{xx} is not zero at the left (and the right) edge. But the total normal force (integral) and the total moment over the cross-section are zero. This means that the nonzero stress is locally in equilibrium. The same is true for the shear stress σ_{xy}. The total shear force is in equilibrium with a total shear force of pl, but the distribution may not be. Based on Saint-Venant's principle, the solution obtained is valid away from the left (and right) edges.

Readers may confirm the satisfaction of the integral equilibrium conditions, Eq. (2.62), as an exercise.

2.8.4 *Comparison with thin beam solution*

If we treat the 2D rectangular solid as a beam (with $b = 1$), its properties are given as follows [7]:

1. The second moment of the cross-section rectangle is $I = \frac{h^3}{12}$.
2. The first moment of the area from the top surface to y on the cross-section is $Q = \frac{h^2}{8} - \frac{y^2}{2}$.

The resulting forces on the cross-section include (see Fig. 2.7):

1. The moment at x: $M = pl(l - x) - \frac{p}{2}(l - x)^2 = \frac{p}{2}(l^2 - x^2)$.
2. The shear force at x: $V = -pl + p(l - x) = -qx$.

In this case, Eq. (2.63) can be rewritten as

$$\begin{aligned}
\sigma_{xx} &= \frac{My}{I} + \frac{py}{h}\left(\frac{4y^2}{h^2} - \frac{3}{5}\right) \\
\sigma_{yy} &= -\frac{p}{2}\left(1 + \frac{y}{h}\right)\left(1 - \frac{2y}{h}\right)^2 \\
\sigma_{xy} &= \frac{VQ}{I}
\end{aligned} \tag{2.65}$$

The two terms in red are the same as the solution for the thin beam based on the Euler–Bernoulli theory. The elasticity theory gives the rest of the terms. The shear stress solution is exactly the same, the normal stress σ_{yy} is entirely additional, and the normal stress σ_{xx} has one additional term. For the σ_{xx} result at $x = 0$ and $y = h/2$, assuming $l = h$ (the beam span is twice its height), the additional term is 1/15 of the leading term (in red). If $l = 2h$, the same ratio becomes 1/60.

```
#               leading term           additional term
#σxx = p*y*(30*L**2-30*x**2    +    20*y**2-3*h**2 )/(5*h**3)
ratio = (20*y**2-3*h**2)/(30*l**2 - 30*x**2)
display(Math(f" \\text{{additional_term/leading_term ($l=h$)}}=\
                 {latex(ratio.subs({x:0, y:h/2}).subs({l:h}))}"))
display(Math(f" \\text{{additional_term/leading_term ($l=h$)}}=\
                 {latex(ratio.subs({x:0, y:h/2}).subs({l:2*h}))}"))
```

$$\text{additional\_term/leading\_term } (l = h) = \frac{1}{15}$$

$$\text{additional\_term/leading\_term } (l = h) = \frac{1}{60}$$

It is clear that when the beam is slender, the thin beam theory gives a better approximation.

2.8.5 *Plot of the stresses*

Let us plot the distribution of these stresses along the cross-section of the rectangle subjected to a distributed force on the top surface.

```
# Plot SymPy expressions (convert sympy functions to numpy ones):

fij = [1, 1, 1, "xl", "yl"]                          # define figure grids
#plt.ioff()
plt.ion()
fig_s = plt.figure(figsize=(4,2.5))

p_val = 1; l_val = 1; h_val = 1
print(f" p = {p_val}, l = {l_val}, h = {h_val}")
dic_ = {p:p_val, l:l_val, h:h_val, x:l_val/2}
np_σxx = sp.lambdify((y), σ[0].subs(dic_),'numpy')   # convert to numpy
np_σyy = sp.lambdify((y), σ[1].subs(dic_),'numpy')
np_σxy = sp.lambdify((y), σ[2].subs(dic_),'numpy')

labels=[r'$σ_{xx}$ at $x=l/2$', r'$σ_{yy}$ at $x=l/2$',
        r'$σ_{xy}$ at $x=l/2$']

y_data = np.linspace(-h_val/2, h_val/2, 400)            # y-axis in plot
f_data = [np_σxx(y_data), np_σyy(y_data), np_σxy(y_data)]

gr.plot_fij(fig_s, y_data, f_data, labels, '$y$', 'Stresses', fij)
plt.savefig('imagesSM/stresses_y.png',dpi=500,bbox_inches='tight')
plt.show()
```

```
p=1,l=1,h=1
```

Distributions of these stresses along the cross-section at $x = l/2$ are plotted in Fig. 2.8. It is seen that the largest stress is the normal stress σ_{xx}. It is not linearly distributed along the cross-section, which is different from the thin beam theory. The linear distribution is an underestimation of the stress level on both surfaces. The shear stress σ_{xy} has a parabolic distribution and is zero on both surfaces, which is the same as the result

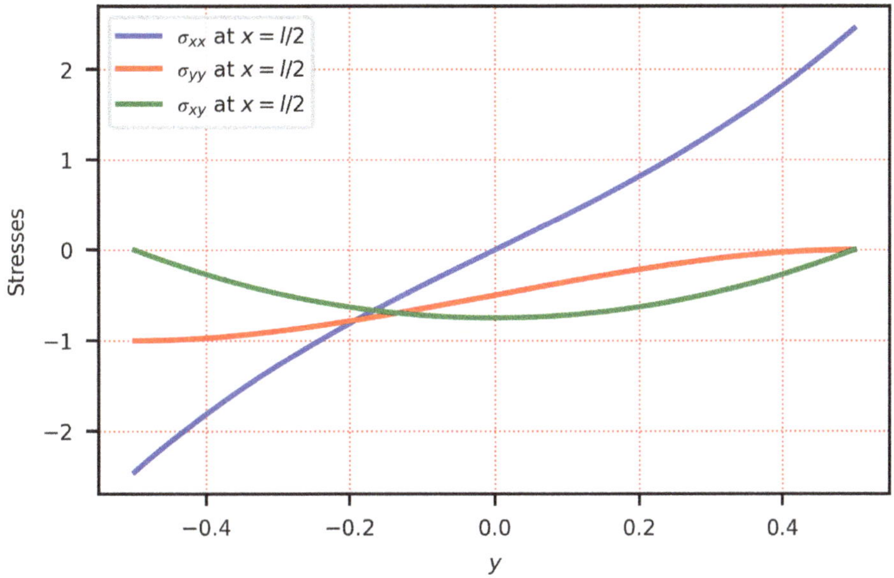

Figure 2.8. Stress distribution along the cross-section at $x = l/2$ in a 2D rectangle solid with dimensions $2l \times h$. It is subjected to a uniform vertical load on the top surface ($y = -h/2$). Parameters used are $p = 1, l = 1, h = 1$.

from the thin beam theory. Both stresses σ_{xy} and σ_{yy} have smaller values compared to σ_{xx}. The largest value of σ_{yy} is the applied vertical stress on the top surface, where the force is applied.

2.8.6 *Contour plot of the stresses*

The contour of the normal stress σ_{xx} in the simply supported beam can be plotted using the `field_plot()` function provided earlier:

```
# Re-define the stresses using the solution obtained:
x, y  = symbols('x, y')                          # define symbolic variables
X = sp.Matrix([x, y])

σ = sp.Matrix([[p*y*(-3*h**2 + 30*l**2 - 30*x**2 + 20*y**2)/(5*h**3)],\
               [p*(-h**3 + 3*h**2*y - 4*y**3)/(2*h**3)], \
               [3*p*x*(-h**2 + 4*y**2)/(2*h**3)]])
display(Math(f" \\text{{ Stresses}}={latex(σ.T)}"))

l_val = 1; h_val = 1
dict_= {p:1, l:l_val, h:h_val}
```

```
# Create mesh grid:
x_mesh, y_mesh = np.mgrid[-l_val:l_val:200j, -h_val/2:h_val/2:100j]
σxxi = sp.nsimplify(σ[0].subs(dict_))
display(Math(f" \\text{{$σ_{{xx}}(x,y)$ to be plotted}}=\
                                        \\scriptsize{latex(σxxi)}"))

plt.figure()
plt.ioff()
#plt.ion()                          # uncomment this to display the plots
field_plot(σxxi, X, x_mesh, y_mesh, title=' Stress $σ_{xx}$', flip='y')
plt.savefig('./imagesSM/SS_Rectangle_Sxx.png',dpi=500,bbox_inches='tight')
#plt.show();                        # uncomment this to display the plots
```

$$\text{Stresses} = \left[\frac{py\left(-3h^2+30l^2-30x^2+20y^2\right)}{5h^3} \quad \frac{p\left(-h^3+3h^2y-4y^3\right)}{2h^3} \quad \frac{3px\left(-h^2+4y^2\right)}{2h^3}\right]$$

$$\sigma_{xx}(x,y) \text{ to be plotted} = \frac{y\left(-30x^2+20y^2+27\right)}{5}$$

```
Value in the domain: Maximum=3.19992; Minimum=-3.19992
```

The contour of the normal stress σ_{xx} in the simply supported rectangle 2D solid with dimensions of $l \times h = 2 \times 1$ is plotted in Fig. 2.9. It is found that the normal stress σ_{xx} is zero at $x = \pm l$ and increases gradually (quadratically) with decreasing $|x|$, as expected. The highest level of normal stresses is on

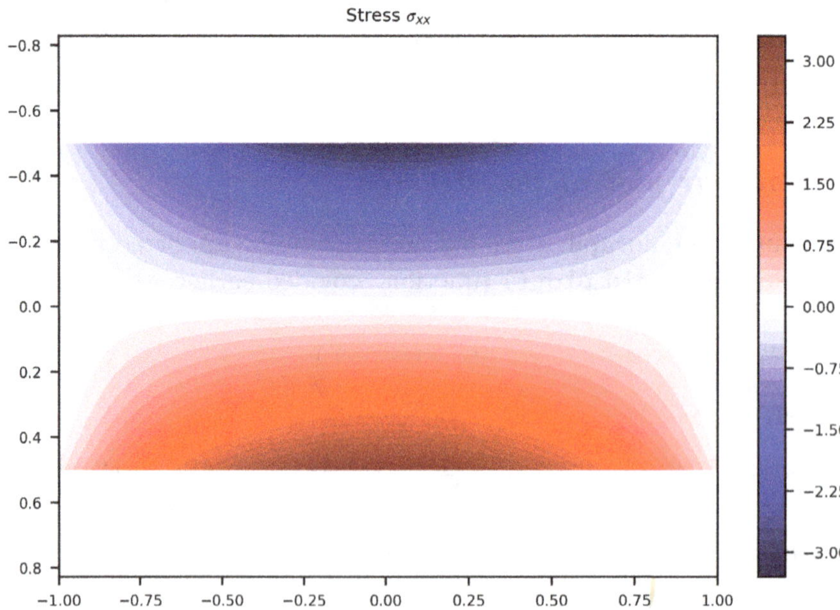

Figure 2.9. The contour of the normal stress solution in the simply supported rectangle 2D solid.

both the upper and lower surfaces. It is also observed that the normal stress σ_{xx} is symmetric with respect to $x = 0$ and antisymmetric with respect to $y = 0$.

Readers may plot the contour of the normal σ_{xy} by changing the code slightly.

2.8.7 *Strain solution*

Using the already obtained stresses and the constitutive Eq. (2.5), the strains in the rectangle solid can be computed using the following snippet:

```
# Redefine stress components as symbolic variables
display(Math(f" \\text{{ Stresses}}={latex(σ.T)}"))

# Formulas for strains (a plane stress problem):
εxx, εyy, γxy = symbols('ε_xx, ε_yy, γ_xy')

εxx = (1/E*(σ[0]-nu*σ[1])).expand().simplify()          # Compute strains
εyy = (1/E*(σ[1]-nu*σ[0])).expand().simplify()
γxy = (2*(1+nu)*σ[2]/E).expand().simplify()
ε = Matrix([εxx, εyy, γxy])
display(Math(f" \\text{{Strains ε}}={latex(ε)}"))
```

$$\text{Stresses} = \left[\frac{py\left(-3h^2+30l^2-30x^2+20y^2\right)}{5h^3} \quad \frac{p\left(-h^3+3h^2y-4y^3\right)}{2h^3} \quad \frac{3px\left(-h^2+4y^2\right)}{2h^3}\right]$$

$$\text{Strains } \varepsilon = \begin{bmatrix} \frac{p\left(5h^3\nu-3h^2y(5\nu+2)+20y\left(3l^2+\nu y^2-3x^2+2y^2\right)\right)}{10Eh^3} \\ \frac{p\left(-5h^3+3h^2y(2\nu+5)+20y\left(-3l^2\nu+3\nu x^2-2\nu y^2-y^2\right)\right)}{10Eh^3} \\ \frac{3px\left(-h^2+4y^2\right)(\nu+1)}{Eh^3} \end{bmatrix}$$

2.8.8 *Displacement solution*

As usual, before computing the displacements, let us check whether the given strains satisfy the compatibility Eq. (2.8), which is a necessary condition for the existence of a displacement field for the given strain field.

```
comptblty = (εxx.diff(y,2)+εyy.diff(x,2)-γxy.diff(x).diff(y)).simplify()
print(f"Is compatibility condition satisfied? {comptblty==0}")
```

```
Is compatibility condition satisfied? True
```

The compatibility condition is satisfied, hence we can now proceed to compute the displacements.

Using the strain–displacement relation Eq. (2.7) we can obtain the displacements via integrations. The procedure is similar to that used in the previous example. Let us integrate the first two equations in Eq. (2.7):

```
# find the displacements via integrations
f = sp.Function('f')(y)                        # prepare an integral function

u_with_f = sp.integrate(ε[0],(x)) + f  # integral constant w.r.t x: f(y)
u_with_f = u_with_f.expand().doit()
display(Math(f" \\text{{ $u$ with f}}={latex(u_with_f)}"))

g = sp.Function('g')(x)

v_with_g = sp.integrate(ε[1],(y)) + g  # integral constant w.r.t y: g(x)
v_with_g = v_with_g.expand().doit()
display(Math(f" \\text{{ $v$ with g}}={latex(v_with_g)}"))
```

$$\text{u with } f = f(y) + \frac{\nu px}{2E} - \frac{3\nu pxy}{2Eh} - \frac{3pxy}{5Eh} + \frac{6l^2pxy}{Eh^3} + \frac{2\nu pxy^3}{Eh^3} - \frac{2px^3y}{Eh^3} + \frac{4pxy^3}{Eh^3}$$

$$\text{v with } g = g(x) - \frac{py}{2E} + \frac{3\nu py^2}{10Eh} + \frac{3py^2}{4Eh} - \frac{3l^2\nu py^2}{Eh^3} + \frac{3\nu px^2y^2}{Eh^3} - \frac{\nu py^4}{Eh^3} - \frac{py^4}{2Eh^3}$$

Two unknown functions are produced. Now, substituting u and v into the third equation in Eq. (2.7) gives

```
expr = (diff(u_with_f,y) + diff(v_with_g,x) - ε[2]).expand()
display(Math(f" \\text{{ Expression from the strain relation}}=\
                                                    {latex(expr)}"))
```

Expression from the strain relation

$$= \frac{d}{dy}f(y) + \frac{d}{dx}g(x) + \frac{3\nu px}{2Eh} + \frac{12px}{5Eh} + \frac{6l^2px}{Eh^3} - \frac{2px^3}{Eh^3}$$

The above equation must be zero. We regroup the terms in it as follows:

$$\underbrace{\frac{d}{dy}f(y)}_{F(y)} + \underbrace{\frac{d}{dx}g(x) + \frac{3\nu px}{2Eh} + \frac{12px}{5Eh} + \frac{6l^2px}{Eh^3} - \frac{2px^3}{Eh^3}}_{G(x)} = 0 \tag{2.66}$$

It is seen that $F(y)$ is a function of only y, and $G(x)$ is a function of only x. However, their sum must be zero. This is not possible unless $F(y)$ and $G(x)$ are constants. We thus denote $F(y) = c_F$ and $G(x) = c_G$. This gives us the following three separate equations:

$$\frac{d}{dx}g(x) + \frac{3\nu px}{2Eh} + \frac{12px}{5Eh} + \frac{6l^2px}{Eh^3} - \frac{2px^3}{Eh^3} = c_G \tag{2.67}$$

$$\frac{d}{dy}f(y) = c_F \tag{2.68}$$

$$c_F + c_G = 0 \tag{2.69}$$

Equations (2.67) and (2.68) are 1D DEs, each with only one unknown function, and Eq. (2.69) is a simple algebraic equation. We first integrate these two DEs. We begin by identifying the terms in the expression:

```
# Use sp.args() to locate the terms in the expression:
display(Math(f" \\text{{ Terms in the expression:}}{latex(expr.args)}"))
```

Terms in the expression: $\left(-\frac{2px^3}{Eh^3}, \frac{12px}{5Eh}, \frac{6l^2px}{Eh^3}, \frac{3\nu px}{2Eh}, \frac{d}{dy}f(y), \frac{d}{dx}g(x)\right)$

Integration of Eq. (2.67) with respect to x:

```
c0, c1, cF, cG = symbols('c0, c1, c_F, c_G')          #integral constants

# find g(x) via integration:
integrand = cG-expr.args[0]-expr.args[1]-expr.args[2]-expr.args[3]
g = sp.integrate(integrand, (x)) + c0          # add an integral constant
g = g.expand().doit()
display(Math(f" \\text{{ Function $g(x)$}}={latex(g)}"))
```

Function $g(x) = c_0 + c_G x - \frac{3\nu px^2}{4Eh} - \frac{6px^2}{5Eh} - \frac{3l^2px^2}{Eh^3} + \frac{px^4}{2Eh^3}$

Integration of Eq. (2.68) with respect to y:

```
# find f(y) via integration:
integrand = cF
f = sp.integrate(integrand,(y)) + c1
f = f.expand().doit()
display(Math(f" \\text{{ Function $f(x)$}}={latex(f)}"))
```

Function $f(x) = c_1 + c_F y$

Now, form the general expression for displacement u:

```
u = sp.integrate(ε[0], (x)) + f
display(Math(f" \\text{{ $u$}}=\\scriptsize {latex(u)}"))
```

$$u = c_1 + c_F y - \frac{2px^3y}{Eh^3} + \frac{x\left(5h^3\nu p - 15h^2\nu py - 6h^2py + 60l^2py + 20\nu py^3 + 40py^3\right)}{10Eh^3}$$

Form the expression for displacement v:

```
v = sp.integrate(ε[1], (y)) + g
display(Math(f" \\text{{ $v$}}=\\scriptsize {latex(v)}"))
```

$$v = c_0 + c_G x - \frac{py}{2E} - \frac{3\nu px^2}{4Eh} - \frac{6px^2}{5Eh} - \frac{3l^2px^2}{Eh^3} + \frac{px^4}{2Eh^3} + \frac{y^4\left(-2\nu p - p\right)}{2Eh^3} + \frac{y^2 \cdot \left(6h^2\nu p + 15h^2p - 60l^2\nu p + 60\nu px^2\right)}{20Eh^3}$$

Determine displacements using BCs

We have obtained the general solution for displacements u and v, with the integral constants c_0 and c_1 and the unknown constants c_F and c_G. There are a total of four unknown constants. Thus, we need DBCs to determine these constants.

Based on the setting of our problem and Fig. 2.7, the boundary conditions are

$$u\big|_{x=-l} + u\big|_{x=l} = 0; \quad v\big|_{x=-l} = 0; \quad v\big|_{x=l} = 0 \tag{2.70}$$

This gives us three conditions. Along with Eq. (2.69), we have the right number of equations to determine these four constants. Let us write a SymPy code to so:

```
cs=sp.solve([u.subs({x:l,y:0})+u.subs({x:-l,y:0}), v.subs({x:l,y:0}),
                          v.subs({x:-l,y:0}),cF+cG], [c0, c1, cF, cG])
display(Math(f" \\text{{ Constants found}}: {latex(cs)}"))
```

Constants found: $\left\{c_0 : \frac{15h^2l^2\nu p + 24h^2l^2p + 50l^4p}{20Eh^3},\ c_1 : 0,\ c_F : 0,\ c_G : 0\right\}$

Since c_0 is the constant term for the deflection v, we can assign it to a variable v_0 to shorten the expressions.

```
v0 = sp.symbols('v0')
v0_ = cs[c0]                          # use v0_ to store the value for v0
cs[c0] = v0
display(Math(f" \\text{{ Constants found}}: {latex(cs)}"))
```

Constants found: $\{c_0 : v_0,\ c_1 : 0,\ c_F : 0,\ c_G : 0\}$

```
# Substitute the constants back to the displacement expression:
u_final = u.subs(cs).expand().simplify()
display(Math(f" \\text{{ Solution $u(x,y)$}}=\
                                        {latex(u_final)}"))
print(f'Is εxx=∂u/∂x? {(ε[0]-diff(u_final,x)).simplify()==0}')  # check

v_final = v.subs(cs).expand().simplify()
display(Math(f" \\text{{ Solution $v(x,y)$}}={latex(v_final)}"))
print(f'Is εyy=∂v/∂y? {(εyy-diff(v_final,y)).simplify()==0}')   # check
print(f'γ_xy=∂u/∂y+∂v/∂x? ' \
    f'{(ε[2].expand()-diff(u_final,y)-diff(v_final,x)).simplify()==0}')
```

$$\text{Solution } u(x,y) = \frac{px\left(5h^3\nu - 3h^2y\left(5\nu+2\right) + 20y\left(3l^2+\nu y^2 - x^2 + 2y^2\right)\right)}{10Eh^3}$$

```
Is εxx=∂u/∂x? True
```

$$\text{Solution } v(x,y) = \frac{Eh^3v_0 - \frac{h^3py}{2} + \frac{3h^2p\left(-5\nu x^2+2\nu y^2-8x^2+5y^2\right)}{20} + \frac{p\left(-6l^2\nu y^2-6l^2x^2+6\nu x^2y^2-2\nu y^4+x^4-y^4\right)}{2}}{Eh^3}$$

```
Is εyy=∂v/∂y? True
γ_xy=∂u/∂y+∂v/∂x? True
```

```
# For expression simplification only:
u_numer = sp.numer(u_final)
display(Math(f"\\text{{Numer of $u$ }}=\\scriptsize {latex(u_numer)}"))
v_numer = sp.numer(v_final)
display(Math(f"\\text{{20×$v$Numer}}=\\scriptsize {latex(20*v_numer)}"))
```

$$\text{Numer of } u \ = px\left(5h^3\nu - 3h^2y\left(5\nu+2\right) + 20y\left(3l^2+\nu y^2-x^2+2y^2\right)\right)$$

$$20\times v\text{Numer} = 20Eh^3v_0 - 10h^3py + 3h^2p\left(-5\nu x^2+2\nu y^2-8x^2+5y^2\right) + 10p\left(-6l^2\nu y^2-6l^2x^2+6\nu x^2y^2-2\nu y^4+x^4-y^4\right)$$

The final solution for the displacements can now be rearranged as follows:

$$\begin{aligned} u(x,y) &= \frac{px}{10Eh^3}\left(5h^3\nu - 3h^2y\left(5\nu+2\right) + 20y\left(3l^2+\nu y^2-x^2+2y^2\right)\right) \\ v(x,y) &= v_0 - \frac{p}{20Eh^3}\Bigg(-10h^3py + 3h^2\left(-5\nu x^2+2\nu y^2-8x^2+5y^2\right) \\ &\quad + \left(-60l^2\nu y^2-60l^2x^2+60\nu x^2y^2-10\nu y^4+20x^4-20y^4\right)\Bigg) \end{aligned} \tag{2.71}$$

We have obtained all the solutions for stresses, strains, and displacements. These are functions of x and y. Let us check the solutions at some critical locations:

```
# Along neutral axis, $y=0$, the deflection (displacement v):
display(Math(f" \\text{{ $v(x,0)$ along $y=0$ is}}: \
                    {latex(v_final.subs(y,0).subs(v0,v0_).expand())}"))
# At $x=0, y=0$, the deflection:
display(Math(f" \\text{{ $v(0,0)$ at tip of the beam is}}:\
      \\small {latex(v_final.subs({x:0,y:0}).subs(v0,v0_).expand())}"))
```

$$v(x,0) \text{ along } y=0 \text{ is: } \frac{3l^2\nu p}{4Eh}+\frac{6l^2p}{5Eh}-\frac{3\nu px^2}{4Eh}-\frac{6px^2}{5Eh}+\frac{5l^4p}{2Eh^3}-\frac{3l^2px^2}{Eh^3}+\frac{px^4}{2Eh^3}$$

$$v(0,0) \text{ at tip of the beam is: } \frac{3l^2\nu p}{4Eh}+\frac{6l^2p}{5Eh}+\frac{5l^4p}{2Eh^3}$$

One may recognize that the term $\frac{5l^4p}{2Eh^3}$ in the tip deflection is, in fact, the same as that for the thin beam based on the Euler–Bernoulli assumption. The displacements along the right edge are given as follows:

```
# at x = l, the displacement u:
display(Math(f" \\text{{ $u(l,y)$ along beam's root edge}}:\
                                    {latex(u_final.subs({x:l}))}"))
# at x = l, the displacement v:
display(Math(f" \\text{{ $v(l,y)$ along beam's root edge}}:\
                                    {latex(v_final.subs({x:l}))}"))
```

$u(l,y)$ along beam's root edge:

$$\frac{lp\left(5h^3\nu-3h^2y\left(5\nu+2\right)+20y\left(2l^2+\nu y^2+2y^2\right)\right)}{10Eh^3}$$

$v(l,y)$ along beam's root edge:

$$\frac{Eh^3v_0-\frac{h^3py}{2}+\frac{3h^2p\left(-5l^2\nu-8l^2+2\nu y^2+5y^2\right)}{20}+\frac{p\left(-5l^4-2\nu y^4-y^4\right)}{2}}{Eh^3}$$

Readers may plot the results using the same codes given in the previous section.

2.9 A general stress method for rectangular 2D solids

In the previous sections, basis functions for the stress functions are carefully chosen. When a good choice can be found, it can drastically reduce the labor in determining these unknown constants and deriving necessary formulas.

It is particularly helpful when all these need to be done manually. On the other hand, making such a choice can be difficult at times. Therefore, the question is naturally: can we simply use all the BBFs below a certain order and then find the solution? The answer is yes, and it can be a preferred method, using symbolic computations.

In addition, each time we create expressions for a condition that needs to be satisfied, we need to make sure the coefficients of all orders of monomials (of x or y) must be zero, and manually create these expressions. This can also be automated. Python functions will be written for this purpose in this section.

We will demonstrate this **general method** by considering a more challenging problem: a dam subjected to water pressure and gravity forces.

2.9.1 *Setting of the problem*

Consider a dam with a rectangular shape cross-section, which is constrained at the bottom. The dam is subjected to water pressure $p(x)$ on the left surface and the gravity force ρg over the entire body, as shown in Fig. 2.10. The top, left, and right surfaces are assumed free. Since the dam is very long along the z-axis, this problem should be treated as a plane strain problem.

2.9.2 *Stress function construction*

Since the body force is constant, the stress function needs to satisfy the biharmonic Eq. (2.29). For the problem of a cantilever beam subjected to a concentrated load at the tip, we used BBFs up to the fourth order. Since the water pressure on the dam is linear and distributed, we shall choose higher-order basis functions up to the sixth order (total order). In this general method, we use all these terms up to the sixth order given in the following without hand-picking, which gives a total of 19 basis functions as follows:

$$\begin{aligned}
&1,\ x,\ y,\ xy,\ x^2,\ y^2,\ x^3,\ yx^2,\ xy^2,\ y^3 && \leq \text{third-order}\\
&(3x^2y^2 - x^4),\ x^3y,\ xy^3,\ (3x^2y^2 - y^4),\ (x^4 - y^4) && \text{fourth-order}\\
&(5x^2y^3 - y^5),\ (x^3y^2 - xy^4),\ (5x^4y - y^5),\ (x^5 - 5xy^4) && \text{fifth-order}\\
&(5x^3y^3 - 3x^5y),\ (x^6 - 10x^4y^2 + 5x^2y^4),\ (xy^5 - x^5y), &&\\
&\quad (2x^6 - 15x^4y^2 + y^6) && \text{sixth-order}
\end{aligned} \tag{2.72}$$

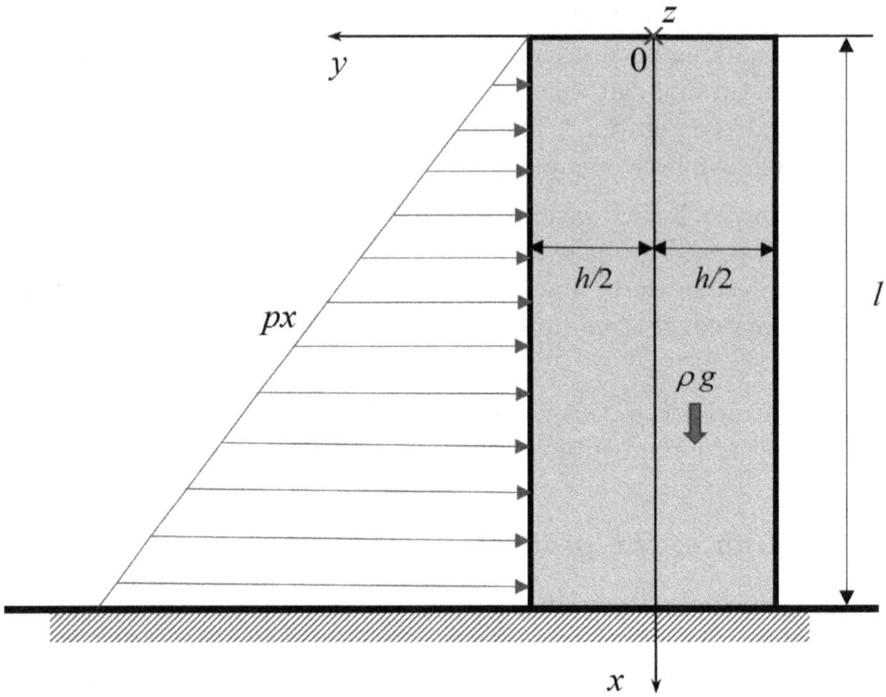

Figure 2.10. A 2D rectangular dam with dimensions $l \times h$, subjected to water pressure and gravity force over the body.

Note that we excluded all terms up to linear order because these do not contribute to the stresses. The stress function is expressed as the linear combination of all these 19 BBFs:

$$\phi(x, y) = a_0 xy + a_1 x^2 + \cdots + a_{18}\left(x^6 - 15x^2y^4 + 2y^6\right) \tag{2.73}$$

where a_i $(i = 0, 1, \ldots, 18)$ are unknown constants. These are 19 constants that need to be determined. Note that in this general approach, we can use more BBFs, but not fewer. On the other hand, too many extra terms can lead to much extra work.

Since we are using only BBFs, ϕ given in Eq. (2.73) automatically satisfies Eq. (2.29). The stress function is defined automatically without the need of typing using the following snippet. Note that the code cells given in this section need to be run sequentially because of the recursive nature of some of the variables. This mean that when a code cell needs a rerun, we shall rerun some of the code cells above it in this section.

```
# define symbolic variables involved in this problem:
E, G, I, h, l, nu, ρg = sp.symbols('E,G,I,h,l,nu,ρg', positive=True)
x, y, p = symbols('x, y, p', real=True)
X = sp.Matrix([x, y])

# unknown constants for BBFs:
a = sp.symbols("a0:19")                              # unknown constants
order = 6                                     # order of biharmonic basis
n_terms = 7 + (order-3)*4                          # total terms of BBFs
ai = Matrix([a[i] for i in range(n_terms)])     # form coeffs. in vector
n_ai = len(ai)
print(f"Total number of unknown constants = {n_ai}")

# Automatically create BBFs up to order 6:
BBFs = Matrix([x*y, x**2, y**2, x**3, x**2*y, x*y**2, y**3])
for i in range(3+1, order+1):
    bbfs = gr.create_nbf(i, X, dfn=4, prnt=False)   # BBFs with order-i.
    BBFs = BBFs.col_join(bbfs)                     # Joining side by side

ϕ0 = ai.T@BBFs                      # initially assumed stress function: ϕ0
d4ϕ = gr.nabla_f(ϕ0, X, dfn=4, smplfy=2)[0]       # Check if ∇^4 ϕ = 0.
display(Math(f" \\text{{Is $∇^4 ϕ$ = 0? }} {latex(d4ϕ==0)}"))

ϕ0 = ϕ0[0]                                             # get the function
#display(Math(f"\\text{{Initial ϕ}}={latex(ϕ0)}"))               # Lengthy
```

```
Total number of unknown constants = 19
```

Is $\nabla^4 \phi = 0$? True

2.9.3 *General stress solution*

Using the stress function given in Eq. (2.73) and Eq. (2.22), the general expressions for stresses can be computed using

```
σxx, σxy, σyy = symbols('σ_xx, σ_xy, σ_yy')
σxx =    ϕ0.diff(y,2) - ρg*x
σyy =    ϕ0.diff(x,2) - ρg*x
σxy = -(ϕ0.diff(x,1)).diff(y,1)

σ0 = Matrix([σxx, σyy, σxy])       # Initial stresses with constants a_i
#display(Math(f" \\text{{ Vector of stresses}}={latex(σ0)}")) # Lengthy
```

Note that we included the stresses from the gravity body force, which is constant and independent of the stress function and which were already obtained in Section 2.3.

Next, we shall use available conditions of only mechanics behavior to determine these constants. This is done step-by-step for different types of conditions, including stress conditions on each of the boundary surfaces, equilibrium conditions, and the assurance of the biharmonic conditions for the stress function.

Step-1: Stress boundary conditions for σ_{xx} on the top edge

For our cantilever dam, the top edge ($x = 0$) is free of normal stress σ_{xx}:

$$\sigma_{xx}\big|_{x=0} \quad \text{for all } y \tag{2.74}$$

Let us compute σ_{xx} along $x = 0$:

```
# SBC for σxx:
σxx_x0 = σ0[0].subs(x,0).expand()
display(Math(f" \\text{{ At $x=0$,  $\\;σ_{{xx}}$}}=\
                                        {latex(σxx_x0)}"))  #σ_{xx} at x=0
```

At $x = 0$, $\sigma_{xx} = -12a_{10}y^2 - 4a_{11}y^3 - 4a_{13}y^3 + 6a_{16}y^4 + 60a_{18}y^4 + 2a_2 + 6a_6y - 4a_8y^2$

Since the expression of $\sigma_{xx}\big|_{x=0}$ is a polynomial function of y, we purposely group them according to the order of the monomials of y:

```
# collect the coefficients of the monomials in the expression:
y_monomials = [y**4, y**3, y**2, y]                            # monomials
coef_y =[σxx_x0.coeff(m) for m in y_monomials]                # get coeffs.
coef_y.append(σxx_x0.coeff(y,0))

display(Math(f" \\text{{ $c_{{y4}}$}}={latex(coef_y[0])}; \
                \\text{{ $c_{{y3}}$}}={latex(coef_y[1])}; \
                \\text{{ $c_{{y2}}$}}={latex(coef_y[2])}"))
display(Math(f" \\text{{ $c_{{y1}}$}}={latex(coef_y[3])}; \
                \\text{{ $c_{{y0}}$}}={latex(coef_y[4])}"))
```

$c_{y4} = 6a_{16} + 60a_{18}; c_{y3} = -4a_{11} - 4a_{13}; c_{y2} = -12a_{10} - 4a_8$

$c_{y1} = 6a_6; c_{y0} = 2a_2$

In order to satisfy Eq. (2.74), the coefficients of all these monomials of y must be zero. This gives us five conditions to determine the constants. The snippet is as follows:

```
# coef_y is now a list of conditions and can be use to solve of ai:
print(f" Number of conditions = {len(coef_y)}")
dic_ai = sp.solve(coef_y, ai)         # solve for some unknown constants
                                  # dic_ai: a dictionary holds the current ai
n_ai = n_ai - len(coef_y)                    # update number of unknowns
print(f" Current number of unknowns = {n_ai}")

σ_ai = σ0.subs(dic_ai)                  # update stresses with current ai
#display(Math(f" \\text{{ Updated stresses}}={latex(σ_ai)}")) # lengthy

ϕ_ai = ϕ0.subs(dic_ai).simplify()            # update ϕ with current ai
#display(Math(f" \\text{{ Updated ϕ}}={latex(ϕ_ai)}"))          # lengthy

# Check whether ϕ_ai satisfies ∇^4 ϕ = 0:
d4ϕ_ai = gr.nabla_f(ϕ_ai, X, dfn=4, smplfy=2).simplify()
display(Math(f" \\text{{ $ ^4ϕ$ = 0?}}={latex(d4ϕ_ai == 0)}"))
```

```
Number of conditions = 5
Current number of unknowns = 14
```

$\nabla^4\phi = 0? = \text{True}$

We purposely commented out the code lines for displaying the new updated stresses and stress function because they are too long. Readers may uncomment these code lines to view them. Note that it is a good practice to track the biharmonic conditions for ϕ because during the process of determining the constants, the terms are frequently altered, leading to the possibility of violation of this condition. We shall reinforce it when it happens.

These stresses still have 14 unknown constants to determine. We move to the next steps using other conditions.

Step-2: Stress boundary conditions on the left and right edges

We now use the SBCs for boundaries along the left and right edges of the dam (or surfaces), which are expressed as

$$\begin{aligned} \sigma_{yy}\big|_{y=h/2} &= -px; \qquad \sigma_{xy}\big|_{y=h/2} = 0 \qquad \text{on left edge} \\ \sigma_{yy}\big|_{y=-h/2} &= 0; \qquad \sigma_{xy}\big|_{y=-h/2} = 0 \qquad \text{on right edge} \end{aligned} \tag{2.75}$$

This gives another set of four conditions, which can be used to further reduce the constants.

2.9.4 *Python code for finding unknowns*

Since the procedure is the same as that in Step-1, we can write a Python function to do this, which can then be used for future steps. The snippet of this function is self-explanatory:

```
def unknownFinder(conditions, ai, dic_ai, ϕ0, σ0, prnt=True):
    '''Find unknowns for a given set of, `conditions` (list);
       updated unknowns, `dic_ai` (dictionary); compute the current
       stresses, `σ_ai` (sympy Matrix); and stress function, `ϕ_ai`.
       (sympy function)
       return: ϕ_ai, σ_ai, d4ϕ_ai, dic_ai
    '''
    sln_ai = sp.solve(conditions, ai)
    if len(sln_ai) == 0: print('Alert: no solution found!')

    dic_ai.update(sln_ai)          # update the dictionary of constants
    dic_ai={k:v.subs(dic_ai).simplify() for k,v in list(dic_ai.items())}
    σ_ai = sp.simplify(σ0.subs(dic_ai))       # update stresses with ai
    ϕ_ai = ϕ0.subs(dic_ai).simplify()                # update ϕ with ai

    if prnt:                          # print the following if needed
        display(Math(f"\\text{{Solutions for $a_i$}}={latex(dic_ai)}"))
        display(Math(f"\\text{{Current σ}}={latex(σ_ai)}"))
        display(Math(f"\\text{{Current ϕ}}={latex(ϕ_ai)}"))

    # Check whether ϕ_ai satisfies ∇^4 ϕ = 0:
    d4ϕ_ai = gr.nabla_f(ϕ_ai, X, dfn=4, smplfy=2).simplify()
    display(Math(f" \\text{{ $∇^4ϕ$ = 0?}}={latex(d4ϕ_ai == 0)}"))
    display(Math(f" \\text{{ $∇^4ϕ$}}={latex(d4ϕ_ai)}"))
    return ϕ_ai, σ_ai, d4ϕ_ai, dic_ai
```

We can now create any set of conditions and run this Python function `unknownFinder()`. This automatically updates all the unknown constants, stresses, and the stress function. In addition, we write the following Python function, `getExprs()`, to form the expressions of coefficients of a monomial with order less than or equal to a certain value in a given set of equations. This can produce a large number of expressions, many of which can be zero. Zero expression removal is also performed in `getExprs()`.

```
def getExprs(eqns, x, order):
    '''Get the expressions for a given list of 1D sympy `eqns`.
    return: exprs - expressions of the coefficients of all orders of x.
    '''
    exprs = []
    for eqn in eqns:
        for i in range(order+1):
            exprs.append(eqn.expand().coeff(x,i))

    exprs = gr.remove_zero_exprs(exprs)
    return exprs
```

```
# SBCs at Left and Right edges:
bc1 = σ_ai[1].subs(y, h/2) + p*x
bc2 = σ_ai[1].subs(y,-h/2)
bc3 = σ_ai[2].subs(y, h/2)
bc4 = σ_ai[2].subs(y,-h/2)

eqns = [bc1, bc2, bc3, bc4]                          # put is in a list
SBC_exprs = getExprs(eqns, x, order)        # get expressions upto order

ϕ_ai, σ_ai, d4ϕ_ai, dic_ai = \
            unknownFinder(SBC_exprs, ai, dic_ai, ϕ0, σ0, prnt=False)

display(Math(f" \\text{{$ϕ$}}=\small {latex(ϕ_ai)}"))
display(Math(f" \\text{{Current $σ$}}={latex(σ_ai)}"))
```

```
Total number of expressions = 28
Number of nonzero expressions = 20
```

$\nabla^4\phi = 0? = \text{True}$

$\nabla^4\phi = 0$

$$\phi = \frac{x(h^3 \cdot (240a_7y^3 - 15hy(12a_7h - p) + x^2(-20p + 40\rho g)) - 60h^2px^2y + 16py^3 \cdot (5x^2 - 3y^2))}{240h^3}$$

$$\text{Current } \sigma = \begin{bmatrix} = \dfrac{x\left(h^3 \cdot (6a_7y - \rho g) + 2py\left(x^2 - 2y^2\right)\right)}{h^3} \\ \dfrac{px\left(-h^3 - 3h^2y + 4y^3\right)}{2h^3} \\ \dfrac{h^3\left(-48a_7y^2 + h\left(12a_7h - p\right)\right) + 12h^2px^2 - 16py^2 \cdot \left(3x^2 - y^2\right)}{16h^3} \end{bmatrix}$$

Step-3: Restore biharmonic condition (if needed)

However, when this set of conditions is imposed, the stress function may fail to satisfy the biharmonic equation. If this happens, we need to reinforce the biharmonic equation again. In this example, the biharmonic condition is satisfied, and nothing needs to be done.

Step-4: Impose integral equilibrium conditions (FBCs)

These stresses still contain one constant, a_7. This time, we shall use the (integral) equilibrium conditions, or FBCs. Consider a free-body diagram of the top part of the dam cut at any x location, as shown in Fig. 2.11. The surface at the bottom cut of the free body has a distributed normal stress σ_{xx}, which contributes to the force in the x-direction and a moment with respect to the z-axis. The distributed shear stress σ_{xy} contributes to the force in the y-direction.

The equilibrium requires all the forces in the x-direction being zero, all the forces in the y-direction being zero, and all the moments with respect to

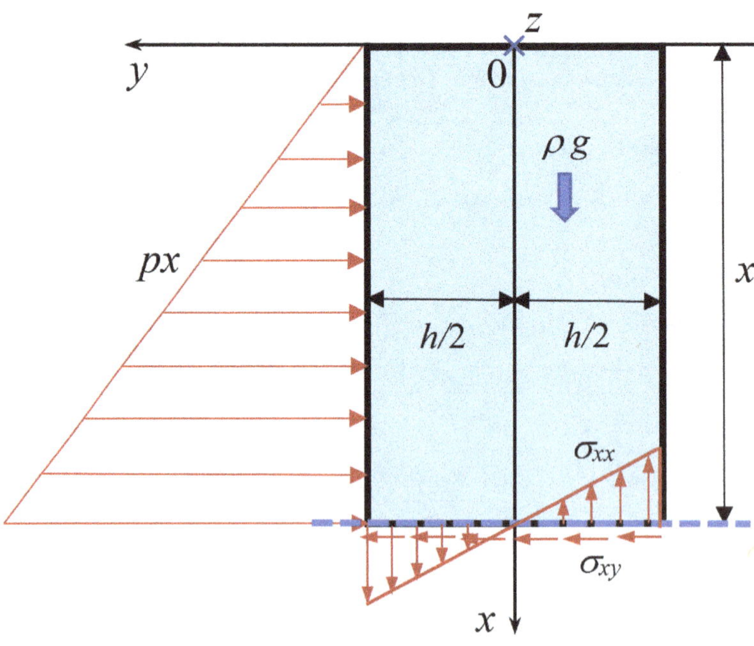

Figure 2.11. A free-body diagram (FBD) of the upper part of the 2D rectangular dam with height x. It is subjected to the water pressure and gravity force. The distribution of stresses is schematically shown on the bottom cut surface.

the z-axis at point $(x, 0)$ being zero:

$$\begin{aligned}
\sum F_x = 0: &\quad \int_{-h/2}^{h/2} \sigma_{xx} dy + hx\rho g = 0 \\
\sum F_y = 0: &\quad \int_{-h/2}^{h/2} \sigma_{xy} dy - \frac{px^2}{2} = 0 \\
\sum M_z = 0: &\quad \int_{-h/2}^{h/2} \sigma_{xx} y dy - \frac{px^3}{6} = 0
\end{aligned} \tag{2.76}$$

This gives us another set of three conditions, which can be used to determine the remaining constant. We will again use **unknownFinder()** to solve for this:

```
# The set of equilibrium conditions:
bcFx = sp.integrate(σ_ai[0],   (y, -h/2, h/2)) + h*x*ρg
bcFy = sp.integrate(σ_ai[2],   (y, -h/2, h/2)) - p*x**2/2
bcMz = sp.integrate(σ_ai[0]*y, (y, -h/2, h/2)) - p*x**3/6
eqns = [bcFx, bcFy, bcMz]                          # put to a list
FBC_exprs = getExprs(eqns, x, 4)

ϕ_ai, σ_ai, d4ϕ_ai, dic_ai = \
            unknownFinder(FBC_exprs, ai, dic_ai, ϕ0, σ0, prnt=False)
display(Math(f" \\text{{ Final ϕ}}:{latex(ϕ_ai)}"))
display(Math(f" \\text{{Final σ}}={latex(σ_ai.expand())}"))
```

```
Total number of expressions = 15
Number of nonzero expressions = 2
```

$\nabla^4 \phi = 0? = \text{True}$

$\nabla^4 \phi = 0$

Final ϕ:

$$\frac{x(h^3(-3hpy + 20x^2(-p + 2\rho g)) + 12h^2py(-5x^2 + 2y^2) + 16py^3 \cdot (5x^2 - 3y^2))}{240h^3}$$

$$\text{Final } \sigma = \begin{bmatrix} -x\rho g + \frac{3pxy}{5h} + \frac{2px^3y}{h^3} - \frac{4pxy^3}{h^3} \\ -\frac{px}{2} - \frac{3pxy}{2h} + \frac{2pxy^3}{h^3} \\ \frac{hp}{80} + \frac{3px^2}{4h} - \frac{3py^2}{10h} - \frac{3px^2y^2}{h^3} + \frac{py^4}{h^3} \end{bmatrix}$$

Finally, we have obtained the stress function and all corresponding stresses without any unknown constants. All the relevant nonzero constants a_i are now determined. The nonzero values are given as follows:

```
dic_ai = {k: v.subs(dic_ai).simplify() for k, v in list(dic_ai.items())}
dic_a = {key:value for key, value in list(dic_ai.items()) if value != 0}

display(Math(f" \\text{{Final nonzero constants}}:{latex(dic_a)}"))
```

Final nonzero constants:

$$\left\{a_0 : -\frac{hp}{80},\ a_{15} : \frac{p}{3h^3},\ a_3 : -\frac{p}{12} + \frac{\rho g}{6},\ a_7 : \frac{p}{10h},\ a_9 : -\frac{p}{4h}\right\}$$

2.9.5 *Final stress solution*

The final stress solution has been found and is rearranged as follows:

$$\begin{bmatrix}\sigma_{xx}\\ \sigma_{yy}\\ \sigma_{xy}\end{bmatrix} = \begin{bmatrix} -\rho g x + \frac{2px^3y}{h^3} + \frac{2p}{h^3}\left(\frac{3h^2xy}{10} - 2xy^3\right)\\ px\left(-\frac{3y}{2h} + \frac{2y^3}{h^3}\right) - \frac{px}{2}\\ \frac{3p}{10h}\left(\frac{h^2}{4} - y^2\right) + \frac{3px^2}{h^3}\left(\frac{h^2}{4} - y^2\right) - \frac{p}{h^3}\left(\frac{h^4}{16} - y^4\right)\end{bmatrix} \tag{2.77}$$

where ρ is the density of the dam material, g is the gravitational acceleration, and q is the weight per unit volume of the fluid. This is the expression presented in Ref. [11]. Now, we understand how it is derived using our general method.

It is straightforward to verify that the SBCs are all satisfied, except for the shear stress on the top of the dam. The nonzero shear stress there is not fully satisfied and is given by

$$\sigma_{xy}\big|_{x=0} = \frac{3p}{10h}\left(\frac{h^2}{4} - y^2\right) - \frac{p}{h^3}\left(\frac{h^4}{16} - y^4\right) \tag{2.78}$$

It is easy to confirm:

$$\int_{-h/2}^{h/2} \sigma_{xy}\big|_{x=0} dy = 0 \tag{2.79}$$

In addition, the values of the shear stress are quite small compared to the others, especially when h is small. However, caution is required when applying these formulas in actual design.

Note that if the pressure on the left edge of the rectangle is a constant p, we can impose a free shear stress condition on the top surface: $\sigma_{xy}\big|_{x=0} = 0$. In this case, $\sigma_{xx}\big|_{x=0}$ should remain undetermined in order to find the analytical closed-form solution. The sum of the normal force on the top surface will be zero: $\int_{-h/2}^{h/2} \sigma_{xx}\big|_{x=0}\, dy = 0$. Additionally, its values will be small when h is small. Readers are encouraged to explore this further.

Based on the author's experience, these formulas for stresses and displacements (to be derived in this section) are highly valuable for benchmarking numerical methods, as they represent an exact solution. For such benchmarking purposes, we can treat the nonzero shear stress on the top of the dam as externally applied distributed tractions. At the bottom of the dam, we can impose the DBC using the formulas to be derived below. With this setup, one can objectively evaluate the numerical solutions against the exact solution.

Now, let us redefine both the stress function and the stresses as known functions, so that we can proceed with our study without the need to rerun all the previous code cells.

```
# Re-define stress function explicitly for future use:

ϕ_ai_ = x*(h**3*(-3*h*p*y + 20*x**2*(-p + 2*ρg)) + 12*h**2*p*y*\
          (-5*x**2 + 2*y**2) + 16*p*y**3*(5*x**2 - 3*y**2))/(240*h**3)

# Check whether ϕ_ai satisfies ∇^4 ϕ = 0:
d4ϕ_ai = gr.nabla_f(ϕ_ai_, X, dfn=4, smplfy=2).simplify()
display(Math(f" \\text{{ $∇^4ϕ$ = 0?}}={latex(d4ϕ_ai==0)}"))
```

$\nabla^4\phi = 0? = \text{True}$

```
# Re-define symbolic variables involved in this problem:
E, G, I, h, l, nu, ρg = sp.symbols('E,G,I,h,l,nu,ρg', positive=True)
x, y, p = symbols('x, y, p', real=True)
X = sp.Matrix([x, y])

# Re-define the final stresses as known functions:
σxx = x*(-5*h**3*ρg + 3*h**2*p*y + 10*p*y*(x**2 - 2*y**2))/(5*h**3)
σyy = p*x*(-h**3 - 3*h**2*y + 4*y**3)/(2*h**3)
σxy = p*(h**4+h**2*(60*x**2-24*y**2)+y**2*(-240*x**2+80*y**2))/(80*h**3)
σ = Matrix([σxx, σyy, σxy])
```

We can use the following code to verify whether the obtained stresses satisfy the stress equilibrium and compatibility equations:

```
# Check stress equilibrium equation (self-weight body force):
print(f"div σ_x+ρg = 0? {(diff(σxx,x)+diff(σxy,y)).simplify()+ρg==0}")
print(f"div σ_y = 0?  {(diff(σxy,x) + diff(σyy,y)).simplify()==0}")

# Check compatibility equation:
compat = diff(diff(σxx,x),x)+diff(diff(σyy,y),y)+2*diff(diff(σxy,x),y)
print(f"Compatable? {compat.simplify()==0}")
```

```
div σ_x+ρg = 0? True
div σ_y = 0?  True
Compatible? True
```

We can also verify all these boundary conditions and the equilibrium conditions for the free-body diagram:

```
# Stresses on the left and right edges:
print(f"SBCs, left:{σ[1].subs(y,h/2),σ[1].subs(y,-h/2)}, right:",\
      f"{σ[2].subs(y, h/2).simplify(), σ[2].subs(y,-h/2).simplify()}")

σxy_x0 = σ[2].subs(x, 0).simplify().collect(y)     # on the top surface
display(Math(f" \\text{{At $x=0$, $σ_{{xy}}$ distribution}}=\
                                        {latex(σxy_x0)}"))

print(f"The integrals of the shear stress on the top surface: "
      f"{sp.integrate(σxy_x0,   (y, -h/2, h/2)).simplify()}, "
      f"{sp.integrate(σxy_x0*y, (y, -h/2, h/2)).simplify()}")

print(f"FBCs:")
print([sp.integrate(σ[0],   (y, -h/2, h/2)),\
       sp.integrate(σ[2],   (y, -h/2, h/2)),\
       sp.integrate(σ[0]*y, (y, -h/2, h/2))])
# should be: -h*x*ρg # p*x**2/2 #p*x**3/6
```

```
SBCs, left:(-p*x, 0), right: (0, 0)
```

$$\text{At } x=0,\ \sigma_{xy} \text{ distribution} = \frac{p\left(h^4 - 24h^2y^2 + 80y^4\right)}{80h^3}$$

```
The integrals of the shear stress on the top surface: 0, 0
FBCs:
[-h*x*ρg, p*x**2/2, p*x**3/6]
```

We observe that all the conditions are satisfied, except for the shear stress on the top surface of the dam. While it is nonzero at that location, both its integral and its first moment are zero.

We are now ready to determine the strain solutions.

2.9.6 *Final strain solution*

Using the stress solution, we can calculate the strain solutions by applying the constitutive equation for a plane strain problem, as shown in Eq. (2.6):

```
# Formulas for strains (a plane strain problem):
εxx, εyy, γxy = symbols('ε_xx, ε_yy, γ_xy')

εxx = (1/E*((1-nu**2)*σ[0]-nu*(1+nu)*σ[1])).expand().simplify()
εyy = (1/E*((1-nu**2)*σ[1]-nu*(1+nu)*σ[0])).expand().simplify()
γxy = (2*(1+nu)*σ[2]/E).expand().simplify()
ε = Matrix([εxx, εyy, γxy])

display(Math(f"\\text{{ε}}={latex(ε)}"))                    # strains

# Check the (strain) compatibility equation:
cmpt=(ε[0].diff(y,2) + ε[1].diff(x,2) - ε[2].diff(x).diff(y)).simplify()
display(Math(f" \\text{{ Strains are compatible? }}{latex(cmpt==0)}"))
```

$$\varepsilon = \begin{bmatrix} \frac{x\left(5h^3\left(\nu^2p+2\nu^2\rho g+\nu p-2\rho g\right)+3h^2py\left(3\nu^2+5\nu+2\right)+20py\left(-\nu^2x^2+\nu^2y^2-\nu y^2+x^2-2y^2\right)\right)}{10Eh^3} \\ \frac{x\left(5h^3\left(\nu^2p+2\nu^2\rho g+2\nu\rho g-p\right)+3h^2py\left(3\nu^2-2\nu-5\right)+20py\left(-\nu^2x^2+\nu^2y^2-\nu x^2+2\nu y^2+y^2\right)\right)}{10Eh^3} \\ \frac{p\left(h^4(\nu+1)+12h^2\cdot\left(5\nu x^2-2\nu y^2+5x^2-2y^2\right)+80y^2\left(-3\nu x^2+\nu y^2-3x^2+y^2\right)\right)}{40Eh^3} \end{bmatrix}$$

Strains are compatible? True

The strain compatibility condition is also satisfied, confirming that the stress and strain solutions are correct. This is a necessary step for finding the displacements based on this set of strains.

2.9.7 *General displacement solution*

With the strain solution obtained and the strain–displacement relations from Eq. (2.7), the displacements can be derived through integration. This process yields unknown functions, as shown earlier. The first two equations in Eq. (2.7) provide the displacements with these unknown functions:

```
# Derive formulas for displacements via integration:

f = sp.Function('f')(y)                              # unknown function of y
u_with_f = (sp.integrate(ε[0],(x)) + f).expand().doit()
#display(Math(f"\\text{{$u$ with $f(y)$}}={latex(u_with_f)}")) # lengthy

g = sp.Function('g')(x)                              # unknown function of x
v_with_g = (sp.integrate(ε[1],(y)) + g).expand().doit()
#display(Math(f"\\text{{$v$ with $g(x)$}}={latex(v_with_g)}")) # lengthy
```

2.9.8 *Final displacement solution*

2.9.8.1 *Determining the unknown functions*

Next, we substitute the displacement functions u and v into the strain-compatibility condition, given by the third equation in Eq. (2.7). The resulting expression is quite lengthy, so we simplify it using the following snippet:

```
expr3 = (diff(u_with_f, y)+diff(v_with_g, x) - ε[2]).expand()
expr_ = expr3.collect(x).collect(y)          # factor out the monomials
#display(Math(f" \\text{{Final compatibility Eq.}}={latex(expr_)}"))

cx4, cx2, cy4, cy2, cy=symbols('c_x4,c_x2,c_y4,c_y2,c_y') # new coeffs.
c = [cx4, cx2, cy4, cy2, cy]
monomials = [x**4, x**2, y**4, y**2, y]   # monomials in the expression
coeff_c=[expr3.collect(m).coeff(m) for m in monomials]#get coefficients

dic_0_3 =dict(list(zip(c[:3], coeff_c[:3])))  # dictionarys for coeffs.
dic_3_  =dict(list(zip(c[3:], coeff_c[3:])))
display(Math(f" \\small{latex(dic_0_3)}"))
display(Math(f" \\small{latex(dic_3_ )}"))

exprS = expr_.subs(list(zip(coeff_c,c)))        # Simplified expression
display(Math(f" \\text{{Compatibility Eq.}}=\\small {latex(exprS)}"))
```

$$\left\{c_{x2} : \frac{9\nu^2 p}{20Eh} - \frac{3\nu p}{4Eh} - \frac{6p}{5Eh},\ c_{x4} : -\frac{\nu^2 p}{2Eh^3} + \frac{p}{2Eh^3},\ c_{y4} : \frac{\nu^2 p}{2Eh^3} - \frac{\nu p}{Eh^3} - \frac{3p}{2Eh^3}\right\}$$

$$\left\{c_y : \frac{\nu^2 p}{2E} + \frac{\nu^2 \rho g}{E} + \frac{\nu \rho g}{E} - \frac{p}{2E},\ c_{y2} : \frac{9\nu^2 p}{20Eh} + \frac{3\nu p}{10Eh} - \frac{3p}{20Eh}\right\}$$

$$\text{Compatibility Eq.} = c_{x2}x^2 + c_{x4}x^4 + c_y y + c_{y2}y^2 + c_{y4}y^4 + \frac{d}{dy}f(y) + \frac{d}{dx}g(x) - \frac{h\nu p}{40E} - \frac{hp}{40E}$$

We obtained a significantly simplified expression. This strain-compatibility expression must equal zero. Following the same procedure as in the cantilever rectangle problem, we separate this expression into two sub-expressions: one that is a function of x, and the other that is a function of y. These terms can be identified using the `args` attribute:

```
exprS.args
```

$$\left(c_{x2}x^2,\ c_{x4}x^4,\ c_y y,\ c_{y2}y^2,\ c_{y4}y^4,\ -\frac{hp}{40E},\ -\frac{h\nu p}{40E},\ \frac{d}{dy}f(y),\ \frac{d}{dx}g(x)\right)$$

This expression is grouped into two parts: one that is a function of x only, denoted as $G(x)$, and the other that is a function of y only, denoted as $F(y)$. Each of these functions is a 1D differential equation and can be solved through integration. This process introduces new integral constants:

```
cF, cG, c0, c1 =symbols('cF, cG, c0, c1')  # variable for new constants

# The expression is grouped into two parts: G(x) is only function of x:
G = sp.integrate(cF-exprS.args[0]-exprS.args[1],(x)) + c0
G = G.expand().doit()
display(Math(f" \\text{{ Function of $x$ only $G(x)$}}={latex(G)}"))

# F(y) is only function of y:
F = sp.integrate(cG-exprS.args[2]-exprS.args[3]-exprS.args[4],(y)) + c1
F = F.expand().doit()
display(Math(f" \\text{{ Function of $y$ only $F(y)$}}={latex(F)}"))

# The expression for these two constants:
K = exprS.args[5]+exprS.args[6]
display(Math(f" \\text{{ Constants $K$}}={latex(K)}"))
```

$$\text{Function of } x \text{ only } G(x) = c_0 + cFx - \frac{c_{x2}x^3}{3} - \frac{c_{x4}x^5}{5}$$

$$\text{Function of } y \text{ only } F(y) = c_1 + cGy - \frac{c_y y^2}{2} - \frac{c_{y2}y^3}{3} - \frac{c_{y4}y^5}{5}$$

$$\text{Constants } K = -\frac{h\nu p}{40E} - \frac{hp}{40E}$$

Using $G(x)$ and $F(y)$, we form the general formulas for displacements:

```
# Compute the general formulas for displacements with new constants:
u_general_cs = (sp.integrate(ε[0],(x)) + F).expand().simplify()
#display(u_general_cs)                                        # Lengthy

v_general_cs = (sp.integrate(ε[1],(y)) + G).expand().simplify()
#display(v_general_cs)                                        # Lengthy
```

2.9.8.2 *Determine the constants*

We have now obtained the general solution for the displacements u and v, with integral constants c_0 and c_1, as well as the constants c_F and c_G. In total, there are four unknown constants, similar to the cantilever problem. To determine these constants, we apply DBCs.

Assume that at the point $(x = l, y = 0)$, as shown in Fig. 2.10, we set

$$u = 0; \quad v = 0; \quad \frac{\partial v}{\partial x} = 0 \tag{2.80}$$

This provides three DBCs. Together with the equation $K = -\frac{h\nu q}{40E} - \frac{hq}{40E}$, obtained earlier, we now have the correct number of equations to determine these four constants. We can proceed by writing the following snippet to solve for them:

```
cs=sp.solve([u_general_cs.subs({x:l,y:0}),v_general_cs.subs({x:l,y:0}),
          diff(v_general_cs,x).subs({x:l,y:0}),cF+cG+K], [c0,c1,cF,cG])

cs = sp.nsimplify(cs,tolerance=1e-10,rational=True)
display(Math(f" \\text{{ Constants $c_0$}}={latex(cs[c0])}"))
display(Math(f" \\text{{ Constants $c_1$}}={latex(cs[c1])}"))
display(Math(f" \\text{{ Constants $c_F$}}={latex(cs[cF])}"))
display(Math(f" \\text{{ Constants $c_G$}}={latex(cs[cG])}"))
```

$$\text{Constants } c_0 = -\frac{2c_{x2}l^3}{3} - \frac{4c_{x4}l^5}{5}$$

$$\text{Constants } c_1 = \frac{-l^2\nu^2 p - 2l^2\nu^2\rho g - l^2\nu p + 2l^2\rho g}{4E}$$

$$\text{Constants } c_F = c_{x2}l^2 + c_{x4}l^4$$

$$\text{Constants } c_G = \frac{-40Ec_{x2}l^2 - 40Ec_{x4}l^4 + h\nu p + hp}{40E}$$

2.9.8.3 *Expression simplification*

Finally, we substitute the constants back into the general expressions for the displacements to obtain the final solutions. The corresponding code snippet is as follows:

```
u_general = u_general_cs.subs(cs).expand() #General expression for u & v
v_general = v_general_cs.subs(cs).expand()
#display(Math(f" \\text{{ $u$}}={latex(u_general)}"))          # Lengthy
#display(Math(f" \\text{{ $v$}}={latex(v_general)}"))          # Lengthy
```

These expressions for the displacements are lengthy and challenging to analyze. Therefore, we rewrite them in terms of monomials with coefficients that represent the parameters of the problem. The corresponding code snippet is as follows:

```python
def get_coeffs(monomials, expression, prnt=True):
    '''Collect the coefficients of each of the `monomials` in descending
       order in the given `expression`. All in sympy expressions.
       return: coeff - the list of coefficients.
    '''
    coeff = []
    expr = expression
    for i, m in enumerate(monomials):
        coeff.append(expr.collect(m).coeff(m))        #get coefficients
        expr = (expr - coeff[i]*m).simplify().expand()
    coeff.append(expr)                            # The collected coefficients

    if prnt:
        for i in range(len(coeff)):
            display(Math(f" \\text{{$c$}}{i}={latex(coeff[i])}"))
    return coeff                                   # the list of coefficients
```

```python
# monomial bases in the expression of displacement u:
ax2y3, ax2y, ax4y, ax2, ay5, ay3, ay2, ay, a0 = symbols(\
 'a_{x2y3}, a_{x2y}, a_{x4y}, a_{x2}, a_{y5}, a_{y3}, a_{y2}, a_y, a_0')
a = [ax2y3, ax2y, ax4y, ax2, ay5, ay3, ay2, ay, a0]  # coefficient list

monomial_u = [x**2*y**3, x**4*y, x**2*y, x**2, y**5, y**3, y**2, y]
coeff_u = get_coeffs(monomial_u, u_general, prnt=False)

dic_u0_3 =dict(list(zip(a[:3 ], coeff_u[ :3]))) # dictionarys for coeffs.
dic_u3_7 =dict(list(zip(a[3:7], coeff_u[3:7])))
dic_u7_  =dict(list(zip(a[7: ], coeff_u[7: ])))
display(Math(f" {latex(dic_u0_3)}"))
display(Math(f" {latex(dic_u3_7)}"))
display(Math(f" {latex(dic_u7_ )}"))

# Re-construct the displacement u using these coefficients:
u_g = 0
for i, m in enumerate(monomial_u):
    u_g += a[i]*m                    # coefficient times the monomial basis
u_g += a[-1]                                 # add in the constant term, d0
display(Math(f" \\text{{Simplified $u$}}={latex(u_g)}"))

# To check: re-construct the displacement u using these coefficients:
u_g = 0
for i, m in enumerate(monomial_u):
    u_g += (coeff_u[i])*m            # coefficient times the monomial basis
u_g += coeff_u[-1]                           # add in the constant term, d0
#display(Math(f" \\text{{Simplified u}}={latex(u_g)}"))

u_final = u_g.subs(list(zip(a, coeff_u))).expand()       # re-construct
print(f" Simplified formula correct? {(u_final-u_general.subs(cs))==0}")
```

$$\left\{ a_{x2y3} : \frac{\nu^2 p}{Eh^3} - \frac{\nu p}{Eh^3} - \frac{2p}{Eh^3},\ a_{x2y} : -\frac{\nu^2 p}{2Eh^3} + \frac{p}{2Eh^3},\right.$$
$$\left. a_{x4y} : \frac{9\nu^2 p}{20Eh} + \frac{3\nu p}{4Eh} + \frac{3p}{10Eh} \right\}$$

$$\left\{ a_{x2} : \frac{\nu^2 p}{4E} + \frac{\nu^2 \rho g}{2E} + \frac{\nu p}{4E} - \frac{\rho g}{2E},\ a_{y2} : -\frac{c_y}{2},\ a_{y3} : -\frac{c_{y2}}{3},\ a_{y5} : -\frac{c_{y4}}{5} \right\}$$

$$\left\{ a_0 : -\frac{l^2\nu^2 p}{4E} - \frac{l^2\nu^2 \rho g}{2E} - \frac{l^2 \nu p}{4E} + \frac{l^2 \rho g}{2E},\ a_y : -c_{x2}l^2 - c_{x4}l^4 + \frac{h\nu p}{40E} + \frac{hp}{40E} \right\}$$

Simplified $u = a_0 + a_y y + a_{x2y3}x^2y^3 + a_{x2y}x^4y + a_{x2}x^2 + a_{x4y}x^2y + a_{y2}y^2 + a_{y3}y^3 + a_{y5}y^5$

```
Simplified formula correct? True
```

The following snippets replicate the process we performed earlier, but for the displacement component v.

```
# new variables of coefficients for the displacement v:
bx3y2, bxy4, bxy2, bxy, bx5, bx3, bx, b0 =\
      symbols('b_x3y2, b_xy4, b_xy2, b_xy, b_x5, b_x3, b_x, b_0 ')
b = [bx3y2, bxy4, bxy2, bxy, bx5, bx3, bx, b0]  # coefficient list

# monomial bases in the displacement expression:
monomial_v = [x**3*y**2, x*y**4, x*y**2, x*y, x**5, x**3, x]

coeff_v = get_coeffs(monomial_v, v_general, prnt=False)
dic_v0_3 =dict(list(zip(b[:3 ], coeff_v[ :3]))) # dictionary for coefs
dic_v3_7 =dict(list(zip(b[3:7], coeff_v[3:7])))
dic_v7_  =dict(list(zip(b[7: ], coeff_v[7: ])))
display(Math(f" {latex(dic_v0_3)}"))
display(Math(f" {latex(dic_v3_7)}"))
display(Math(f" {latex(dic_v7_ )}"))

# Re-construct the displacement v using these coefficients:
v_g = 0
for i, m in enumerate(monomial_v):
    v_g += b[i]*m                   # coefficient times the monomial basis
v_g += b[-1]                                # add in the constant term, d0
display(Math(f" \\text{{Simplified $v$}}={latex(v_g)}"))

v_final = v_g.subs(list(zip(b, coeff_v))).expand()       # re-construct
print(f" Simplified formula correct? {(v_final-v_general.subs(cs))==0}")
```

$$\left\{b_{x3y2}: -\frac{\nu^2 p}{Eh^3} - \frac{\nu p}{Eh^3},\ b_{xy2}: \frac{9\nu^2 p}{20Eh} - \frac{3\nu p}{10Eh} - \frac{3p}{4Eh},\right.$$
$$\left. b_{xy4}: \frac{\nu^2 p}{2Eh^3} + \frac{\nu p}{Eh^3} + \frac{p}{2Eh^3}\right\}$$

$$\left\{b_x: c_{x2}l^2 + c_{x4}l^4,\ b_{x3}: -\frac{c_{x2}}{3},\ b_{x5}: -\frac{c_{x4}}{5},\ b_{xy}: \frac{\nu^2 p}{2E} + \frac{\nu^2 \rho g}{E} + \frac{\nu \rho g}{E} - \frac{p}{2E}\right\}$$

$$\left\{b_0: -\frac{2c_{x2}l^3}{3} - \frac{4c_{x4}l^5}{5}\right\}$$

Simplified $v = b_0 + b_x x + b_{x3}x^3 + b_{x3y2}x^3y^2 + b_{x5}x^5 + b_{xy}xy + b_{xy2}xy^2 + b_{xy4}xy^4$

```
Simplified formula correct? True
```

2.9.8.4 *Final expression*

The final solutions for the displacements, produced in the previous code cell, are rearranged as follows:

$$\begin{aligned} u &= a_0 + a_{x2}x^2 + a_{x2y}x^4y + a_{x2y3}x^2y^3 \\ &\quad + a_{x4y}x^2y + a_y y + a_{y2}y^2 + a_{y3}y^3 + a_{y5}y^5 \\ v &= b_0 + b_x x + b_{x3}x^3 + b_{x3y2}x^3y^2 + b_{x5}x^5 \\ &\quad + b_{xy}xy + b_{xy2}xy^2 + b_{xy4}xy^4 \end{aligned} \tag{2.81}$$

The displacements are fifth-order polynomials. All of these coefficients are provided earlier in the dictionaries generated by the codes, expressed in terms of the given parameters for the dam problem.

These are the exact solutions for the displacements of the dam subjected to water pressure and gravity load. Although the expressions are lengthy, they remain manageable due to the simplifications made earlier. To the best of the author's knowledge, this is the first time explicit closed-form solutions for the displacements for this problem have been obtained.

Let us now check if the displacement solutions satisfy the strain–displacement relations:

```
expr1 = (ε[0]-diff(u_final,x)).subs(list(zip(a, coeff_u))).simplify()
print(f'Is εxx=∂u/∂x? {expr1==0}')                              # check
expr2 = (ε[1]-diff(v_final,y)).subs(list(zip(b, coeff_v))).simplify()
print(f'Is εyy=∂v/∂y? {expr2==0}')                              # check
expr3 = (ε[2]-(diff(u_final,y)+diff(v_final,x)))\
        .subs(list(zip(c,coeff_c))).subs(list(zip(a, coeff_u)))\
        .subs(list(zip(b, coeff_v))).simplify()
print(f'γ_xy=∂u/∂y+∂v/∂x? {expr3==0}')                          # check
```

```
Is εxx=∂u/∂x? True
Is εyy=∂v/∂y? True
γ_xy=∂u/∂y+∂v/∂x? True
```

2.9.9 *Solutions at special locations*

Let us now compute the solutions at specific locations. Along the line $y = 0$, which runs vertically through the center of the dam, the displacement in the horizontal y-direction becomes

```
u_y0 = nsimplify(u_final.subs({y:0})).simplify().factor()
display(Math(f" \\text{{ $u(x,0)$ along the vertical mid line}}=\
                                                    {latex(u_y0)}"))
```

$u(x,0)$ along the vertical mid line

$$= \frac{(-l+x)(l+x)(\nu+1)(\nu p+2\nu\rho g-2\rho g)}{4E}$$

As observed, the distribution of $u(x,0)$ is simply parabolic. The displacement $v(x,0)$, on the other hand, is more complicated. To simplify the expression, we set Poisson's ratio to $\frac{1}{3}$ by substitution. This needs to be done for all related parameters and coefficients, so we use a dictionary to accomplish this. The snippet is as follows:

```
# Set Poisson's ratio to 1/3 to simplify the expressions:

dict_ν = {nu:1/3}
v_y0 = nsimplify(v_final.subs({y:0}).subs(list(zip(c,coeff_c)))\
                                                        .subs(dict_ν))
display(Math(f" \\text{{ $v(x,0)$ along vertical mid line}}=\
                                                        {latex(v_y0)}"))
```

$$v(x,0) \text{ along vertical mid line} = \frac{14l^3p}{15Eh} - \frac{7l^2px}{5Eh} + \frac{7px^3}{15Eh} - \frac{16l^5p}{45Eh^3} + \frac{4l^4px}{9Eh^3} - \frac{4px^5}{45Eh^3}$$

As observed, the displacement $v(x, 0)$ along the middle vertical line is a fifth-order polynomial in x.

At $x = 0$ and $y = 0$, the midpoint at the top of the dam, the displacement in the y-direction becomes

```
dict_ = {x:0,y:0, nu:1/3}
u_tip = nsimplify(u_final.subs(dict_)).simplify()
display(Math(f" \\text{{ Tip displacement, $u(0,0)$}}={latex(u_tip)}"))

v_tip = nsimplify(v_final.subs(list(zip(c,coeff_c))).subs(dict_))\
                                                        .simplify()
display(Math(f" \\text{{ Tip displacement, $v(0,0)$}}={latex(v_tip)}"))
```

$$\text{Tip displacement, } u(0,0) = \frac{l^2\left(-p+4\rho g\right)}{9E}$$

$$\text{Tip displacement, } v(0,0) = \frac{2l^3p\left(21h^2-8l^2\right)}{45Eh^3}$$

2.9.10 *Contour plots of solutions*

To gain a better understanding of the solutions across the entire domain of the dam, we generate contour plots that visualize the displacement fields. These plots will provide a clear overview of the behavior of the displacements throughout the structure.

```
# Create mesh grid for plots:
l_val = 2; h_val = 1
x_mesh, y_mesh = np.mgrid[0.:l_val:500j, -h_val/2:h_val/2:250j]

# Set the parameters to specific values:
dic = {p:1, E:1, nu: 1/3, l:l_val, h:h_val, ρg:2}

vi = sp.nsimplify(v_final.subs(list(zip(c,coeff_c))).subs(dict_v)\
                        .subs(dic), tolerance=1e-10,rational=True)
display(Math(f" \\text{{$v(x,y)$ to be plotted}}={latex(vi)}"))

plt.figure()
plt.ioff()
#plt.ion()                        # uncomment this to display the plots
gr.field_plot(vi, X, x_mesh, y_mesh, title='Displacement $v$')
plt.xlabel(r"$x$")
plt.ylabel(r"$y$")
plt.savefig('imagesSM/Dam_v.png',dpi=500,bbox_inches='tight')
#plt.show()                       # uncomment this to display the plots
```

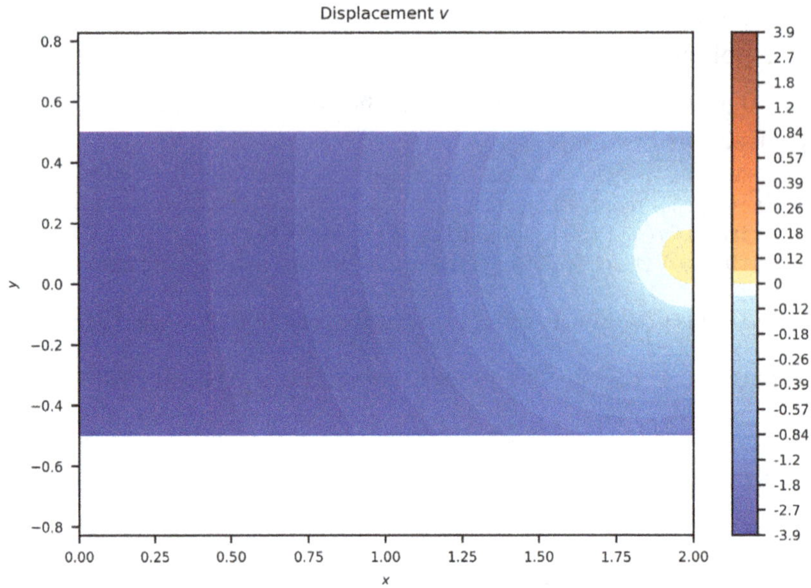

Figure 2.12. The contour of the displacement in the y-direction v in the dam.

$$v(x,y) \text{ to be plotted} = -\frac{4x^5}{45} - \frac{4x^3y^2}{9} + \frac{7x^3}{15} + \frac{8xy^4}{9} - \frac{4xy^2}{5} + \frac{4xy}{9} + \frac{68x}{45} - \frac{176}{45}$$

```
Minimum value in the domain: -3.91111
Maximum value in the domain: 0.0384129
```

The contour of the displacement in the y-direction v in the dam is plotted in Fig. 2.12. The distribution is a fifth-order polynomial, with the highest value of v at the top-left point of the dam (the plot is rotated 90° counter-clockwise).

Readers may plot the contour of the displacement in the x-direction u by simply modifying the code slightly. The following code will plot the contour of the normal stress σ_{xx} in the dam:

```
σxxi = sp.nsimplify(σxx.subs(dic))
display(Math(f"\\text{{$σ_{{xx}}(x,y)$ to be plotted}}={latex(σxxi)}"))

#plt.ioff()
plt.ion()
gr.field_plot(σxxi, X, x_mesh, y_mesh, title='Stress $σ_{xx}$')
plt.xlabel(r"$x$")
plt.ylabel(r"$y$")
plt.savefig('imagesSM/Dam_Sxx.png',dpi=500,bbox_inches='tight')
plt.show()
```

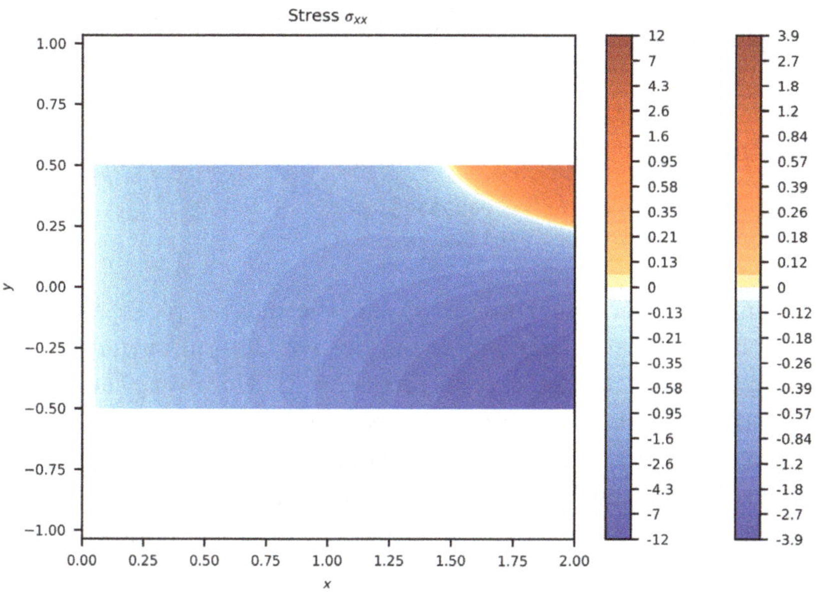

Figure 2.13. The contour of the normal stress solution in the dam.

$$\sigma_{xx}(x, y) \text{ to be plotted} = \frac{x\left(10y\left(x^2 - 2y^2\right) + 3y - 10\right)}{5}$$

```
Minimum value in the domain: -11.6
Maximum value in the domain: 3.6
```

The contour of the normal stress σ_{xx} in the dam is plotted in Fig. 2.13. It is found that the normal stress σ_{xx} is zero at $x = 0$ and increases gradually (quadratically) with x, as expected. The highest levels of normal stress are observed on both the upper and lower surfaces.

Let us now compute and plot the contour of the shear stress σ_{xy}:

```
σxyi = sp.nsimplify(σxy.subs(dic), tolerance=1e-10)
display(Math(f"\\text{{$σ_{{xy}}(x,y)$ to be plotted}}={latex(σxyi)}"))
σxy00 = σxyi.subs({x:0,y:0})                          # stress at (0,0)
display(Math(f" \\text{{$σ_{{xy}}(0,0)$ }}= {latex(σxy00)}"))

#plt.ioff()
plt.ion()
gr.field_plot(σxyi, X, x_mesh, y_mesh, title='Stress $σ_{xx}$')
plt.xlabel(r"$x$")
plt.ylabel(r"$y$")
plt.savefig('imagesSM/Dam_Sxy.png',dpi=500,bbox_inches='tight')
plt.show()
```

$$\sigma_{xy}(x,y) \text{ to be plotted} = \frac{3x^2}{4} + \frac{y^2\left(-240x^2+80y^2\right)}{80} - \frac{3y^2}{10} + \frac{1}{80}$$

$$\sigma_{xy}(0,0) \ = \frac{1}{80}$$

```
Minimum value in the domain: -0.00999996
Maximum value in the domain: 3.01245
```

The contour of the shear stress σ_{xy} in the dam is plotted in Fig. 2.14. It is seen that the free shear stress on the two lateral surfaces $y = \pm h/2$ is satisfied, but not on the top surface $x = 0$. However, the shear stress is very small, only about 0.42% (approximately $\frac{1}{80} \times \frac{1}{3.01245}$ for $h = 1$) of the maximum shear stress. It is interesting to observe how the shear stress builds up to satisfy the boundary conditions on the left and right surfaces. The highest level of normal stresses occurs around the middle point on the dam's bottom.

Readers may plot the contour of the normal stress σ_{yy} by changing the code slightly.

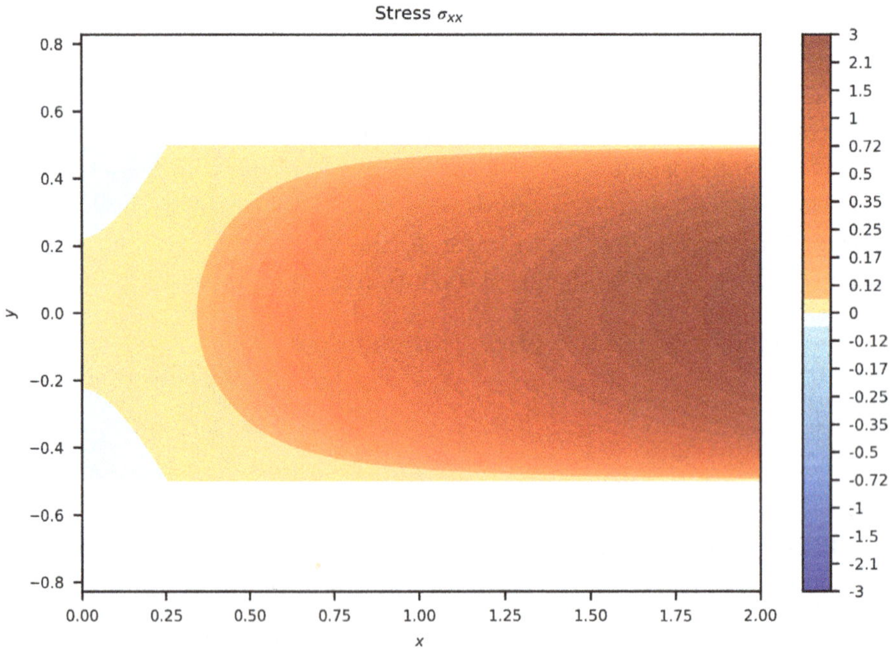

Figure 2.14. The contour of the shear stress solution in the dam.

2.10 Review questions

1. Solve the problem of cantilever rectangle 2D solid studied in Section 2.6 for stresses and displacement, assuming it is a plane strain problem.
2. Apply the general method to the problem of cantilever rectangle 2D solid studied in Section 2.7, by using all BBFs up to fourth order.
3. Solve the dam problem of cantilever rectangle 2D solid studied in Section 2.9 for stresses and displacement, assuming the pressure on the left edge is uniform p. Check the stress solution against all the necessary conditions. [Hint: Use up to fifth-order BBFs, and enforce shear (not the normal) stress on the top surface zero.]
4. Consider a thin rectangular 2D solid shown in Fig. 2.7. It is subjected to a parabolically distributed load p on the top surface:

$$p(x) = q(l^2 - x^2) \tag{2.82}$$

   Derive the formulas for stresses, strains, and the displacements in the rectangle. [Hint: Use up to seventh-order BBFs.]
5. Consider an idealized wedge-shaped dam shown in Fig. 2.15. It is subjected to water pressure py on the left edge, and gravity force ρg over the dam body. The right inclined surface is assumed free. Since the dam is

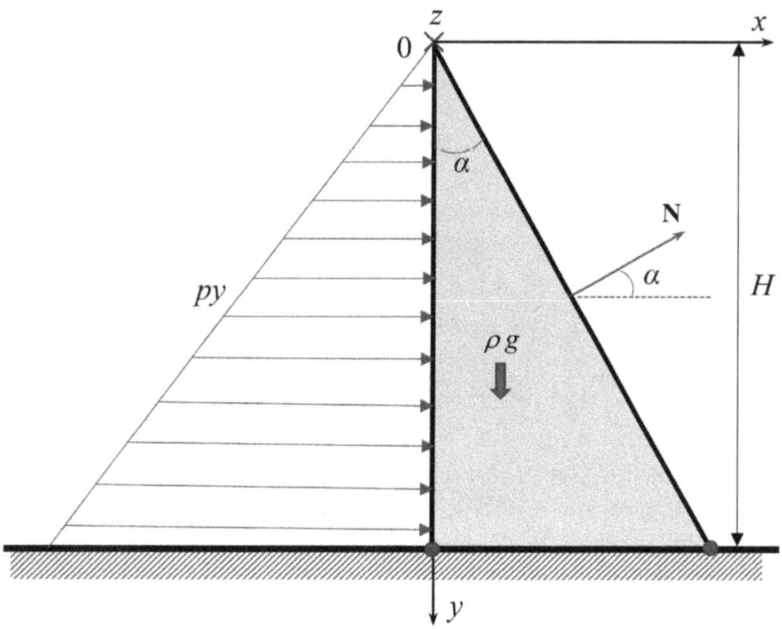

Figure 2.15. An idealized 2D wedge dam with a tip angle of α, subjected to water pressure and gravity force over the dam body.

very long along the z-axis, this problem can be treated as a plane strain problem. At point $(y = l, x = 0)$, we have the following conditions for the displacements:

$$u = 0; \quad v = 0; \quad \frac{\partial v}{\partial x} = 0 \tag{2.83}$$

Use a stress function and derive the formulas for stresses, strains, and the displacements in the dam. [Hint: The wedge has only two boundaries. The externally applied force is linear in y. It is likely that the stresses will be linear functions. Since the stresses are obtained using second-order derivatives of the stress function, the order of the stress function is likely to be a third-order polynomial. Thus, we can use third-order BBFs, and use Eq. (2.2) for the boundary condition along the right slope surface.]

2.11 Concluding remarks

This chapter discusses in detail the stress method for solving 2D solid mechanics problems. Python codes are developed, and a number of problems are solved for solutions in stresses, then in strains, and finally in displacements. We close this chapter with the following remarks.

1. The basic solution strategy is to convert the 2D problem into 1D-like ones, which can then be solved.
2. The stress function is used to convert three functions of stresses into a single function.
3. A stress function can be constructed using a linear combination of biharmonic basis functions (BBFs), which can be generated using the Python function `create_nbf()` developed in this chapter.
4. Careful initial selection of BBFs can help ease the solution procedure. This is particularly helpful for manual derivations.
5. A general method has been proposed to solve a given 2D solid mechanics problem by an automatic construction of BBFs and using sets of conditions to determine the unknown constants. Initial selection of BBFs is not required.
6. The solutions in stress distributions are applicable away from the boundaries where the SBCs are not fully satisfied in terms of detailed distribution.
7. To the best of the author's knowledge, this is the first time explicit closed-form solutions are found for the displacements in the rectangular dam problem. These displacements are fifth-order polynomials.

The next chapter presents displacement methods for 2D problems.

References

[1] S. Timoshenko and J.N. Goodier, *Theory of Elasticity*, McGraw-Hill, New York, 1970. Available online http://books.google.com/books?id=yFISAAAAIAAJ&dq=theory+of+elasticity&ei=ICiKSsr3G4jwkQSbxMyPCg.

[2] Z.L. Xu, *Elasticity*, Vol. 1 & 2, People's Publisher, China, 1979.

[3] J.D. Achenbach, *Wave Propagation in Elastic Solids*, North-Holland, Amsterdam, 1984.

[4] Y.C. Fung, *Foundations of Solid Mechanics*, Prentice-Hall of India, New Delhi, 1968.

[5] A.P. Boresi, K. Chong and J.D. Lee, *Elasticity in Engineering Mechanics*, John Wiley & Sons, New York, 2010.

[6] A.P. Boresi, R.J. Schmidt and O.M. Sidebottom, *Advanced Mechanics of Materials*, John Wiley & Sons, New York, 1985.

[7] G.R. Liu, *Mechanics of Materials: Formulations and Solutions with Python*, World Scientific, New Jersey, 2024.

[8] G.R. Liu, *Calculus: A Practical Course with Python*, World Scientific, New Jersey, 2025.

[9] G.R. Liu, *Vector Calculus: Formulation, Applications, and Python Codes*, World Scientific, New Jersey, 2025.

[10] G.R. Liu, *Numbers and Functions: Theory, Formulation, and Python Codes*, World Scientific, New Jersey, 2024.

[11] S. Timoshenko and J.N. Goodier, *Theory of Elasticity*, McGraw-Hill, New York, 1970. Available online http://books.google.com/books?id=yFISAAAAIAAJ&dq=theory+of+elasticity&ei=ICiKSsr3G4jwkQSbxMyPCg.

[12] G.R. Liu, *Solid Mechanics with Python: Non-Cartesian Coordinates*, World Scientific, New Jersey, 2026.

Chapter 3

Displacement Methods for 2D Solids

```
# Place cursor in this cell,and press Ctrl+Enter to import dependencies.
import sys                          # For accessing the computer system
sys.path.append('../grbin/')   # Change to the directory in your system

from commonImports import *       # Import dependencies from '../grbin/'
import grcodes as gr                    # Import the module of the author
importlib.reload(gr)              # When grcodes is modified, reload it

from continuum_mechanics.solids import sym_grad
init_printing(use_unicode=True)         # For latex-like quality printing
np.set_printoptions(precision=4,suppress=True,
                    formatter={'float_kind': '{:.4e}'.format})
```

In the previous chapter, we discussed in detail stress methods for solving 2D solid mechanics problems in Cartesian coordinate systems, where the stresses are treated as primary variable functions. This chapter discusses an alternative method: the displacement method.

A **displacement method** treats displacements as primary variable functions. It usually starts by approximating the displacements using some kind of basis functions [1], equipped with adjustable constants for these bases. The key advantage is that all the other variables for a solid mechanics problem, such as strain and stresses, can be obtained using the assumed displacements via only differentiations, which can always be done deterministically for all elementary basis functions used in the approximation [2]. The most difficult part is choosing proper sets of basis functions for the displacements. This chapter details this process for problems defined in the Cartesian coordinate system. The problem domain is confined to three boundary segments in a wedge shape or four segments in a rectangular shape. Our discussion will focus on the use of monomial basis functions for assuming the displacements. This means that the displacements to be found belong to a polynomial space. If we somehow know that the displacement has some bases, we simply add these in Ref. [1].

The equations to be solved in a displacement method for a given solid mechanics problem include:

- Displacement boundary conditions (DBCs), Eqs. (2.3) and (2.4), which impose constraints on the displacements.
- Equilibrium equations in differential form, Eq. (2.1), which relate the body forces to stresses inside the solid.
- Stress boundary conditions (SBCs), Eq. (2.2), which relate the tractions on the boundary to the stresses on the boundary.
- Integral equilibrium equations, or FBCs, which relate the forces on the boundary to the stresses inside the solid.
- Compatibility equation, Eq. (2.8), written in terms of strains, or Eq. (2.18), written in terms of stresses, which imposes constraints on the strain or stress components.

In contrast to the stress method, a displacement method usually does not need to consider the compatibility equations. This is because when the strains (hence the stresses) are obtained from displacements, compatibility is always ensured.

3.1 Procedure of displacement methods

The general solution procedure in a displacement method is as follows:

1. Select a set of monomials as basis functions for each of the displacement components, u and v. The orders of these sets may differ. The choice of monomial bases should be based on the problem setup and prior experience.
2. Use known conditions, such as symmetry conditions, to exclude unnecessary bases. This can help reduce burdens in subsequent derivations.
3. Construct displacement functions for each component via linear combinations of the bases with unknown constants for these monomials.
4. Use the given displacement boundary conditions (DBCs) to determine some of the unknown constants.

5. Use the strain–displacement relations to compute the strains, which involve only differentiations. The strains are then expressed in terms of these constants.
6. Use the constitutive equations to compute the stresses, which involve only algebraic operations and are therefore always feasible. The stresses are then expressed in terms of the remaining unknown constants. This step can be combined with step 5 by directly using the stress–displacement relations.
7. Determine the remaining unknown constants using the stress boundary conditions (SBCs) and integral force boundary conditions (FBCs), which may also be applied to a free-body diagram (FBD) cut.
8. Check whether all necessary conditions — equilibrium, stress, and DBC — are satisfied. If not, reinforce them as needed.

In contrast to a stress method, the displacement method does not require indefinite integrations, eliminating concerns about integral functions and integral constants. The process flows naturally and straightforwardly, with all operations being manageable. The most difficult part is selecting proper sets of basis monomials.

The following sections detail the procedure using example problems.

3.2 Cantilever rectangle 2D solid

Consider first the 2D solid cantilever problem in Cartesian coordinates, as shown in Fig. 3.1.

The detailed setup and use of Cartesian coordinates for this problem are given in Section 2.7, with DBCs that initially fix the displacements and some of their slopes at the origin.

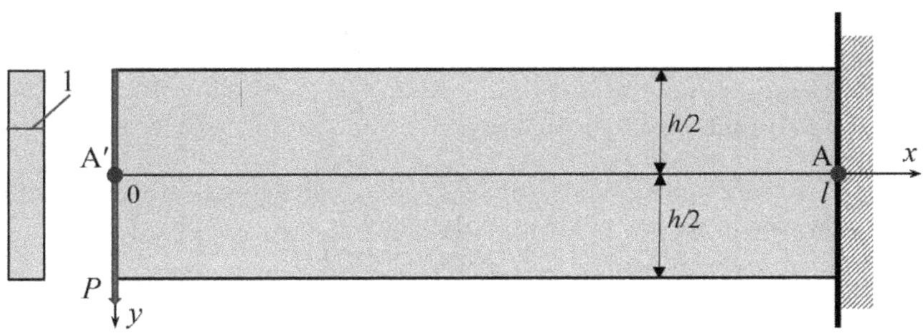

Figure 3.1. A 2D rectangular thin cantilever with dimensions of $l \times h$ and unit thickness $b = 1$, subjected to a downward force P at the left end.

3.2.1 *General expressions of assumed displacements*

A general expression for the displacement components $u(x, y)$ and $v(x, y)$ should be a linear combination of polynomial basis functions. One may also use or add other special basis functions, such as trigonometric and harmonic bases, for special types of problems if needed. The only requirement is that these basis functions be linearly independent, which is crucial to prevent breakdowns in the solution process.

First, we need to define monomial basis functions, generate symbolic constants, and construct these two displacement functions. In this process, we can also make use of the known symmetry properties for the 2D cantilever rectangular beam problem:

1. The displacement function $u(x, y)$ is antisymmetric in y.
2. The displacement function $v(x, y)$ is symmetric in y.

These conditions are used to filter out unnecessary basis functions. The snippet is as follows:

```
# Define necessary symbolic variables:
x, y, z = sp.symbols("x, y, z ", real=True)
h, l, P, E, G, ν, I = sp.symbols("h, l, P, E, G, ν, I",positive=True)
σxx, σyy, σxy = sp.symbols("σ_xx, σ_yy, σ_xy")
εxx, εyy, εxy   = sp.symbols("ε_xx, ε_yy, ε_xy")

order = 3                           # use complete monomial upto `order`
# Generate the complete set of the basis functions (bfs):
bfs_list = [x**i*y**j for i in range(order + 1)
                        for j in range(order + 1) if i+j <= order]
bfs = sp.Matrix(bfs_list)           # form sympy vector with these bases
display(Math(f"\\text{{Complete 2D monomials upto order }}{order}:"))
display(Math(f"\\text{{bfs_all}} = {latex(bfs_list)}"))

def is_even_in_y(poly):
    return poly.subs(y, -y) == poly
def is_odd_in_y(poly):
    return -poly.subs(y, -y) == poly

# Select basis functions (bfs) for u(x,y,z), antisymmteric conditions:
u_bfs = [poly for poly in bfs if -poly.subs(y, -y) == poly]
display(Math(f" \\text{{u_bfs}}={latex(u_bfs)}"))          # Bases for u

# Select bfs for v(x,y,z) using symmteric conditions:
v_bfs = [poly for poly in bfs if  poly.subs(y, -y) == poly]
display(Math(f" \\text{{v_bfs}}={latex(v_bfs)}"))          # Bases for v
```

Complete 2D monomials upto order 3:

bfs_all = $\left[1,\ y,\ y^2,\ y^3,\ x,\ xy,\ xy^2,\ x^2,\ x^2y,\ x^3\right]$

u_bfs = $\left[y,\ y^3,\ xy,\ x^2y\right]$

v_bfs = $\left[1,\ y^2,\ x,\ xy^2,\ x^2,\ x^3\right]$

Based on the number of basis functions used, we generate the necessary number of symbolic unknown constants. This must be done separately for each displacement component. The general expressions for these displacement components are given by the dot products of the basis functions and their corresponding constants.

```
# unknown constants for displacements:
ai = sp.symbols(f'a:{len(u_bfs)}')                        # unknown constants
display(Math(f" \\text{{Unknown constants for $u$}}={latex(ai)}"))
ai = Matrix([ai[i] for i in range(len(u_bfs))]) # form coeffs. in vector

bi = sp.symbols(f'b:{len(v_bfs)}')                       # unknown constants
display(Math(f" \\text{{Unknown constants for $v$}}={latex(bi)}"))
bi = Matrix([bi[i] for i in range(len(v_bfs))])             # form a vector

cs = ai.col_join(bi)                     # form a single vector of constants

ug = (ai.T@Matrix(u_bfs))[0]             # Form the general displacements
vg = (bi.T@Matrix(v_bfs))[0]
display(Math(f"\\text{{Assumed displacements, $u(x,y,z)$}}={latex(ug)}"))
display(Math(f"\\hspace{{4cm}} \\text{{ $v(x,y,z)$}}={latex(vg)}"))
print(f" Total number of unknowns {len(cs)}")
```

Unknown constants for $u = (a_0,\ a_1,\ a_2,\ a_3)$

Unknown constants for $v = (b_0,\ b_1,\ b_2,\ b_3,\ b_4,\ b_5)$

Assumed displacements, $u(x, y, z) = a_0y + a_1y^3 + a_2xy + a_3x^2y$
$$v(x, y, z) = b_0 + b_1y^2 + b_2x + b_3xy^2 + b_4x^2 + b_5x^3$$

```
Total number of unknowns 10
```

3.2.2 *Enforcing displacement boundary conditions*

We assume that the displacements at the beam root $(l, 0)$ are constrained as follows:

$$\begin{aligned} u &= v = 0 \\ \frac{\partial v}{\partial x} &= 0 \end{aligned} \tag{3.1}$$

This provides three DBCs, which are used to determine some of the unknowns, as shown in the following snippet:

```
dic0 = {x:l, y:0}                                    # for DBCs at the beam root
DBCs = [ug.subs(dic0), vg.subs(dic0), vg.diff(x).subs(dic0)]
sln_DBC = sp.solve(DBCs, cs)                # solve for unknowns
display(Math(f" \\text{{Conditions for fixed displacements}}=\
                                                  {latex(sln_DBC)}"))
# update the solutions for the unknown constants:
sln_cs = {**sln_DBC}

# update for the displacements:
u = ug.subs(sln_cs).simplify()
v = vg.subs(sln_cs).simplify()
display(Math(f" \\text{{General displacement $u$}}={latex(u)}"))
display(Math(f" \\text{{General displacement $v$}}={latex(v)}"))
display(Math(f" \\text{{Constants found so far}}:{latex(sln_cs)}"))
```

Conditions for fixed displacements $= \left\{b_0\colon b_4 l^2 + 2b_5 l^3,\ b_2\colon -2b_4 l - 3b_5 l^2\right\}$

General displacement $u = y\left(a_0 + a_1 y^2 + a_2 x + a_3 x^2\right)$

General displacement $v = b_1 y^2 + b_3 x y^2 + b_4 l^2 + b_4 x^2 + 2b_5 l^3 + b_5 x^3 - lx\left(2b_4 + 3b_5 l\right)$

Constants found so far: $\left\{b_0\colon b_4 l^2 + 2b_5 l^3,\ b_2\colon -2b_4 l - 3b_5 l^2\right\}$

We have found two of the constants (in terms of the other constants) using the DBCs. Note that DBCs can also be enforced at the last step. We choose to do it here to slightly reduce the length of the expressions.

The displacement expressions can now be used to compute the stresses, which are then used to enforce the SBCs.

3.2.3 *Enforcing stress boundary conditions*

The general expressions for the three stress components are derived using the constitutive equations and strain–displacement relations. Since this 2D cantilever is considered a plane stress problem, the constitutive equations are the inverse of Eq. (2.5). The snippet for this computation is as follows:

```
# Compute general stresses using the current displacement functions:

σxxg = (2*G/(1-ν)*(ug.diff(x)+ν*vg.diff(y))).expand().collect(G*x)
σyyg = (2*G/(1-ν)*(vg.diff(y)+ν*ug.diff(x))).expand().collect(G*x)
σxyg = (G*(vg.diff(x)+ug.diff(y))).expand().collect(G*y)

display(Math(f" \\text{{General expressions for stresses}}:"))
display(Math(f" \\text{{$σ_{{xx}}$}}=\\small {latex(σxxg)}"))
display(Math(f" \\text{{$σ_{{yy}}$}}=\\small {latex(σyyg)}"))
display(Math(f" \\text{{$σ_{{xy}}$}}=\\small {latex(σxyg)}"))
```

General expressions for stresses:

$$\sigma_{xx} = \frac{2Ga_2y}{1-\nu} + \frac{4Gb_1y\nu}{1-\nu} + Gx\left(\frac{4a_3y}{1-\nu} + \frac{4b_3y\nu}{1-\nu}\right)$$

$$\sigma_{yy} = \frac{2Ga_2y\nu}{1-\nu} + \frac{4Gb_1y}{1-\nu} + Gx\left(\frac{4a_3y\nu}{1-\nu} + \frac{4b_3y}{1-\nu}\right)$$

$$\sigma_{xy} = Ga_0 + Ga_2x + Ga_3x^2 + Gb_2 + 2Gb_4x + 3Gb_5x^2 + Gy\left(3a_1y + b_3y\right)$$

We now apply the SBCs for stresses in the x–y plane on the top and bottom surfaces of the rectangular beam:

$$\text{On the top and bottom, } y = \pm h\text{:} \quad \sigma_{yy} = 0,\ \sigma_{xy} = 0 \tag{3.2}$$

These conditions must hold for all x. Thus, the coefficients of all monomial terms x^2, x^1, and x^0 in the stress expressions on these boundaries must vanish. This results in a new set of five conditions, which can be used to determine some of the remaining constants.

```
display(Math(f" \\text{{On top surface, $σ_{{yy}}$}}=\
                                    \\small {latex(σyyg.subs(y,h/2))}"))
display(Math(f" \\text{{On top surface, $σ_{{xy}}$}}=\
                                    \\small {latex(σxyg.subs(y,h/2))}"))
SBCs = []
SBCs.append(σyyg.subs(y,h/2).coeff(x,1))          # expressions of SBCs
SBCs.append(σyyg.subs(y,h/2).coeff(x,0))
SBCs.append(σxyg.subs(y,h/2).coeff(x,2))
SBCs.append(σxyg.subs(y,h/2).coeff(x,1))
SBCs.append(σxyg.subs(y,h/2).coeff(x,0))

sln_SBC = sp.solve(SBCs, cs)
display(Math(f" \\text{{Constants found}}:\\small {latex(sln_SBC)}"))
sln_cs = {**sln_DBC, **sln_SBC}          # update the unknown constants
sln_cs = {key: value.subs(sln_cs).simplify() for key, value in
                                                    sln_cs.items()}
```

```
# Using the current constants to update the stresses:
σxx = σxxg.subs(sln_cs).simplify(); σyy = σyyg.subs(sln_cs).simplify()
σxy = σxyg.subs(sln_cs).simplify()
display(Math(f"σ_{{xx}}={latex(σxx)} \hspace{{2cm}} σ_{{yy}}=\\small \
                                                        {latex(σyy)}"))
display(Math(f"σ_{{xy}} =\\small {latex(σxy)}"))

# Using the current constants to update the displacements:
u = ug.subs(sln_cs).simplify(); v = vg.subs(sln_cs).simplify()
display(Math(f"\\text{{General displacement $u$}}=\\small{latex(u)}"))
display(Math(f"\\text{{General displacement $v$}}=\\small{latex(v)}"))
```

On top surface, $\sigma_{yy} = \frac{Ga_2 h\nu}{1-\nu} + \frac{2Gb_1 h}{1-\nu} + Gx\left(\frac{2a_3 h\nu}{1-\nu} + \frac{2b_3 h}{1-\nu}\right)$

On top surface, $\sigma_{xy} = Ga_0 + Ga_2 x + Ga_3 x^2 + Gb_2 + 2Gb_4 x + 3Gb_5 x^2 + \frac{Gh\left(\frac{3a_1 h}{2} + \frac{b_3 h}{2}\right)}{2}$

Constants found: $\left\{a_0\colon -\frac{3a_1 h^2}{4} - b_2 - \frac{3b_5 h^2 \nu}{4},\ a_2\colon -2b_4,\ a_3\colon -3b_5,\ b_1\colon b_4\nu,\ b_3\colon 3b_5\nu\right\}$

$\sigma_{xx} = 4Gy\left(-b_4\nu - b_4 - 3b_5 x\nu - 3b_5 x\right)$ $\qquad \sigma_{yy} = 0$

$\sigma_{xy} = \frac{3G\left(-a_1 h^2 + 4a_1 y^2 - b_5 h^2\nu + 4b_5 y^2\nu\right)}{4}$

General displacement: $u = \frac{y\left(-3a_1 h^2 + 4a_1 y^2 + 8b_4 l - 8b_4 x - 3b_5 h^2\nu + 12b_5 l^2 - 12b_5 x^2\right)}{4}$

General displacement: $v = b_4 x^2 + b_4 y^2\nu + b_5 x^3 + 3b_5 xy^2\nu + l^2\left(b_4 + 2b_5 l\right) - lx\left(2b_4 + 3b_5 l\right)$

All the stresses and displacements have been updated using the constants found so far. Note that we did not enforce the SBCs on the bottom surface, as we previously used the symmetry condition for the displacements. Thus, when the SBCs on the top surface are enforced, the SBCs on the bottom surface are automatically satisfied. We found that the stress component $\sigma_{yy} = 0$ throughout the entire domain, even though we only applied zero SBC conditions on the top (and bottom) surfaces. Therefore, we now focus only on the two stress components σ_{xx} and σ_{xy} for this problem.

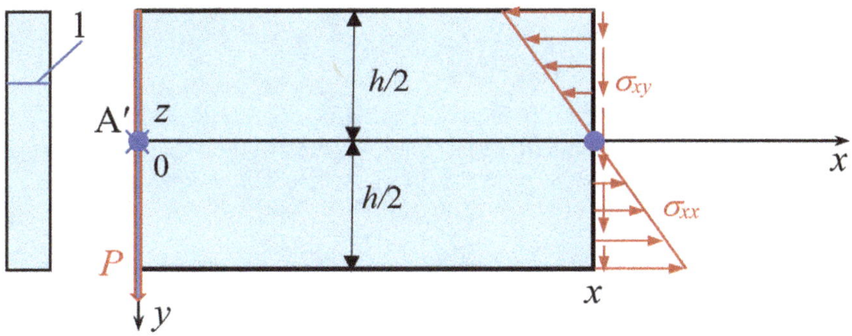

Figure 3.2. A free-body diagram (FBD) of the left part of the 2D rectangular beam with length x. It is subjected to a downward force on the left edge. The distribution of stresses is schematically shown on the bottom cut surface.

3.2.4 *Enforcing integral equilibrium conditions*

Next, we apply the following integral equilibrium conditions for stresses on the cross-section of the rectangular beam at an arbitrary location x, as shown in Fig. 3.2.

The cross-sectional surface of the free body has a distributed normal stress σ_{xx} in the x-direction and a moment about the z-axis. The distributed shear stress σ_{xy} generates a force in the y-direction. All these forces and moments must be in equilibrium. The formulas are as follows:

$$\begin{aligned} &\sum F_x = 0\text{:}\ \int_A \sigma_{xx} dy = 0 \\ &\sum M_z = 0\text{:}\ -\int_A \sigma_{xx} y dy - Px = 0 \\ &\sum F_y = 0\text{:}\ \int_A \sigma_{xy} dy + P = 0 \end{aligned} \tag{3.3}$$

Since σ_{xx} is odd in y, the first equation is automatically satisfied. Therefore, we only need to enforce the last two equations. Since this set of equations is expressed in terms of forces, it is also referred to as the FBCs. The boundary can be either the actual boundary of the problem domain or a FBD boundary cut from the domain, as in the current case.

Note that, in general, the equations can be functions of x. In this problem, the second equation is clearly linear in x. Therefore, all coefficient expressions for the monomials x^1 and x^0 must be zero. We use `gr.getExprs()`, developed in the previous chapter, to generate these expressions. The code snippet is as follows:

```
sln_cs = {**sln_DBC, **sln_SBC}                          # current constants

Int = sp.integrate
FBCs = []
FBCs.append(Int(σxxg.subs(sln_cs),    (y,-h/2,h/2)))
FBCs.append(Int(σxxg.subs(sln_cs)*y, (y,-h/2,h/2)) + P*x)
FBCs.append(Int(σxyg.subs(sln_cs),    (y,-h/2,h/2)) + P)

# display these FBCs:
display(Math(f"\\text{{∑Fx=0 expression: }}\small {latex(FBCs[0])} "))
display(Math(f"\\text{{∑Mz=0 expression: }}\small {latex(FBCs[1])} "))
display(Math(f"\\text{{∑Fy=0 expression: }}\small {latex(FBCs[2])} "))
```

$$\sum Fx = 0 \text{ expression: } 0$$

$$\sum Mz = 0 \text{ expression: } Px + \frac{h^3\left(-\frac{4Gb_4\nu}{3} - \frac{4Gb_4}{3} - 4Gb_5x\nu - 4Gb_5x\right)}{4}$$

$$\sum Fy = 0 \text{ expression: } P + \frac{h^3\left(Ga_1 + Gb_5\nu\right)}{4} + h\left(-\frac{3Ga_1h^2}{4} - \frac{3Gb_5h^2\nu}{4}\right)$$

Note that, in general, these FBCs can be functions of x. In this problem, the second equation is clearly linear in x, as shown in the output above. Therefore, all coefficient expressions for the monomials x^1 and x^0 must be zero. We use **gr.getExprs()**, developed in the previous chapter, to generate these expressions and solve them. The code snippet is as follows:

```
FBC_exprs = gr.getExprs(FBCs, x, 1)

sln_FBC = sp.solve(FBC_exprs, cs)
display(Math(f" \\text{{Constants found}}:\\small {latex(sln_FBC)}"))

sln_cs = {**sln_DBC, **sln_SBC, **sln_FBC}                       # update
sln_cs = {key: value.subs(sln_cs) for key, value in sln_cs.items()}
```

3.2.5 *Final solutions and verification*

We can now substitute the obtained constants into the current expressions for stresses and displacements to obtain the final set of solutions. The snippet is as follows:

```
# Using the current constants to update the stresses:
σxx_final = σxxg.subs(sln_cs).simplify()
σyy_final = σyyg.subs(sln_cs).simplify()
σxy_final = σxyg.subs(sln_cs).simplify()
display(Math(f"\\text{{Final stresses: $ σ_{{xx}}$}}= \
                    {latex(σxx_final)}; σ_{{yy}} ={latex(σyy_final)}; \
                                        σ_{{xy}} ={latex(σxy_final)}"))

# Using the current constants to update the displacements:
u_final = u.subs(sln_cs).simplify()
v_final = v.subs(sln_cs).simplify()
display(Math(f" \\text{{Final displacement $u$}}= {latex(u_final)}"))
display(Math(f" \\text{{Final displacement $v$}}= {latex(v_final)}"))
```

```
Total number of expressions = 6
Number of nonzero expressions = 3
```

Constants found: $\left\{a_1: \frac{P\nu + 2P}{Gh^3\nu + Gh^3},\ b_4: 0,\ b_5: \frac{P}{Gh^3\nu + Gh^3}\right\}$

Final stresses: $\sigma_{xx} = -\frac{12Pxy}{h^3}; \sigma_{yy} = 0; \sigma_{xy} = \frac{3P\left(-h^2 + 4y^2\right)}{2h^3}$

Final displacement: $u = \frac{Py\left(-3h^2\nu - 3h^2 + 6l^2 - 6x^2 + 2y^2\nu + 4y^2\right)}{2Gh^3(\nu + 1)}$

Final displacement: $v = \frac{P\left(2l^3 - 3l^2x + x^3 + 3xy^2\nu\right)}{Gh^3(\nu + 1)}$

We have finally obtained the full set of solutions for both displacements and stresses, which are the same as those obtained in the previous chapter using the stress method.

Since, in our displacement method, we did not use the differential form of the equilibrium equations (with body forces being free), let us now check whether these equilibrium equations are satisfied by the stresses obtained.

```
display(Math(f" \\text{{Is equilibrium Eq. 1 = 0?}}=\
          {latex((σxx_final.diff(x)+σxy_final.diff(y)).simplify())}"))
display(Math(f" \\text{{Is equilibrium Eq. 2 = 0?}}=\
          {latex((σxy_final.diff(x)+σyy_final.diff(y)).simplify())}"))
```

Is equilibrium Eq. 1 = 0? $= 0$

Is equilibrium Eq. 2 = 0? $= 0$

It can be seen that the equilibrium equations are all satisfied. This is because: (1) the body force is zero, and (2) the FBCs are enforced at all locations of x. Therefore, equilibrium is satisfied at every point in the solid. It is always good practice to verify this. Next, let's check whether the strains satisfy the compatibility equations.

```
comptblty = ((σxx_final+σyy_final).diff(x,2)+\
                      (σxx_final+σyy_final).diff(y,2)).simplify()  #∇^2
print(f"Is compatibility condition satisfied? {comptblty==0}")
```

```
Is compatibility condition satisfied? True
```

It can be seen that the compatibility equation is also satisfied. Readers can verify all the other conditions for this set of solutions.

The constants for the initial monomials of the assumed displacements are determined as follows:

```
sln_cs = {key: value.subs(sln_cs).simplify() for key, value in
                                                    sln_cs.items()}
n = 3
sln_cs_= gr.split_dict_n(sln_cs, n) # split into n smaller dictionaries
display(sln_cs_[0])
display(sln_cs_[1])
display(sln_cs_[2])
```

$$\left\{a_0: \frac{3P\left(-h^2(\nu+1)+2l^2\right)}{2Gh^3(\nu+1)},\ b_0: \frac{2Pl^3}{Gh^3(\nu+1)},\ b_2: -\frac{3Pl^2}{Gh^3(\nu+1)}\right\}$$

$$\left\{a_2: 0,\ a_3: -\frac{3P}{Gh^3(\nu+1)},\ b_1: 0\right\}$$

$$\left\{a_1: \frac{P(\nu+2)}{Gh^3(\nu+1)},\ b_3: \frac{3P\nu}{Gh^3(\nu+1)},\ b_4: 0,\ b_5: \frac{P}{Gh^3(\nu+1)}\right\}$$

It can be seen that three of these constants are found to be zero.

This concludes our study of the 2D cantilever problem using a displacement method. Now, let's apply the same approach to a problem with three boundary segments.

3.3 A wedge dam

3.3.1 *Setting of the problem*

Consider the idealized wedge-shaped dam shown in Fig. 3.3. It has a height of H, and the angle of the slope surface with respect to the y-axis is α. It is subjected to water pressure py on the left vertical edge and gravity

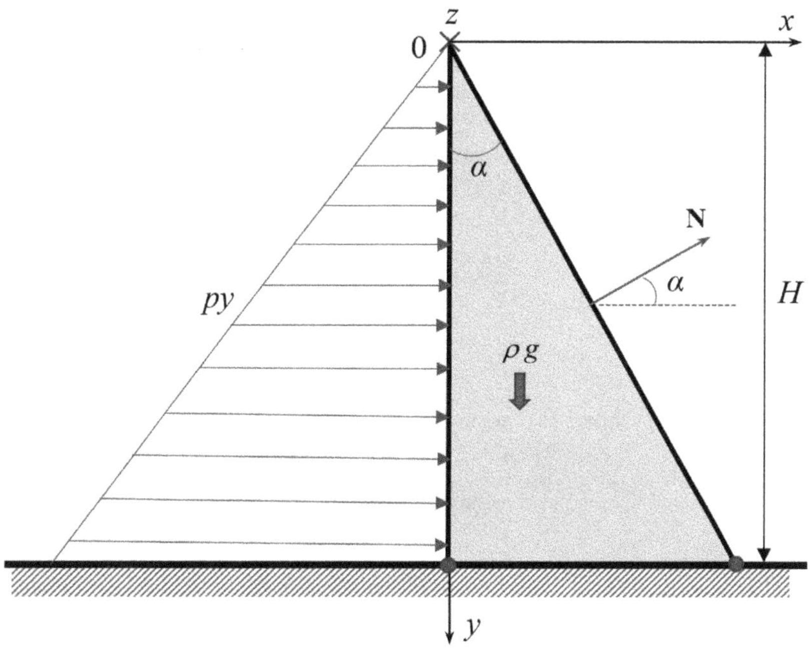

Figure 3.3. An idealized 2D wedge dam with a height of H and a tip angle of α. It is subjected to water pressure along the vertical edge and gravity force over the dam body.

force ρg over the dam body. The right slope surface is assumed to be free, and the bottom edge is constrained. At the point $(y = H, x = 0)$, we assume the following displacement conditions:

$$u = 0; \quad v = 0; \quad \frac{\partial u}{\partial y} = 0 \tag{3.4}$$

Since the dam is very long along the z-axis, this problem can be treated as a plane strain problem. Our task is to derive the formulas for the stresses and displacements in the wedge dam using a displacement method.

3.3.2 *General expression of assumed displacements*

The wedge has only three boundaries. The externally applied water pressure is linear in y, and the geometric dimension also changes linearly. It is likely that the stresses will be linear functions. Since the stresses are related to the first derivatives of the displacement functions, the displacement functions are likely to be second-order polynomials. Thus, we can use monomial basis functions of complete second order [1] and follow the procedure for a displacement method given earlier to solve the problem.

For the wedge, we need to define its angle and use it to compute the direction cosines of the slope surface so that we can enforce the SBCs on that surface. We shall use the following codes to get this done step-by-step:

```
# Define necessary symbolic variables for coordinates and parameters:
x, y = sp.symbols("x, y", real=True)
H, p, ρg, E, G, ν = sp.symbols("H, p, ρg, E, G, ν ", positive=True)

# Symbolic variables for the orientation of the slope surface:
n1, n2, α, tα = sp.symbols("n1, n2, α, tα", real=True)
n1_=cos(α); n2_=-sin(α)                                # direction cosines
tα_= sin(α)/cos(α)

σxx, σyy, σxy = sp.symbols("σ_xx, σ_yy, σ_xy")
εxx, εyy, εxy   = sp.symbols("ε_xx, ε_yy, ε_xy")

order = 2                  # Set to use upto complete 2nd order monomials:
# Generate the complete set of monomial basis functions in 2D:
bfs_list = [x**i*y**j for i in range(order + 1)
                              for j in range(order + 1) if i+j <= order]

bfs = sp.Matrix(bfs_list)           # form sympy vector with these bases
display(Math(f" \\text{{Complete 2nd order monomials in 2D}}: \
                                    \\text{{bfs}} = {latex(bfs_list)}"))
u_bfs = bfs                                                # Bases for u
v_bfs = bfs                                   # Use the same bases for v
```

Complete second-order monomials in 2D: bfs $= \left[1,\ y,\ y^2,\ x,\ xy,\ x^2\right]$.

Based on the number of basis functions used, the general expressions for these displacement components are formed using the dot product of the bases and the corresponding constants.

```
# unknown constants for displacements:
ai = sp.symbols(f'a:{len(u_bfs)}')                     # unknown constants
display(Math(f" \\text{{Unknown constants for $u$}}={latex(ai)}"))
ai = Matrix([ai[i] for i in range(len(u_bfs))])           # form a vector

bi = sp.symbols(f'b:{len(v_bfs)}')                     # unknown constants
display(Math(f" \\text{{Unknown constants for $v$}}={latex(bi)}"))
bi = Matrix([bi[i] for i in range(len(v_bfs))])           # form a vector

cs = ai.col_join(bi)       # form a single vector of using all constants

ug = (ai.T@Matrix(u_bfs))[0] # Form general expressions of displacements
vg = (bi.T@Matrix(v_bfs))[0]
display(Math(f"\\text{{Assumed displacements,$\;u(x,y)$}}={latex(ug)}"))
display(Math(f"\hspace{{4cm}} \\text{{ $v(x,y)$}}={latex(vg)}"))
```

Unknown constants for $u = (a_0,\ a_1,\ a_2,\ a_3,\ a_4,\ a_5)$

Unknown constants for $v = (b_0,\ b_1,\ b_2,\ b_3,\ b_4,\ b_5)$

Assumed displacements, $u(x,y) = a_0 + a_1 y + a_2 y^2 + a_3 x + a_4 xy + a_5 x^2$

$$v(x,y) = b_0 + b_1 y + b_2 y^2 + b_3 x + b_4 xy + b_5 x^2$$

3.3.3 *Enforcing displacement boundary conditions*

We assume that the displacements at $(0, H)$ are fixed as follows:

$$\begin{aligned} u = v = 0 \\ \frac{\partial u}{\partial y} = 0 \end{aligned} \tag{3.5}$$

This gives three DBCs, which can be used to find some of the unknown constants using the following snippet:

```
dic0 = {x:0, y:H}                                          # for DBCs
DBCs = [ug.subs(dic0), vg.subs(dic0), ug.diff(y).subs(dic0)]
sln_DBC = sp.solve(DBCs, cs)
display(Math(f" \\text{{Conditions for fixed displacements}}=\
                                                {latex(sln_DBC)}"))
# update the solutions for the unknown constants:
sln_cs = {**sln_DBC}
u = ug.subs(sln_cs).simplify(); v = vg.subs(sln_cs).simplify()
display(Math(f" \\text{{Current displacement $u$}}={latex(u)}"))
display(Math(f" \\text{{Current displacement $v$}}={latex(v)}"))
```

Conditions for fixed displacements $= \{a_0\colon H^2 a_2,\ a_1\colon\ -2Ha_2,$
$b_0\colon\ -H^2 b_2 - Hb_1\}$

Current displacement: $u = H^2 a_2 - 2Ha_2 y + a_2 y^2 + a_3 x + a_4 xy + a_5 x^2$

Current displacement: $v = -H^2 b_2 - Hb_1 + b_1 y + b_2 y^2 + b_3 x + b_4 xy + b_5 x^2$

We have found three of the constants (in terms of the other constants) using these DBCs. The displacement expressions can now be used to compute the stresses, which can then be used to enforce the SBCs for stresses.

3.3.4 *Enforcing stress boundary conditions*

The general expressions for the three stress components are obtained using the constitutive equations and strain–displacement relations. Since this wedge dam is considered as a plane strain problem, the constitutive equations are given in Eq. (2.6) (in its inverse form). The snippet for this computation is as follows:

```
# Compute stresses using displacement-function, plane strain:

σxxg = (2*G/(1-2*ν)*((1-ν)*u.diff(x)+ν*v.diff(y))).simplify()
σyyg = (2*G/(1-2*ν)*(ν*u.diff(x)+(1-ν)*v.diff(y))).simplify()
σxyg = (G*(v.diff(x)+u.diff(y))).expand().collect(G)

display(Math(f" \\text{{General expressions for stresses}}:"))
display(Math(f" \\text{{$σ_{{xx}}$}}={latex(σxxg)}"))
display(Math(f" \\text{{$σ_{{yy}}$}}={latex(σyyg)}"))
display(Math(f" \\text{{$σ_{{xy}}$}}={latex(σxyg)}"))
```

General expressions for stresses:

$$\sigma_{xx} = -\frac{2G\left(\nu\left(b_1 + 2b_2y + b_4x\right) - (\nu - 1)\left(a_3 + a_4y + 2a_5x\right)\right)}{2\nu - 1}$$

$$\sigma_{yy} = -\frac{2G\left(\nu\left(a_3 + a_4y + 2a_5x\right) - (\nu - 1)\left(b_1 + 2b_2y + b_4x\right)\right)}{2\nu - 1}$$

$$\sigma_{xy} = G\left(-2Ha_2 + 2a_2y + a_4x + b_3 + b_4y + 2b_5x\right)$$

We can now apply SBCs for stresses on these two edges of the wedge dam. On the left vertical surface, where $x = 0$, we have

$$\sigma_{xx} = -py, \; \sigma_{xy} = 0 \tag{3.6}$$

On the slope surface of the wedge dam, where $x = y\tan\alpha$, we have

$$\begin{aligned} n_1\sigma_{xx}\Big|_{x=y\tan\alpha} + n_2\sigma_{xy}\Big|_{x=y\tan\alpha} &= 0 \\ n_1\sigma_{yy}\Big|_{x=y\tan\alpha} + n_2\sigma_{xy}\Big|_{x=y\tan\alpha} &= 0 \end{aligned} \tag{3.7}$$

where n_1 and n_2 are the direction cosines of the outward normal with respect to the x- and y-axes. Thus, $n_1 = \cos\alpha$, $n_2 = -\cos\alpha$. These conditions must be satisfied at any y. Therefore, all the coefficients for the monomials y^1 and y^0 in these expressions must vanish. This gives a new set of seven conditions, which can be used to solve for some of the remaining constants.

```
SBCs = []
SBCs.append((σxxg.subs(x,0)+p*y))                          # SBC equations
SBCs.append((σxxg.subs(x,0)+p*y))
SBCs.append(σxyg.subs(x,0))
SBCs.append(σxyg.subs(x,0))
SBCs.append((n1*σxxg.subs(x,y*tα)+n2*σxyg.subs(x,y*tα)))
SBCs.append((n1*σxxg.subs(x,y*tα)+n2*σxyg.subs(x,y*tα)))
SBCs.append((n2*σyyg.subs(x,y*tα)+n1*σxyg.subs(x,y*tα)))
SBCs.append((n2*σyyg.subs(x,y*tα)+n1*σxyg.subs(x,y*tα)))
```

```
SBC_exprs = gr.getExprs(SBCs, y, 1)        # get expressions for SBC eqns

sln_SBC = sp.solve(SBC_exprs, cs)
#display(Math(f" \\text{{Constants found}}:{latex(sln_7)}")) # lengthy
sln_cs = {**sln_DBC, **sln_SBC}            # update the unknown constants

# Using the current constants to update the stresses:
σxx = σxxg.subs(sln_cs).simplify(); σyy = σyyg.subs(sln_cs).simplify()
σxy = σxyg.subs(sln_cs).simplify()
#display(Math(f"σ_{{xx}}={latex(σxx)};\;\;\;\; σ_{{yy}}={latex(σyy)}"))
display(Math(f"σ_{{xy}} ={latex(σxy)}"))       # display one component

# Using the current constants to update the displacements:
u = u.subs(sln_cs).simplify(); v = v.subs(sln_cs).simplify()
#display(Math(f" \\text{{General displacement $u$}}={latex(u)}"))
#display(Math(f" \\text{{General displacement $v$}}={latex(v)}"))
```

```
Total number of expressions = 16
Number of nonzero expressions = 16
```

$$\sigma_{xy} = \frac{n_1 n_2 x \left(2Gb_4 t\alpha\nu - 4Gb_5\nu + 4Gb_5 + 2p\nu - p\right)}{n_1^2 t\alpha\nu^2 - n_1^2 t\alpha\nu - 2n_1 n_2\nu + 2n_1 n_2 + n_2^2 t\alpha\nu^2}$$

We obtained the updated stresses and displacements, which are quite lengthy. We will focus on one of these expressions. The stresses are linear in the coordinates, and they still contain unknown constants. We need to proceed further.

3.3.5 *Enforcing the equilibrium equations*

Using the updated stresses, we can now compute the equilibrium equation (2.1). For the wedge dam, its weight acts as the body force. Therefore, we have

$$\frac{\partial \sigma_{xx}(x,y)}{\partial x} + \frac{\partial \sigma_{yx}(x,y)}{\partial y} + 0 = 0 \tag{3.8a}$$

$$\frac{\partial \sigma_{xy}(x,y)}{\partial x} + \frac{\partial \sigma_{yy}(x,y)}{\partial y} + \rho g = 0 \tag{3.8b}$$

Since the stresses are linear, these two equations become constants, which can be imposed using the following snippet:

```
sln_cs = {**sln_DBC, **sln_SBC}
expr1 = (σxx.diff(x)+σxy.diff(y)).expand()
expr2 = (σxy.diff(x)+σyy.diff(y)).expand() + ρg
sln_eq = sp.solve([expr1,expr2],cs)

sln_cs = {**sln_DBC, **sln_SBC, **sln_eq}        # update the constants
sln_cs = {key: value.subs(sln_cs) for key, value in sln_cs.items()}

# Using the current constants to update the stresses:
dicα = {n1:n1_, n2:n2_, tα:tα_}
σxx_final = σxxg.subs(sln_cs).subs(dicα).simplify()
σyy_final = σyyg.subs(sln_cs).subs(dicα).simplify().expand().simplify()
σxy_final = σxyg.subs(sln_cs).subs(dicα).simplify()

print(f"σxx={σxx_final}, \nσyy={σyy_final}, \nσxy={σxy_final}")
#display(Math(f"\\text{{Stresses found: $σ_{{xx}}$}}={latex(σxx_final)};\
#     \; σ_{{yy}} ={latex(σyy_final)};\; σ_{{xy}} ={latex(σxy_final)}"))
```

```
σxx=-p*y,
σyy=-2*p*x/tan(α)**3 + p*y/tan(α)**2 + x*ρg/tan(α) - y*ρg,
σxy=-p*x/tan(α)**2
```

3.3.6 *Final stress solutions*

The final solutions for the stresses are found and rearranged as follows:

$$\begin{aligned}\sigma_{xx} &= -py \\ \sigma_{yy} &= -\frac{2px}{\tan^3(\alpha)} + \frac{py}{\tan^2(\alpha)} + \frac{x\rho g}{\tan(\alpha)} - y\rho g \\ \sigma_{xy} &= -\frac{px}{\tan^2(\alpha)}\end{aligned} \tag{3.9}$$

Clearly, the SBCs on the left vertical surface are satisfied. We can further verify the SBCs on the slope surface and check the equilibrium equations using the following snippet:

```
dicn = {**{x:y*tα_}, **dicα}      # for checking BCs on the slope surface
slopeBC1 = (n1*σxx_final+n2*σxy_final).subs(dicn).trigsimp().simplify()
slopeBC1 = (n2*σyy_final+n1*σxy_final).subs(dicn).trigsimp().simplify()
display(Math(f" \\text{{Is BCs at the slope surface satisfied? }} \
            {latex(slopeBC1==0)} \\text{{ and }} {latex(slopeBC1==0)}"))

#σyy = (ρg/sp.tan(α)-2*p/sp.tan(α)**3)*x + (p/sp.tan(α)**2-ρg)*y
display(Math(f" \\text{{Is equilibrium Eq. 1 = 0?}}=\
            {latex((σxx_final.diff(x)+σxy_final.diff(y)).simplify())}"))
display(Math(f" \\text{{Is equilibrium Eq. 2 = 0?}}=\
         {latex((σxy_final.diff(x)+σyy_final.diff(y)+ρg).simplify())}"))
```

Is BCs at the slope surface satisfied? True and True

Is equilibrium Eq. 1 = 0? = 0

Is equilibrium Eq. 2 = 0? = 0

3.3.7 *Final displacement solutions*

The final displacements can be computed using

```
# Using the current constants to update the displacements:
u_final = u.subs(sln_cs).subs(dicα).simplify().expand().trigsimp()
v_final = v.subs(sln_cs).subs(dicα).expand().simplify()
#display(u_final, v_final)                                # Lengthy
```

Since the expressions are lengthy and complicated, we write the following utility code function to group the terms based on their contribution to the force terms and to simplify the expressions.

```
def reform_expr(expr, terms):
    '''Re-write the expr by grouping based on the terms given.'''
    expr = expr.simplify()
    de_expr=sp.denom(expr)              # get the denominator of the expr
    nu_expr=sp.numer(expr)                # get the numerator of the expr
    reformed_term = []
    for term in terms:      # group the items of a term in the numerator
        reformed_term.append((nu_expr.expand().coeff(term)/de_expr)\
                                                          .simplify())
    return reformed_term
```

```
# Define symbolic variables for function of force contributors:
fp, fg =sp.symbols("fp, fg")

contributors = [p,  ρg]                         # contributors to results
f_contributors=[fp, fg]                       # functions of contributors

# get the functions of contributors to u:
u_terms = reform_expr(u_final, contributors)
u_terms = list(zip(f_contributors, u_terms))
display(Math(f"\\text{{Contributions to $u$}}:"))
display(Math(f"\hspace{{0.1cm}} \\scriptsize {latex(u_terms[0])}"))
display(Math(f"\hspace{{0.1cm}} \\scriptsize {latex(u_terms[1])}"))

# get the functions of contributors to v:
v_terms = reform_expr(v_final, contributors)
v_terms = list(zip(f_contributors, v_terms))
display(Math(f"\\text{{Contributions to $v$}}:"))
display(Math(r"\;\scriptsize{"+latex(v_terms[0]).replace('- 2 x^{2}',\
                                          ' \\\\ - 2 x^{2}')+"}"))
display(Math(f"\hspace{{0.1cm}} \\scriptsize {latex(v_terms[1])}"))
```

Contributions to u:

$$\left(fp, \frac{-H^2\nu + H^2 + 2Hy\nu - 2Hy + x^2\nu + xy\nu\tan^3(\alpha) - xy\nu\tan(\alpha) - xy\tan^3(\alpha) - y^2\nu + y^2}{2G\tan^3(\alpha)}\right)$$

$$\left(fg, \frac{H^2\nu - H^2 - 2Hy\nu + 2Hy - x^2\nu + 2xy\nu\tan(\alpha) + y^2\nu - y^2}{4G\tan(\alpha)}\right)$$

Contributions to v:

$$\left(fp, \frac{-H^2\nu\tan^3(\alpha) + H^2\nu\tan(\alpha) - H^2\tan(\alpha) - 4Hx\nu + 4Hx - x^2\nu\tan^3(\alpha) + x^2\nu\tan(\alpha) + x^2\tan^3(\alpha) - 2x^2\tan(\alpha) + 4xy\nu - 4xy + y^2\nu\tan^3(\alpha) - y^2\nu\tan(\alpha) + y^2\tan(\alpha)}{4G\tan^3(\alpha)}\right)$$

$$\left(fg, \frac{-H^2\nu + H^2 + \frac{2Hx\nu}{\tan(\alpha)} - \frac{2Hx}{\tan(\alpha)} - x^2\nu - \frac{2xy\nu}{\tan(\alpha)} + \frac{2xy}{\tan(\alpha)} + y^2\nu - y^2}{4G}\right)$$

The simplified formulas for the final displacement solutions are written in the following unified form:

$$u \text{ or } v = pf_p + \rho g f_g \tag{3.10}$$

where f_p represents the function of the contribution from the pressure force term p and f_g represents the function of the contribution from the gravity force term ρg. All these contribution functions are provided in the outputs from the previous code cell for both u and v.

The displacement solution can also be rearranged manually as follows:

$$\begin{aligned}
u =& \frac{-p}{2G\tan^3\alpha}\left[H(H-2y)(\nu-1) - x^2\nu + xy\left(\nu + (1-\nu)\tan^2\alpha\right)\right. \\
& \left.\times \tan\alpha + y^2(\nu-1)\right] \\
& + \frac{\rho g}{4G\tan\alpha}\left[H(H-2y)(\nu-1) - x^2\nu + 2xy\nu\tan\alpha + y^2(\nu-1)\right] \\
v =& \frac{-p}{4G\tan^3\alpha}\Big[4x(H-y)(\nu-1) + \left[H^2(\nu\tan^2\alpha - \nu + 1)\right. \\
& \left. + x^2\left((\nu-1)\tan^2\alpha + 2 - \nu\right) + y^2(\nu - 1 - \nu\tan^2\alpha)\right]\tan\alpha\Big] \\
& - \frac{\rho g}{4G\tan\alpha}\left[2x(-H+y)(\nu-1) + \left(H^2(\nu-1) + x^2\nu + y^2(1-\nu)\right)\tan\alpha\right]
\end{aligned} \tag{3.11}$$

The displacement is quadratic in both x and y. To the best of the author's knowledge, this may be the first time a closed-form solution for the displacements of wedges has been obtained. The largest displacements occur at the tip of the dam.

```
#u_final.subs(dic0),v_final.subs(dic0),u_final.diff(y).subs(dic0) #check

dic_tip = {x:0,y:0}                    # the largest displacements at the tip
u_tip = u_final.subs(dic_tip).collect(H**2).collect(tan(α)).simplify()
v_tip = v_final.subs(dic_tip).collect(H**2).collect(tan(α)**2)

display(Math(f"\\text{{At the dam tip, }} u= {latex(u_tip)}"))
display(Math(f"\\text{{At the dam tip, }} v= {latex(v_tip)}"))
u_tip_ = H**2*(-2*p*(ν - 1) - ρg*(1 - ν)*tan(α)**2)/(4*G*tan(α)**3)
v_tip_ = H**2*(-p*ν - ν*ρg + ρg + (p*ν - p)/tan(α)**2)/(4*G)

# Other interest locations:
#dicx0 = {x:0}  # for computing the displacement along the vertical edge
#u_final.subs(dicx0).simplify(),v_final.subs(dicx0).simplify()
#dic_slope = {x:y*tα_}          # for the displacement along the slope edge
#u_final.subs(dic_slope),v_final.subs(dic_slope).subs(dic_slope)
```

At the dam tip, $u = \dfrac{H^2\left(-2p\left(\nu-1\right)-\rho g\left(1-\nu\right)\tan^2\left(\alpha\right)\right)}{4G\tan^3\left(\alpha\right)}$

At the dam tip, $v = \dfrac{H^2\left(-p\nu-\nu\rho g+\rho g+\frac{p\nu-p}{\tan^2\left(\alpha\right)}\right)}{4G}$

3.3.8 *Checking integral equilibrium conditions*

Consider an FBD of the top part of the wedge dam, cut at any y-location, as shown in Fig. 3.4. The surface at the bottom of the FBD has a distributed normal stress σ_{yy}, which contributes to the force in the y-direction and the moment with respect to the z-axis at point $(y,0)$. The distributed shear stress σ_{xy} contributes to the force in the x-direction.

The integral equilibrium requires all the forces in the x-direction be zero, all the forces in the y-direction be zero, and all the moments with respect to the z-axis at point $(y,0)$ be zero:

$$\sum F_x = 0\colon \int_0^{y\tan\alpha} \sigma_{xy}dx + \frac{py^2}{2} = 0$$

$$\sum F_y = 0\colon \int_0^{y\tan\alpha} \sigma_{yy}dx + \frac{\rho g y^2\tan(\alpha)}{2} = 0$$

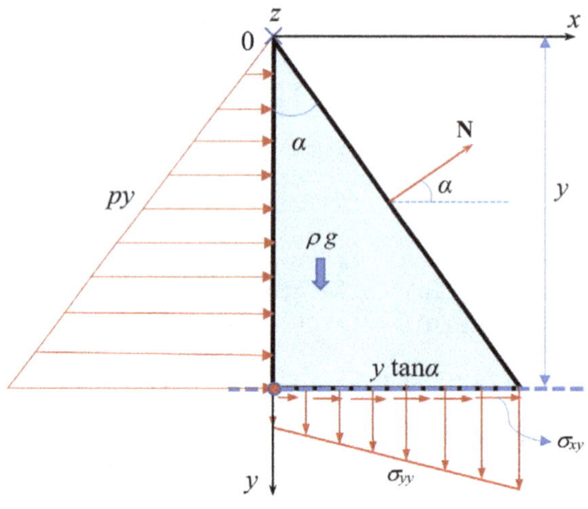

Figure 3.4. An FBD of the upper part of the wedge dam with height y. It is subjected to water pressure and gravity force. The distribution of stresses is schematically shown on the bottom cut surface.

$$\sum M_z = 0\text{:} \quad \int_0^{y\tan\alpha} \sigma_{yy} x dx + \frac{py^3}{6} + \frac{\rho g y^3 \tan^2(\alpha)}{6} = 0 \qquad (3.12)$$

Using the stresses obtained, we can confirm that these integral equilibrium conditions are satisfied.

```
Int = sp.integrate
FBD_ΣFx = (Int(σxy_final,    (x, 0, y*tα_)).simplify() + p*y**2/2)
FBD_ΣFy = (Int(σyy_final,    (x, 0, y*tα_)) + ρg*y*y*tα_/2).simplify()
FBD_ΣMz = (Int(σyy_final*x, (x, 0, y*tα_))
                              + p*y**3/6 + ρg*y**3*tα_**2/6).simplify()
display(Math(f" \\text{{Equilibrium of the free body diagram? }}\
        {latex(FBD_ΣFx==0)}; {latex(FBD_ΣFy==0)};{latex(FBD_ΣMz==0)}"))
```

Equilibrium of the free body diagram? True; True; True

The solution given in Eq. (3.9) is the same as that presented in Ref. [3]. As observed, all these stresses are linear in both x and y. The normal stress σ_{xx} and shear stress σ_{xy} result solely from the water pressure p. The normal stress σ_{yy} arises from both p and gravity ρg. The largest stresses occur along the base at $y = H$. Due to the linear distributions, the critical stresses are found at either $(0, H)$ or $(H \tan\alpha, H)$. The formulas are as follows:

```
display(Math(f" \\text{{At $(0,H)$, $σ_{{xx}}$ }}=\
                          {latex(σxx_final.subs({x:0,y:H}))}"))
display(Math(f" \\text{{At ($H$ tanα, $H$), $σ_{{xx}}$ }}=\
                       {latex(σxx_final.subs({x:y*tα_,y:H}))}"))
display(Math(f" \\text{{At $(0,H)$, $σ_{{yy}}$ }}=\
                          {latex(σyy_final.subs({x:0,y:H}))}"))
display(Math(f" \\text{{At ($H$ tanα, $H$), $σ_{{yy}}$ }}=\
                       {latex(σyy_final.subs({x:y*tα_,y:H}))}"))
display(Math(f" \\text{{At ($H$ tanα, $H$), $σ_{{xy}}$ }}=\
                       {latex(σxy_final.subs({x:y*tα_,y:H}))}"))
```

$$\text{At } (0, H),\ \sigma_{xx} = -Hp$$

$$\text{At } (H \tan\alpha,\ H),\ \sigma_{xx} = -Hp$$

$$\text{At } (0, H),\ \sigma_{yy} = \frac{Hp}{\tan^2(\alpha)} - H\rho g$$

$$\text{At } (H \tan\alpha,\ H),\ \sigma_{yy} = -\frac{2Hp\sin(\alpha)}{\cos(\alpha)\tan^3(\alpha)} + \frac{Hp}{\tan^2(\alpha)} + \frac{H\rho g\sin(\alpha)}{\cos(\alpha)\tan(\alpha)} - H\rho g$$

$$\text{At } (H \tan\alpha,\ H),\ \sigma_{xy} = -\frac{Hp\sin(\alpha)}{\cos(\alpha)\tan^2(\alpha)}$$

3.4 A general displacement method

This section introduces a general displacement method for solving solid mechanics problems defined in 2D rectangular domains. The method is automated and can be applied as long as the problem is well-posed. We use a 2D rectangular cantilever subjected to higher-order distributed forces as an example due to its wide range of applications.

3.4.1 *Problem setup and coordinates*

Consider a 2D rectangular cantilever of length L, as shown in Fig. 3.5. It has a height h and is subjected to a distributed force q along the bottom edge, a force P and moment M on the right edge, and gravity force ρg over the cantilever body. Due to its relatively long length in the x-direction and small thickness (assumed to be unit thickness), this problem can be treated as a plane stress problem. At the root point $(y = 0, x = 0)$, the following displacement conditions are assumed:

$$u = 0, \quad v = 0, \quad \frac{\partial v}{\partial x} = 0 \tag{3.13}$$

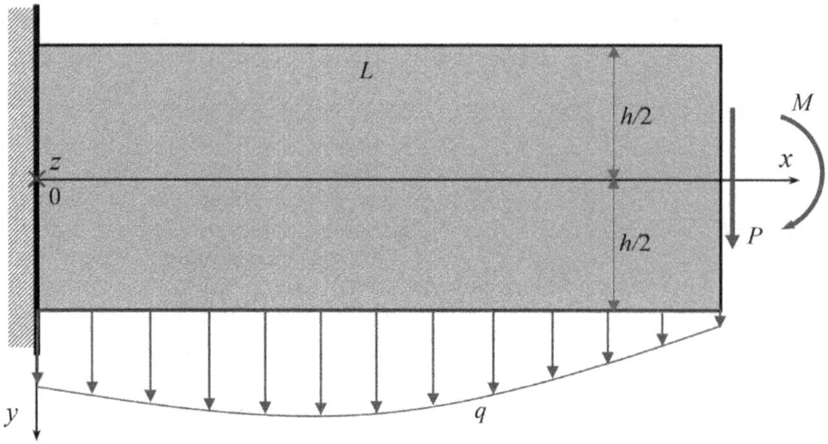

Figure 3.5. A 2D cantilever of length L and height h, subjected to multiple loads.

Our goal is to derive the formulas for the stresses and displacements in the cantilever using the displacement method.

3.4.2 *General expression of assumed displacements*

The cantilever has four boundaries. Since it is relatively long in the x-direction, the top and bottom edges are treated as the primary stress boundaries, where the SBCs will be rigorously enforced. For the right edge, we will use the integral form of the FBCs to ensure equilibrium. The left edge is defined by displacement conditions at the root point. The distribution of displacements and stresses along the left edge can be obtained once the problem is solved, following the general procedure for a displacement method.

In this study, we assume that the distributed force q varies quadratically in y (which can be adjusted as needed). Given that the displacement functions should be three orders higher than those in the case of no distributed force, we use monomial basis functions up to the sixth order. Since the loads are multiple, it is challenging to predict the symmetry of the displacements. Therefore, we will utilize all these monomial basis functions.

The following code defines the problem and creates the basis functions for the displacements:

```
# Define necessary symbolic variables for coordinates and parameters:
x, y = symbols("x, y", real=True)
h, L = symbols("h, L", positive=True)                          # Geometry
E, G, ν = symbols("E, G, ν", positive=True)                     # Property
```

```
# Parameters for loads:
q0, q1, q2, P, M, ρg = symbols("q0, q1, q2, P, M, ρg", positive=True)
q = q0 + q1*x + q2*x**2                       # distributed load, quadratic

σxx, σyy, σxy = sp.symbols("σ_xx, σ_yy, σ_xy")
εxx, εyy, εxy   = sp.symbols("ε_xx, ε_yy, ε_xy")

order = 6                     # Set to use upto complete 6th order monomials
# Generate the complete set of monomial basis functions in 2D:
bfs_list = [x**i*y**j for i in range(order + 1)
                          for j in range(order + 1) if i+j <= order]
ij = [(i, j) for i in range(order + 1)
                          for j in range(order + 1) if i+j <= order-2]
bfs = sp.Matrix(bfs_list)           # form sympy vector with these bases

u_bfs = bfs_list # all bfs for u(x,y) do not use any symmetry condition
display(Math(f" \\text{{u_bfs}}={latex(u_bfs[:15])}...")) # Bases for u

v_bfs = bfs_list                                   # all bfs for u(x,y)
display(Math(f" \\text{{v_bfs}}={latex(v_bfs[:15])}...")) # Bases for v

#display(Math(f" \\text{{Complete 6th order monomials in 2D}}:"))
#display(bfs_list); display(ij)
```

$$\text{u\_bfs} = \left[1,\ y,\ y^2,\ y^3,\ y^4,\ y^5,\ y^6,\ x,\ xy,\ xy^2,\ xy^3,\ xy^4,\ xy^5,\ x^2,\ x^2y\right]\ldots$$

$$\text{v\_bfs} = \left[1,\ y,\ y^2,\ y^3,\ y^4,\ y^5,\ y^6,\ x,\ xy,\ xy^2,\ xy^3,\ xy^4,\ xy^5,\ x^2,\ x^2y\right]\ldots$$

Based on the number of basis functions used, we generate the necessary symbolic unknown constants in matrix form using SymPy. This step is repeated for each displacement component. The general expressions for these displacement components are formed as the dot product of the basis functions and their corresponding constants. Since no symmetry conditions are applied, the resulting expressions are expected to be lengthy.

```
# unknown constants for displacements:
ai = sp.symbols(f'a:{len(u_bfs)}')                    # unknown constants
print(f'Number of terms used = {len(u_bfs)}')
print(f'Total number of unknown constants = {2*len(u_bfs)}')

display(Math(f" \\text{{constants for $u$}}={latex(ai[:10])}..."))
ai = Matrix([ai[i] for i in range(len(u_bfs))]) # form coeffs. in vector

bi = sp.symbols(f'b:{len(v_bfs)}')                    # unknown constants
display(Math(f" \\text{{constants for $v$}}={latex(bi[:10])}..."))
bi = Matrix([bi[i] for i in range(len(v_bfs))]) # form coeffs. in vector

cs = ai.col_join(bi)                  # form a single vector of constants
```

```
cs = ai.col_join(bi)                      # form a single vector of constants

ug = (ai.T@Matrix(u_bfs))[0]              # Form the general displacements
vg = (bi.T@Matrix(v_bfs))[0]
display(Math(f"\\text{{Assumed displacements,$\;u(x,y)$}}=\
                                        {latex(ug.args[:7])}..."))
display(Math(f"\hspace{{4cm}}\\text{{$v(x,y)$}}=\
                                        {latex(vg.args[:8])}..."))
```

```
Number of terms used = 28
Total number of unknown constants = 56
```

Constants for $u = (a_0,\ a_1,\ a_2,\ a_3,\ a_4,\ a_5,\ a_6,\ a_7,\ a_8,\ a_9)\ldots$

Constants for $v = (b_0,\ b_1,\ b_2,\ b_3,\ b_4,\ b_5,\ b_6,\ b_7,\ b_8,\ b_9)\ldots$

Assumed displacements,
$u(x,y) = \left(a_0,\ a_1 y,\ a_{13}x^2,\ a_{18}x^3,\ a_2 y^2,\ a_{22}x^4,\ a_{25}x^5\right)\ldots$

$v(x,y) = \left(b_0,\ b_1 y,\ b_{13}x^2,\ b_{18}x^3,\ b_2 y^2,\ b_{22}x^4,\ b_{25}x^5,\ b_{27}x^6\right)\ldots$

3.4.3 *Enforcing displacement boundary conditions*

The DBCs for this problem are given in Eq. (3.13). These can be imposed using the following snippet:

```
# Solve for unknown constants using DBCS and update:
dic0 = {x:0, y:0}
DBCs = [ug.subs(dic0), vg.subs(dic0), ug.diff(y).subs(dic0)]      # DBCs
sln_DBC = sp.solve(DBCs, cs)

sln_cs = {**sln_DBC}   # update the solutions for the unknown constants
print(f" Len of sln_cs (number of constants found) = {len(sln_cs)}")

u = ug.subs(sln_cs).simplify()                  # update the displacements
v = vg.subs(sln_cs).simplify()
display(Math(f" \\text{{Current $u$}}={latex(u.args[:8])}..."))
display(Math(f" \\text{{Current $v$}}={latex(v.args[:8])}..."))
display(Math(f" \\text{{Constants found using displacement BCs}}=\
                                                {latex(sln_cs)}"))
```

```
Len of sln_cs (number of constants found) = 3
```

Current $u = \left(a_{13}x^2,\ a_{18}x^3,\ a_2 y^2,\ a_{22}x^4,\ a_{25}x^5,\ a_{27}x^6,\ a_3 y^3,\ a_4 y^4\right)\ldots$

Current $v = \left(b_1 y,\ b_{13}x^2,\ b_{18}x^3,\ b_2 y^2,\ b_{22}x^4,\ b_{25}x^5,\ b_{27}x^6,\ b_3 y^3\right)\ldots$

Constants found using displacement BCs $= \{a_0\colon 0,\ a_1\colon 0,\ b_0\colon 0\}$

We have found three of the constants (possibly in terms of the other constants) using the DBCs. The displacement expressions can now be used to compute the stresses, which will then allow us to enforce the SBCs.

3.4.4 *Enforcing stress boundary conditions*

The general expressions for the three stress components are obtained using the constitutive equations and strain–displacement relations. Since this cantilever is considered a plane stress problem, the constitutive equations for plane stress should be applied. The snippet for this computation is as follows:

```
# Use the Stress-Displacement relations:
plane_stress = True
if plane_stress:
    # Compute stresses using displacement-function, plane stress:
    σxxg=(2*G/(1-ν)*(ug.diff(x)+ν*vg.diff(y))).simplify()
    σyyg=(2*G/(1-ν)*(vg.diff(y)+ν*ug.diff(x))).simplify()
    σxyg=(G*(vg.diff(x)+ug.diff(y))).expand().collect(G*y)
else:
    # Compute stresses using displacement-function, plane strain:
    σxxg=(2*G/(1-2*ν)*((1-ν)*u.diff(x)+ν*v.diff(y))).simplify()
    σyyg=(2*G/(1-2*ν)*(ν*u.diff(x)+(1-ν)*v.diff(y))).simplify()
    σxyg=(G*(v.diff(x)+u.diff(y))).simplify().cancel().expand()

#display(Math(f"\\text{{Stress general expressions (few terms)}}:"))
#display(Math(f"\\text{{$σ_{{xx}}$}}={latex(σxxg)}"))          # Lengthy
#display(Math(f"\\text{{$σ_{{yy}}$}}={latex(σyyg)}"))
#display(Math(f"\\text{{$σ_{{xy}}$}}={latex(σxyg)}"))
```

The current expressions for these stresses are quite lengthy. Readers may review them by uncommenting the lines for displaying the results.

We can now apply SBCs for stresses on both the top and bottom edges of the cantilever. On the top edge, where $y = -h/2$, both stress components, σ_{yy} and σ_{xy}, must be zero. On the bottom edge, where $y = h/2$, the normal stress σ_{yy} must equal the externally applied downward load q (which is quadratic in x), and the shear stress σ_{xy} should be zero. Thus, the following SBCs are required:

$$\begin{aligned} \sigma_{yy} &= 0,\ \sigma_{xy} = 0 \quad \text{on } y = -h/2 \\ \sigma_{yy} &= q,\ \sigma_{xy} = 0 \quad \text{on } y = h/2 \end{aligned} \tag{3.14}$$

These conditions must be satisfied at any x. Therefore, the coefficients for monomials of all orders in x in these expressions for the SBCs must vanish.

This provides a set of conditions which can be used to solve for some of the remaining constants.

```
# Stress BCs on the top and bottom edges, y = ±h/2:

eqn1 = (σyyg.subs(y, h/2) - q).simplify().cancel().expand()
eqn2 = (σyyg.subs(y,-h/2)).simplify().cancel().expand()
eqn3 = (σxyg.subs(y, h/2)).simplify().cancel().expand()
eqn4 = (σxyg.subs(y,-h/2)).simplify().cancel().expand()

SBCs = []
for i in range(order):
    SBCs.append(eqn1.coeff(x,i))
    SBCs.append(eqn2.coeff(x,i))
    SBCs.append(eqn3.coeff(x,i))
    SBCs.append(eqn4.coeff(x,i))
```

Since these expressions are generated automatically, some of the SBC expressions may turn out to be zero. To account for this, we write the following utility snippet to remove any possible zero expressions. If no zero expressions are found, the code will simply do nothing:

```
def remove_zero_exprs(exprs):
    print(f" Total number of expressions = {len(exprs)}")
    expr0s = [i for i, expr in enumerate(exprs) if expr == 0]
    print(f" The Index of the zero expressions: {expr0s}")
    exprs = [exprs[i] for i in range(len(exprs)) if i not in expr0s]
    print(f" Number of nonzero expressions = {len(exprs)}")
    return exprs
```

```
# Use SBCs to solve for unknown constants and update:

SBCs = remove_zero_exprs(SBCs)

sln_cs = {**sln_DBC} # update the solutions for the unknown constants
print(f" Current length of sln_cs = {len(sln_cs)}")

sln_SBC = sp.solve(SBCs, cs)  # solve for unknown constants and update
if len(sln_SBC)==0: print("Alert: no solution is found!")
sln_SBC = {key: value.subs(sln_cs).simplify().cancel() \
                                    for key, value in sln_SBC.items()}
print(f" Number of stress BCs = {len(SBCs)}")
#display(Math(f" \\text{{Constants found}}:{latex(sln_SBC)}")) #Lengthy
sln_cs = {**sln_DBC, **sln_SBC}        # update the unknown constants
sln_cs = {key: value.subs(sln_cs) for key, value in sln_cs.items()}
```

```
# Using the current constants to update the stresses:
σxx = σxxg.subs(sln_cs).simplify()
σyy = σyyg.subs(sln_cs).simplify()
σxy = σxyg.subs(sln_cs).simplify()
#display(Math(f"σ_{{xx}}={latex(σxx)};\;\;\;\; σ_{{yy}}={latex(σyy)}"))
#display(Math(f"σ_{{xy}} ={latex(σxy)}"))                       # Lengthy
```

```
Total number of expressions = 24
The Index of the zero expressions: []
Number of nonzero expressions = 24
Current length of sln_cs = 3
Number of stress BCs = 24
```

```
# Check SBCs on top and bottom surfaces:
σyy.subs(y, h/2).simplify(), σxy.subs(y, h/2).simplify(),\
σyy.subs(y,-h/2).simplify(), σxy.subs(y,-h/2).simplify()
```

$$\left(q_0 + q_1 x + q_2 x^2,\ 0,\ 0,\ 0\right)$$

We obtained the updated stresses and displacements (which are quite lengthy), and the BCs are satisfied.

3.4.5 *Enforcing equilibrium equations*

Using the updated stresses, we can now compute the differential form of the equilibrium equation Eq. (2.1). For the cantilever, gravity acts as the body force. Therefore, we have

$$\frac{\partial \sigma_{xx}(x,y)}{\partial x} + \frac{\partial \sigma_{yx}(x,y)}{\partial y} + 0 = 0 \tag{3.15a}$$

$$\frac{\partial \sigma_{xy}(x,y)}{\partial x} + \frac{\partial \sigma_{yy}(x,y)}{\partial y} + \rho g = 0 \tag{3.15b}$$

These two equilibrium equations are imposed using the following snippet. As before, since these conditions must be satisfied for any x and y, all the coefficients for each of the monomials, including the mixed ones $x^i y^j$ (recorded earlier when the bases were generated), must vanish. This results in a new set of equations, which can be quite large.

```
# Enforce the differential equilibrium equations:

sln_cs = {**sln_DBC, **sln_SBC}                          # solve and update
sln_cs = {key: value.subs(sln_cs) for key, value in sln_cs.items()}
σxx = σxxg.subs(sln_cs).simplify().cancel()
σyy = σyyg.subs(sln_cs).simplify().cancel().expand().simplify()
σxy = σxyg.subs(sln_cs).simplify().cancel()

# The equilibrium equations:
eqn1 = (σxx.diff(x)+σxy.diff(y)).simplify().cancel().expand()
eqn2 = (σxy.diff(x)+σyy.diff(y)+ρg).simplify().cancel().expand()

print(f" Number of monomials in stress PDEs = {len(ij)}")
Equils = []
for i in range(len(ij)):
    Equils.append(eqn1.coeff(x,ij[i][0]).coeff(y,ij[i][1]))
for i in range(len(ij)):
    Equils.append(eqn2.coeff(x,ij[i][0]).coeff(y,ij[i][1]))

Equils = remove_zero_exprs(Equils)

sln_Equil = sp.solve(Equils, cs) # solve for unknown constants & update
```

```
Number of monomials in stress PDEs = 15
Total number of expressions = 30
The Index of the zero expressions: [27, 29]
Number of nonzero expressions = 28
```

```
# Update the stresses:

sln_cs = {**sln_DBC, **sln_SBC, **sln_Equil}
sln_cs = {key: value.subs(sln_cs) for key, value in sln_cs.items()}
print(f" Current len of sln_cs = {len(sln_cs)}; Total cs = {len(cs)}")

# Using the current constants to update the stresses:
σxx = σxxg.subs(sln_cs).simplify().cancel()
σyy = σyyg.subs(sln_cs).simplify().cancel().expand().simplify()
σxy = σxyg.subs(sln_cs).simplify().cancel()

#display(Math(f"\\text{{Stresses found: $σ_{{xx}}$}}={latex(σxx)};\
#           σ_{{yy}} ={latex(σyy)}; σ_{{xy}} ={latex(σxy)}"))  # lengthy

# Check how many constants are found and how many unknowns remain:
list_keys = list(sln_cs.keys())
cs_Eql = list(set(list(cs))-set(list_keys))
print(f" Constants yet to be found: {cs_Eql}")
```

```
Current len of sln_cs = 52; Total cs = 56
Constants yet to be found: [b13, b1, b18, b7]
```

We have updated all the stresses and displacements using the constants found so far. However, there are still four unknown constants yet to be determined.

3.4.6 *Enforcing integral equilibrium conditions*

Next, we apply the integral equilibrium conditions for the right edge. The distributed normal stress σ_{xx} on the cross-section of the cantilever contributes to the force in the x-direction and to the moment $\boldsymbol{M}$ with respect to the z-axis. The distributed shear stress σ_{xy} on the cross-section contributes to the force $\boldsymbol{P}$ in the y-direction, as shown in Fig. 3.6. Thus, at any arbitrary x, we have

$$\begin{aligned}
&\sum F_x = 0\colon \int_{-h/2}^{h/2} \sigma_{xx} dy = 0 \\
&\sum F_y = 0\colon \int_{-h/2}^{h/2} \sigma_{xy} dy + P + \int_{x}^{L} (q + \rho g h) dt = 0 \qquad (3.16) \\
&\sum M_z = 0\colon \int_{-h/2}^{h/2} \sigma_{xx} y dy + P(L - x) + M + \int_{x}^{L} (q + \rho g h) t dt = 0
\end{aligned}$$

Note that the pressure and the body force must be involved in these FBCs, as shown in the foregoing equations. We need to integrate along x over (x, L), where the integral variable t is used [2]. The equations can, in

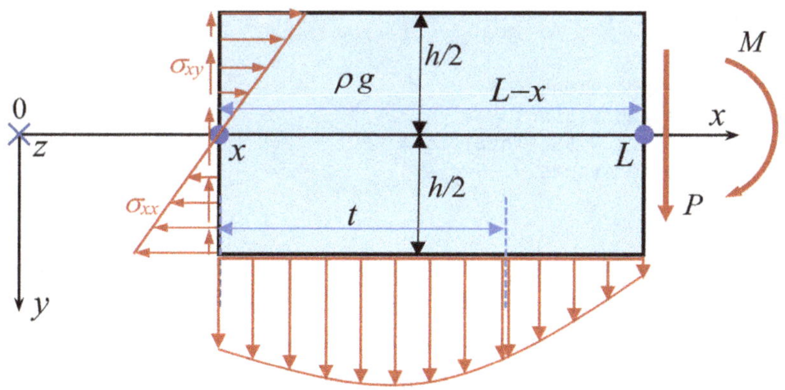

Figure 3.6. An FBD of the left portion of the 2D rectangular beam with a length of x. The beam is subjected to a downward force and moment at the right edge, and the distribution of stresses is schematically illustrated on the cut surface.

general, be a function of x. In our problem, the second equation is clearly linear in x. Therefore, we need all coefficient expressions for monomials x^1 and x^0 to be zero. We should use **gr.getExprs()** developed in the previous chapter to form all these expressions. This gives us another set of three conditions, which is exactly the number needed to determine the three remaining constants. The snippet is as follows:

```
# Integral equilibrium equations on cross-section at x:
t = sp.symbols('t')                                  # variable for integration only
pt = (q + ρg*h).subs(x, x+t)
Int = sp.integrate
interval_y = (y, -h/2, h/2)
interval_x = (t, 0, L-x)
FBC_ΣFx = Int(σxx,   interval_y).simplify() - 0
FBC_ΣFy = (-Int(σxy, interval_y) + Int(pt,   interval_x) + P ).simplify()
FBC_ΣMz = (Int(σxx*y,interval_y) + Int(pt*t,interval_x) \
                                          + P*(L-x) - M).expand()
#display(Math(f"\\text{{Integral equilibrium on cross-section at $x$:\
#              }} {latex(FBC_ΣFx)}; {latex(FBC_ΣFy)}; {latex(FBC_ΣMz)}"))

FBC_exprs = gr.getExprs([FBC_ΣFx, FBC_ΣFy, FBC_ΣMz], x, 1)
sln_FBC = sp.solve(FBC_exprs, cs)
if len(sln_FBC)==0: print("Alert: no solution is found!")

sln_cs = {**sln_DBC, **sln_SBC, **sln_Equil, **sln_FBC}
sln_cs = {key: value.subs(sln_cs) for key, value in sln_cs.items()}
print(f" Current len of sln_cs = {len(sln_cs)}; Total cs = {len(cs)}")
cs_final = {key: value.simplify() \
                          for key, value in sln_cs.items() if value != 0
print(f" Number of nonzero constants = {len(cs_final)}")
#display(Math(f" \\text{{ Nonzero constants}}= {latex(cs_final)}"))
```

```
Total number of expressions = 6
Number of nonzero expressions = 4
Current len of sln_cs = 55; Total cs = 56
Number of nonzero constants = 35
```

We have successfully found all the unknown constants. For this cantilever problem, we started with a total of 56 constants: 35 of them are nonzero, and the rest are zero. Note that all of this is done automatically without much manual operation, which differs significantly from stress methods. This is because the displacement method does not usually use any indefinite integrals, which would introduce additional unknown functions and constants.

3.4.7 *Final solution for the stresses*

We are now ready to produce the full set of solutions. The following snippet produces the final formulas for the displacements and then for the stresses for the 2D cantilever:

```
# Using the obtained constants to produce the displacement formulas:
u_final = ug.subs(sln_cs).simplify()
v_final = vg.subs(sln_cs).simplify()

# Using the obtained constants to produce the stress formulas:
σxx_final = σxxg.subs(sln_cs).simplify()
σyy_final = σyyg.subs(sln_cs).simplify()
σxy_final = σxyg.subs(sln_cs).simplify()
```

The formulas for the stresses are quite lengthy and are expressed in the following unified form:

$$\sigma_{ij} = Mf_M + Pf_P + \rho g f_g + q_0 f_{q0} + q_1 f_{q1} + q_2 f_{q2} \tag{3.17}$$

where $M, P, \rho g, q_0, q_1, q_2$ are the terms defining the forces applied on the cantilever, $i, j = x, y$, and f_k is a function representing the contribution from the force term k.

All these contribution functions are computed using the following snippet for each of the stress components. Since the expressions are quite lengthy and complicated, we provide the following utility function to group the terms based on the contribution of these force terms and simplify the expressions:

```
# Define symbolic variables for function of force contributors:
fM, fP, fg, fq0, fq1, fq2=sp.symbols("f_M, f_P, f_g, f_q0, f_q1, f_q2")

contributors = [ M,  P, ρg,  q0,  q1,  q2]    # contributors to results
f_contributors=[fM, fP, fg, fq0, fq1, fq2]  # functions of contributors

# get the functions of contributors to σxx:
σxx_terms_ = reform_expr(σxx_final, contributors)
σxx_terms = list(zip(f_contributors, σxx_terms_))    # convert to tuble
display(Math(f"\\text{{Contributions to $σ_{{xx}}$}}:"))
display(Math(f"\hspace{{0.1cm}} \small {latex(σxx_terms[:3])}"))
display(Math(f"\hspace{{0.1cm}} \\scriptsize {latex(σxx_terms[3:5])}"))
display(Math(r"\;\small {"+latex(σxx_terms[5]).replace('y^{2} -',\
                                                    'y^{2} -\\\\')+"}"))
```

Contributions to σ_{xx}:

$$\left[\left(f_M,\ \frac{12y}{h^3}\right),\right.$$

$$\left.\left(f_P,\ \frac{12y\left(-L+x\right)}{h^3}\right),\ \left(f_g,\ \frac{y\left(-30L^2+60Lx-3h^2-30x^2+20y^2\right)}{5h^2}\right)\right]$$

$$\left[\left(f_{q0},\ \frac{y\left(-30L^2+60Lx-3h^2-30x^2+20y^2\right)}{5h^3}\right),\right.$$

$$\left.\left(f_{q1},\ \frac{y\left(-20L^3+30L^2x-3h^2x-10x^3+20xy^2\right)}{5h^3}\right)\right]$$

$$\left(f_{q2},\ \frac{\begin{array}{c}-25200L^4y+33600L^3xy+350h^5+261h^4y-4200h^3y^2\\ -5040h^2x^2y-840h^2y^3-8400x^4y+33600x^2y^3-5040y^5\end{array}}{8400h^3}\right)$$

Once we group our solution in this manner, we find the following from these results:

1. The highest (combined) order of the stress is 4 and is contributed by q_1, which is the second-order term in the distributed force q.
2. When the order of the distributed force reduces by one, so does the order of the stresses. For the point forces $\boldsymbol{M}$ and $\boldsymbol{P}$, their contribution is linear.
3. We can expect that if the order of the distributed force increases further, we would need to increase the order of the monomial basis functions when approximating the displacements at the beginning of the solution process.
4. The order of x is at least one degree lower than that of y.

```
# Functions of contributors to σyy:
σyy_terms_ = reform_expr(σyy_final, contributors)
σyy_terms = list(zip(f_contributors, σyy_terms_))
display(Math(f"\\text{{Contributions to $σ_{{yy}}$}}:"))
display(Math(f"\hspace{{0.5cm}}\small {latex(σyy_terms[:-2])}"))
display(Math(f"\hspace{{0.5cm}}\small {latex(σyy_terms[-2:])}"))
```

Contributions to σ_{yy}:

$$\left[(f_M,\ 0),\ (f_P,\ 0),\ \left(f_g,\ \frac{y}{2}-\frac{2y^3}{h^2}\right),\ \left(f_{q0},\ \frac{1}{2}+\frac{3y}{2h}-\frac{2y^3}{h^3}\right)\right]$$

$$\left[\left(f_{q1},\ \frac{x\left(h^{3}+3h^{2}y-4y^{3}\right)}{2h^{3}}\right),\right.$$
$$\left.\left(f_{q2},\ \frac{hy}{40}+\frac{x^{2}}{2}+\frac{3x^{2}y}{2h}-\frac{y^{3}}{5h}-\frac{2x^{2}y^{3}}{h^{3}}+\frac{2y^{5}}{5h^{3}}\right)\right]$$

From this set of results for σ_{yy}, we find the following:

1. Only forces in the y-direction contribute to σ_{yy}.
2. The order of x is at least two degrees lower than that of y.

```
# Functions of contributors to σxy:

σxy_terms_ = reform_expr(σxy_final, contributors)
σxy_terms = list(zip(f_contributors, σxy_terms_))
display(Math(f"\\text{{Contributions to $σ_{{xy}}$}}:"))
display(Math(f"\hspace{{0.1cm}}\\scriptsize{latex(σxy_terms[:4])}"))
display(Math(f"\hspace{{0.1cm}}\small {latex(σxy_terms[4])}"))
display(Math(f"\hspace{{0.1cm}}\small {latex(σxy_terms[5])}"))
```

Contributions to σ_{xy}:

$$\left[(f_{M},\ 0),\ \left(f_{P},\ \frac{3}{2h}-\frac{6y^{2}}{h^{3}}\right),\ \left(f_{g},\ \frac{3L}{2}-\frac{6Ly^{2}}{h^{2}}-\frac{3x}{2}+\frac{6xy^{2}}{h^{2}}\right),\right.$$
$$\left.\left(f_{q0},\ \frac{3\left(Lh^{2}-4Ly^{2}-h^{2}x+4xy^{2}\right)}{2h^{3}}\right)\right]$$

$$\left(f_{q1},\ \frac{60L^{2}h^{2}-240L^{2}y^{2}-h^{4}-60h^{2}x^{2}+24h^{2}y^{2}+240x^{2}y^{2}-80y^{4}}{80h^{3}}\right)$$

$$\left(f_{q2},\ \frac{20L^{3}h^{2}-80L^{3}y^{2}-h^{4}x-20h^{2}x^{3}+24h^{2}xy^{2}+80x^{3}y^{2}-80xy^{4}}{40h^{3}}\right)$$

From this set of results for σ_{xy}, we find the following:

1. The moment contributes nothing to the shear stress σ_{xy}.
2. Other findings are largely similar to those for σ_{xx}.

Let us redefine these stress solutions explicitly in Python for future use, without the need to rerun the code above. Readers can also simply copy these formulas into their code, if needed.

```
# The obtained stress formulas for are defined in Python for future use:
#print("σxx_terms_=", σxx_terms_,"\n")
#print("σyy_terms_=", σyy_terms_,"\n"); print("σxy_terms_=",σxy_terms_)
σxx_terms_=[12*y/h**3, 12*y*(-L+x)/h**3,
            y*(-30*L**2+60*L*x-3*h**2-30*x**2+20*y**2)/(5*h**2),
            y*(-30*L**2+60*L*x-3*h**2-30*x**2+20*y**2)/(5*h**3),
            y*(-20*L**3+30*L**2*x-3*h**2*x-10*x**3+20*x*y**2)/(5*h**3),
            (-25200*L**4*y+33600*L**3*x*y+350*h**5+261*h**4*y-
             4200*h**3*y**2-5040*h**2*x**2*y-840*h**2*y**3-8400*x**4*y+
             33600*x**2*y**3-5040*y**5)/(8400*h**3)]
σxx_ = sp.Matrix(contributors).T @ sp.Matrix(σxx_terms_)
#display(Math(f"\\text{{ The final expression for $σ_{{xx}}$}}=\
#                                            \\small{latex(σxx_ )}"))
σyy_terms_= [0, 0, y/2-2*y**3/h**2, 1/2+3*y/(2*h)-2*y**3/h**3,
           x*(h**3+3*h**2*y-4*y**3)/(2*h**3), h*y/40+x**2/2+
           3*x**2*y/(2*h)-y**3/(5*h)-2*x**2*y**3/h**3+2*y**5/(5*h**3)]
σyy_ = sp.Matrix(contributors).T @ sp.Matrix(σyy_terms_)
#display(Math(f"\\text{{ The final expression for $σ_{{yy}}$}}=\
#                                           \\small{latex(σyy_ )}"))
σxy_terms_= [0, 3/(2*h)-6*y**2/h**3, 3*L/2-6*L*y**2/h**2-3*x/2+
           6*x*y**2/h**2, 3*(L*h**2-4*L*y**2-h**2*x+4*x*y**2)/(2*h**3),
           (60*L**2*h**2-240*L**2*y**2-h**4-60*h**2*x**2+24*h**2*y**2+
           240*x**2*y**2-80*y**4)/(80*h**3),
           (20*L**3*h**2-80*L**3*y**2-h**4*x-20*h**2*x**3+
           24*h**2*x*y**2+80*x**3*y**2-80*x*y**4)/(40*h**3)]
σxy_ = sp.Matrix(contributors).T @ sp.Matrix(σxy_terms_)
#display(Math(f"\\text{{ The final expression for $σ_{{xy}}$}}=\
#                                           \\small{latex(σxy_ )}"))
```

3.4.8 *Final solution for the displacements*

The following code cell produces the formulas for displacements for the 2D cantilever. The displacement expressions are rewritten in the following unified form:

$$u \text{ or } v = Mf_M + Pf_P + \rho g f_g + q_0 f_{q0} + q_1 f_{q1} + q_2 f_{q2} \tag{3.18}$$

where f_k is a function of the contribution from force term k. All these contribution functions are computed using the following snippet for each displacement component, respectively. First, for the displacement component u:

```
# Using the obtained constants produce the displacement formulas, u:
u_final = ug.subs(sln_cs).simplify()     # u
u_terms_ = reform_expr(u_final, contributors)
u_terms = list(zip(f_contributors, u_terms_))
```

```
display(Math(f"\\text{{Contributor functions for $u$}}:"))
display(Math(f"\hspace{{0cm}} \\scriptsize{latex(u_terms[:2])}"))
display(Math(f"\hspace{{0cm}} \\scriptsize{latex(u_terms[2])}"))
display(Math(r"\scriptsize{"+latex(u_terms[3]).replace('L y^{3} -',\
                                            'L y^{3} -\\\\')+"}"))
display(Math(r"\scriptsize{"+latex(u_terms[4]).replace('x^{2} ν',\
                                            'x^{2} ν\\\\')+"}"))
display(Math(r"\scriptsize{"+latex(u_terms[5]).replace('L^{3} y^{3} +',\
        'L^{3} y^{3} +\\\\').replace('x^{3} y +','x^{3} y +\\\\')+"}"))
```

Contributor functions for u:

$$\left[\left(f_M,\ \frac{6xy}{Gh^3(\nu+1)}\right),\ \left(f_P,\ \frac{y\left(-12Lx+3h^2\nu+3h^2+6x^2-2y^2\nu-4y^2\right)}{2Gh^3(\nu+1)}\right)\right]$$

$$\left(f_g,\ \frac{y\begin{pmatrix}-60L^2x+30Lh^2\nu+30Lh^2+60Lx^2-20Ly^2\nu-40Ly^2-5h^2x\nu\\ -6h^2x-20x^3+20xy^2\nu+40xy^2\end{pmatrix}}{20Gh^2(\nu+1)}\right)$$

$$\left(f_{q0},\ \frac{\begin{matrix}-60L^2xy+30Lh^2y\nu+30Lh^2y+60Lx^2y-20Ly^3\nu-40Ly^3\\ -5h^3x\nu-15h^2xy\nu-6h^2xy-20x^3y+20xy^3\nu+40xy^3\end{matrix}}{20Gh^3(\nu+1)}\right)$$

$$\left(f_{q1},\ \frac{\begin{matrix}-160L^3xy+60L^2h^2y\nu+60L^2h^2y+120L^2x^2y-40L^2y^3\nu\\ -80L^2y^3-h^4y\nu-h^4y-10h^3x^2\nu-10h^3y^2-30h^2x^2y\nu-12h^2x^2y\\ +4h^2y^3\nu-2h^2y^3-20x^4y+40x^2y^3\nu+80x^2y^3-8y^5\nu-12y^5\end{matrix}}{80Gh^3(\nu+1)}\right)$$

$$\left(f_{q2},\ \frac{\begin{matrix}-25200L^4xy+8400L^3h^2y\nu+8400L^3h^2y+16800L^3x^2y\\ -5600L^3y^3\nu-11200L^3y^3+350h^5x-210h^4xy\nu+261h^4xy\\ -1400h^3x^3\nu-4200h^3xy^2-4200h^2x^3y\nu-1680h^2x^3y\\ +1680h^2xy^3\nu-840h^2xy^3-1680x^5y+5600x^3y^3\nu+11200x^3y^3\\ -3360xy^5\nu-5040xy^5\end{matrix}}{16800Gh^3(\nu+1)}\right)$$

```
# Next, let get the formulas for displacement component, v:
v_final = vg.subs(sln_cs).simplify()    # v
v_terms_ = reform_expr(v_final, contributors)
v_terms = list(zip(f_contributors, v_terms_))
display(Math(f"\\text{{Contributor functions for $v$}}:"))
display(Math(f"\hspace{{0cm}} \\scriptsize{latex(v_terms[:2])}"))
display(Math(r"\scriptsize{"+latex(v_terms[2]).replace('h^{2} x^{2} +',\
                                          'h^{2} x^{2} +\\\\')+"}"))
display(Math(r"\scriptsize{"+latex(v_terms[3]).replace('x^{2} ν -',\
                                                 'x^{2} ν -\\\\')+"}"))
display(Math(r"\scriptsize{"+latex(v_terms[4]).replace('x^{3} ν -',\
                                                 'x^{3} ν -\\\\')+"}"))
display(Math(r"\scriptsize{"+latex(v_terms[5])\
                 .replace('h^{4} y^{2} ν +', 'h^{4} y^{2} ν +\\\\')\
                 .replace('h^{2} y^{4} ν -','h^{2} y^{4} ν -\\\\')+"}"))
```

Contributor functions for v:

$$\left[\left(f_M,\ \frac{3\left(-x^2-y^2\nu\right)}{Gh^3(\nu+1)}\right),\ \left(f_P,\ \frac{3Lx^2+3Ly^2\nu-x^3-3xy^2\nu}{Gh^3(\nu+1)}\right)\right]$$

$$\left(f_g,\ \frac{\begin{matrix}60L^2x^2+60L^2y^2\nu-40Lx^3-120Lxy^2\nu-25h^2x^2\nu-24h^2x^2\\+6h^2y^2\nu+5h^2y^2+10x^4+60x^2y^2\nu-20y^4\nu-10y^4\end{matrix}}{40Gh^2(\nu+1)}\right)$$

$$\left(f_{q0},\ \frac{\begin{matrix}60L^2x^2+60L^2y^2\nu-40Lx^3-120Lxy^2\nu+10h^3y-15h^2x^2\nu\\-24h^2x^2+6h^2y^2\nu+15h^2y^2+10x^4+60x^2y^2\nu-20y^4\nu-10y^4\end{matrix}}{40Gh^3(\nu+1)}\right)$$

$$\left(f_{q1},\ \frac{\begin{matrix}40L^3x^2+40L^3y^2\nu-20L^2x^3-60L^2xy^2\nu+10h^3xy-5h^2x^3\nu\\-8h^2x^3+6h^2xy^2\nu+15h^2xy^2+2x^5+20x^3y^2\nu-20xy^4\nu-10xy^4\end{matrix}}{40Gh^3(\nu+1)}\right)$$

$$\left(f_{q2},\ \frac{\begin{matrix}25200L^4x^2+25200L^4y^2\nu-11200L^3x^3-33600L^3xy^2\nu-700h^5y\nu\\-210h^4x^2\nu-681h^4x^2-261h^4y^2\nu+210h^4y^2+8400h^3x^2y\\+2800h^3y^3\nu-2100h^2x^4\nu-3360h^2x^4+5040h^2x^2y^2\nu+12600h^2x^2y^2\\+420h^2y^4\nu-840h^2y^4+560x^6+8400x^4y^2\nu-16800x^2y^4\nu\\-8400x^2y^4+1680y^6\nu+1120y^6\end{matrix}}{33600Gh^3(\nu+1)}\right)$$

3.4.9 *Contributors to the displacements*

The expressions for the displacements are significantly more complex and comprehensive. A careful examination reveals the following observations:

1. In the assumed expressions, we used complete sixth-order monomials. While terms like x^6 have vanished, the combined order of the monomials remains six.
2. If $q_2 = 0$, implying that the distributed force is linear, the displacement function is of fifth order, which corresponds to the water dam problem studied in Section 2.9.
3. Further, if $q_1 = 0$ as well, implying that the distributed force or body force is constant, the displacement function is of fourth order, which corresponds to the simply-supported cantilever problem studied in Section 2.8.
4. If $q_0 = 0$ as well, implying that there is no distributed force, the displacement function is of third order, which corresponds to the cantilever loaded at its end studied in Section 2.7.
5. Finally, if $P = 0$ as well, implying that only a pure moment is applied, the displacement function is of second order, which corresponds to the cantilever problem studied in Section 2.6.

Our general displacement method uncovers these important properties of 2D solid mechanics problems defined in a rectangular domain when polynomial basis functions are used. It is easy to predict the order of the displacement if the distributed force is cubic.

To the best of the author's knowledge, this is the first time a closed-form solution for displacements in cantilever problems with multiple loads has been derived.

3.4.10 *Verification of the results*

We shall now verify the results by computing the stresses from the final displacements. We can then substitute these into the equilibrium equations to check if they are satisfied. Since the displacements were initially assumed with unknown constants, and all these constants were determined using equilibrium and boundary conditions on stresses, substituting the determined constants back into the assumed displacement allows for the re-computation of stresses. If these re-computed stresses still satisfy all the conditions, our solutions for both stresses and displacements must be correct. The snippet is as follows:

```
# Compute stresses using final displacements, plane stress:
σxx_=(2*G/(1-ν)*(u_final.diff(x)+ν*v_final.diff(y))).simplify()
σyy_=(2*G/(1-ν)*(v_final.diff(y)+ν*u_final.diff(x))).simplify()
σxy_=(G*(v_final.diff(x)+u_final.diff(y))).expand().collect(G*y)

# Check the differential equilibrium equations:
eqn1 = (σxx_.diff(x) + σxy_.diff(y)).simplify()
eqn2 = (σxy_.diff(x) + σyy_.diff(y) + ρg).simplify()

display(Math(f" \\text{{Is equilibrium Eq. 1 = 0? }}{latex(eqn1==0)}"))
display(Math(f" \\text{{Is equilibrium Eq. 2 = 0? }}{latex(eqn2==0)}"))

# Check SBCs on both top and bottom surfaces:
σyy_.subs(y, h/2).simplify(), σxy_.subs(y, h/2).simplify(),\
display(Math(f" \\text{{SBCs on the bottom surface: }}\
{latex((σyy_.subs(y, h/2).simplify(), σxy_.subs(y, h/2).simplify()))}"))
display(Math(f" \\text{{SBCs on the top surface: }} \
{latex((σyy_.subs(y,-h/2).simplify(), σxy_.subs(y,-h/2).simplify()))}"))

# Check FBCs on any cross-section at x:
FBC_ΣFx = Int(σxx_,   interval_y).simplify() - 0
FBC_ΣFy = (-Int(σxy_, interval_y) + Int(pt,  interval_x) + P).simplify()
FBC_ΣMz = (Int(σxx_*y,interval_y) + Int(pt*t,interval_x)\
                                          + P*(L-x) - M).expand()
display(Math(f" \\text{{FBCs on any cross-section: }}\
                              {latex((FBC_ΣFx, FBC_ΣFy, FBC_ΣMz))}"))
# Check FBCs on the boundary:
FBC_ΣFy_xL = Int(σxy_, interval_y).subs(x,L).simplify()
FBC_ΣMz_xL = Int(σxx_*y,interval_y).subs(x,L).simplify()
display(Math(f" \\text{{FBCs at x=L: }}\
                          {latex((FBC_ΣFx, FBC_ΣFy_xL, FBC_ΣMz_xL))}"))
```

Is equilibrium Eq. 1 = 0? True

Is equilibrium Eq. 2 = 0? True

SBCs on the bottom surface: $\left(q_0 + q_1 x + q_2 x^2,\ 0\right)$

SBCs on the top surface: $(0,\ 0)$

FBCs on any cross-section: $(0,\ 0,\ 0)$

FBCs at $x = L$: $(0,\ P,\ M)$

The results obtained are all correct. Readers may further verify the DBCs. Note that we do not need to check the compatibility equations because our method is displacement-based and thus always ensures compatibility. Readers may explore this further if desired.

3.4.11 *Solutions at special locations*

Let us check solutions at special locations. First, at $x = L$:

```
# Stresses on the cross-section at x=L:

σxx_xL_terms = reform_expr(σxx_final.subs(x,L), contributors)
σxx_xL_terms = list(zip(f_contributors, σxx_xL_terms))
display(Math(f"\\text{{Contributor functions for $σ_{{xx}}$ \
                                                    at $x=L$:}}"))
display(Math(f"\hspace{{0cm}} \\scriptsize {latex(σxx_xL_terms[:4])}"))
display(Math(f"\hspace{{0cm}} \\scriptsize {latex(σxx_xL_terms[4:])}"))

σxy_xL = σxy_final.subs(x,L).simplify().collect(q1).collect(q2)\
                                        .collect(P).collect(L)
display(Math(f"\\text{{ At $x=L$: }}"))
display(Math(f"\\text{{ $σ_{{xy}}$}}=\\small {latex(σxy_xL)}"))
```

Contributor functions for σ_{xx} at $x = L$:

$$\left[\left(f_M,\ \frac{12y}{h^3}\right),\ (f_P,\ 0),\ \left(f_g,\ -\frac{3y}{5}+\frac{4y^3}{h^2}\right),\ \left(f_{q0},\ -\frac{3y}{5h}+\frac{4y^3}{h^3}\right)\right]$$

$$\left[\left(f_{q1},\ \frac{Ly\left(-3h^2+20y^2\right)}{5h^3}\right),\ \left(f_{q2},\ -\frac{3L^2y}{5h}+\frac{4L^2y^3}{h^3}+\frac{h^2}{24}+\frac{87hy}{2800}\right.\right.$$
$$\left.\left.-\frac{y^2}{2}-\frac{y^3}{10h}-\frac{3y^5}{5h^3}\right)\right]$$

At $x = L$:

$$\sigma_{xy} = \frac{Lq_2\left(-2h^4+48h^2y^2-160y^4\right)+P\left(120h^2-480y^2\right)+q_1\left(-h^4+24h^2y^2-80y^4\right)}{80h^3}$$

We find that the normal stress σ_{xx} on the cross-section is of fifth order in y, compared to the linear distribution predicted by the 1D thin beam theory. The shear stress σ_{xy} on the cross-section is of fourth order in y, compared to the quadratic distribution predicted by the 1D thin beam theory. The integral of the shear stress σ_{xy} over the cross-section gives P.

Let us compute the stress solutions at $y = 0$:

```
# Stresses along the axial-axis, y=0:
display(Math(f"\\text{{ $σ_{{xx,y=0}}$}}=\
                        \\small{latex(σxx_final.subs(y, 0).simplify())}"))
display(Math(f"\\text{{ $σ_{{yy,y=0}}$}}=\
                        \\small{latex(σyy_final.subs(y, 0).simplify())}"))

σxy_y0= σxy_final.subs(y, 0).simplify().collect(q0).collect(q1)\
    .collect(q2).collect(ρg).collect(h).collect(120*(L-x)).collect(120)
display(Math(f"\\text{{ $σ_{{xy,y=0}}$}}=\\small {latex(σxy_y0)}"))
```

$$\sigma_{xx,y=0} = \frac{h^2 q_2}{24}$$

$$\sigma_{yy,y=0} = \frac{q_0}{2} + \frac{q_1 x}{2} + \frac{q_2 x^2}{2}$$

$$\sigma_{xy,y=0} = \frac{\begin{array}{l}120P + q_1\left(60L^2 - h^2 - 60x^2\right) + q_2\left(40L^3 - 2h^2x - 40x^3\right)\\ +120\left(L - x\right)\left(h\rho g + q_0\right)\end{array}}{80h}$$

We observe that the normal stress σ_{xx} along the axial axis is not zero but remains constant with respect to x. There is a contribution from the force term q_2. The normal stress σ_{yy} varies quadratically with x in response to the distributed force along the bottom edge. The shear stress σ_{xy} varies cubically with x, also in response to the distributed force along the bottom edge. The term $1.5P/h$ (the first term in $\sigma_{xy,y=0}$) matches the value given by the 1D thin beam theory. In other words, if the distributed force is zero, the stresses along the beam axis are identical to those predicted by the 1D thin beam theory.

```
# Displacements along the axial axis, y=0:
display(Math(f"\\text{{ $u_{{y=0}}$}}=\
                    \\small {latex(u_final.subs(y, 0).simplify())}"))

v0_terms = reform_expr(v_final.subs(y, 0), contributors)
v0_terms = list(zip(f_contributors, v0_terms))
display(Math(f"\\text{{Contributor to $v_{{y=0}}$}}:"))
display(Math(f"\hspace{{0.1cm}} \\scriptsize{latex(v0_terms[:3])}"))
display(Math(f"\hspace{{0.1cm}} \\scriptsize{latex(v0_terms[3:5])}"))
display(Math(f"\hspace{{0.1cm}} \\small {latex(v0_terms[5])}"))
```

$$u_{y=0} = \frac{x\left(h^2 q_2 - 12q_0\nu - 6q_1 x\nu - 4q_2 x^2\nu\right)}{48G\left(\nu + 1\right)}$$

Contributor to $v_{y=0}$:

$$\left[\left(f_M,\ -\frac{3x^2}{Gh^3(\nu+1)}\right),\ \left(f_P,\ \frac{x^2(3L-x)}{Gh^3(\nu+1)}\right),\right.$$

$$\left.\left(f_g,\ \frac{x^2\left(60L^2-40Lx-25h^2\nu-24h^2+10x^2\right)}{40Gh^2(\nu+1)}\right)\right]$$

$$\left[\left(f_{q0},\ \frac{x^2\left(60L^2-40Lx-15h^2\nu-24h^2+10x^2\right)}{40Gh^3(\nu+1)}\right),\right.$$

$$\left.\left(f_{q1},\ \frac{x^2\left(L^3-\frac{L^2x}{2}-\frac{h^2x\nu}{8}-\frac{h^2x}{5}+\frac{x^3}{20}\right)}{Gh^3(\nu+1)}\right)\right]$$

$$\left(f_{q2},\ \frac{x^2\left(25200L^4-11200L^3x-210h^4\nu-681h^4-2100h^2x^2\nu-3360h^2x^2+560x^4\right)}{33600Gh^3(\nu+1)}\right)$$

Note that the terms contributed by M and P are the same as those given in the 1D thin beam theory.

```
# Displacements at the root edge, x=0. Component u:

u0_terms = reform_expr(u_final.subs(x, 0), contributors)
u0_terms = list(zip(f_contributors, u0_terms))
display(Math(f"\\text{{Contributor to $u_{{x=0}}$}}:"))
display(Math(f"\hspace{{0.1cm}} \\scriptsize{latex(u0_terms[:3])}"))
display(Math(f"\hspace{{0.1cm}} \\scriptsize{latex(u0_terms[3:6:2])}"))
display(Math(f"\hspace{{0.1cm}} \\scriptsize {latex(u0_terms[4])}"))
```

Contributor to $u_{x=0}$:

$$\left[(f_M,\ 0),\ \left(f_P,\ \frac{y\left(3h^2\nu+3h^2-2y^2\nu-4y^2\right)}{2Gh^3(\nu+1)}\right),\right.$$

$$\left.\left(f_g,\ \frac{Ly\left(3h^2\nu+3h^2-2y^2\nu-4y^2\right)}{2Gh^2(\nu+1)}\right)\right]$$

$$\left[\left(f_{q0},\ \frac{Ly\left(3h^2\nu+3h^2-2y^2\nu-4y^2\right)}{2Gh^3(\nu+1)}\right),\right.$$

$$\left.\left(f_{q2},\ \frac{L^3y\left(3h^2\nu+3h^2-2y^2\nu-4y^2\right)}{6Gh^3(\nu+1)}\right)\right]$$

$$\left(f_{q1}, \frac{y(60L^2h^2\nu + 60L^2h^2 - 40L^2y^2\nu - 80L^2y^2 - h^4\nu - h^4 - 10h^3y + 4h^2y^2\nu - 2h^2y^2 - 8y^4\nu - 12y^4)}{80Gh^3(\nu+1)} \right)$$

```
# Displacement component v:

v0_terms = reform_expr(v_final.subs(x, 0), contributors)
v0_terms = list(zip(f_contributors, v0_terms))
display(Math(f"\\text{{Contributor to $v_{{x=0}}$}}:"))
display(Math(f"\hspace{{0.1cm}} \\scriptsize{latex(v0_terms[:3])}"))
display(Math(f"\hspace{{0.1cm}} \\scriptsize{latex(v0_terms[3:5])}"))
display(Math(f"\hspace{{0.1cm}} \\scriptsize{latex(v0_terms[5])}"))
```

Contributor to $v_{x=0}$:

$$\left[\left(f_M, -\frac{3y^2\nu}{Gh^3(\nu+1)}\right), \left(f_P, \frac{3Ly^2\nu}{Gh^3(\nu+1)}\right),\right.$$

$$\left.\left(f_g, \frac{y^2\cdot\left(60L^2\nu + 6h^2\nu + 5h^2 - 20y^2\nu - 10y^2\right)}{40Gh^2(\nu+1)}\right)\right]$$

$$\left[\left(f_{q0}, \frac{y\left(60L^2y\nu + 10h^3 + 6h^2y\nu + 15h^2y - 20y^3\nu - 10y^3\right)}{40Gh^3(\nu+1)}\right),\right.$$

$$\left.\left(f_{q1}, \frac{L^3y^2\nu}{Gh^3(\nu+1)}\right)\right]$$

$$\left(f_{q2}, \frac{y(25200L^4y\nu - 700h^5\nu - 261h^4y\nu + 210h^4y + 2800h^3y^2\nu + 420h^2y^3\nu - 840h^2y^3 + 1680y^5\nu + 1120y^5)}{33600Gh^3(\nu+1)} \right)$$

The displacement distributions along the root of the cantilever are useful when the solutions are used to validate a numerical model. These displacements can be used as a DBC, and the stresses on the left edge are used as external loads in a numerical model, allowing the results of stresses to be compared [4].

The displacement expressions are too long. It would be easier for readers to run the code in this section to obtain these formulas.

3.5 Review questions

1. Solve the half-space problem studied in Section 2.5, assuming it is a plane stress problem, and the displacements are set to zero at $y = d$.
2. Solve the pure bending problem studied in Section 2.6, assuming the displacements at point $(y = 0, x = 0)$ satisfy the following conditions:

$$u = 0; \quad v = 0; \quad \frac{\partial v}{\partial x} = 0 \tag{3.19}$$

3. Solve the simply-supported beam problem studied in Section 2.8, using all the conditions given there, including the same displacement conditions.
4. Solve the water dam problem studied in Section 2.9, using all the conditions given there, including the same displacement conditions.

3.6 Concluding remarks

This chapter discusses in detail the displacement method for solving 2D solid mechanics problems defined in Cartesian coordinates. A number of problems are solved, providing full sets of solutions for stresses, strains, and displacements. We close this chapter with the following remarks:

1. The basic solution strategy is to treat the displacements as the primary unknown functions and assume them as a linear combination of monomials. The differential equilibrium equations, along with various types of boundary conditions — DBCs, SBCs, and FBCs — can then be used to determine these unknown constants.
2. The operations in our displacement method involve differentiation, definite integration, and algebraic manipulation. Hence, the method is always workable, straightforward, and powerful for problems for which a proper set of basis functions can be found.
3. A set of Python scripts is provided for these types of problems to automatically solve them.

Note that the sequence of imposing various types of boundary conditions — DBCs, SBCs, and FBCs — can vary depending on convenience. Additionally, the differential equilibrium equations can also be used. The bottom line is to ensure all these conditions are satisfied.

Note also that the rigid body movement of a 2D solid can always be added to the displacement solutions if desired. The formula is as follows:

$$u_0 = a - by; \quad v_0 = c - bx \tag{3.20}$$

where u_0 and v_0 are the two components of the rigid body displacements and a, b, and c are three arbitrary constants representing the three rigid body movements. Using the strain–displacement relations, one can easily confirm that these displacements will not contribute to the strains and hence the stress solutions. They only affect the reference frame in which the displacements are measured.

The next chapter presents stress methods for 3D problems.

References

[1] G.R. Liu, *Numbers and Functions: Theory, Formulation, and Python Codes*, World Scientific, New Jersey, 2024.
[2] G.R. Liu, *Calculus: A Practical Course with Python*, World Scientific, New Jersey, 2025.
[3] Z.L. Xu, *Elasticity*, Vol. 1 & 2, People's Publisher, China, 1979.
[4] G.R. Liu and T.T. Nguyen, *Smoothed Finite Element Methods*, Taylor & Francis Group, New York, 2010.

Chapter 4

Stress Methods for 3D Solids

```
# Place cursor in this cell, and press Ctrl+Enter to import dependencies.
import sys                          # For accessing the computer system
sys.path.append('../grbin/')   # Change to the directory in your system

from commonImports import *        # Import dependencies from '../grbin/'
import grcodes as gr                   # Import the module of the author
importlib.reload(gr)               # When grcodes is modified, reload it

from continuum_mechanics.solids import sym_grad
init_printing(use_unicode=True)        # For latex-like quality printing
np.set_printoptions(precision=4,suppress=True,
                    formatter={'float_kind': '{:.4e}'.format})
```

This chapter discusses three-dimensional (3D) solid mechanics problems in Cartesian coordinates using stress methods. As discussed in Chapter 2, a typical stress method treats stresses as primary variables and tries to find solutions for these first. It often uses stress functions whose differentiations give the stress components. This chapter extends such a stress method to 3D solids, and the so-called **stream functions**, also known as stress functions, are often used in the derivation process. This chapter is written with reference to [1–4].

4.1 Basic equations for 3D solids

Let us list the necessary basic equations and conditions that we need to solve. The derivation of these equations can be found in Ref. [5]. We use these equations directly, considering only static cases.

4.1.1 *Equilibrium equations*

In Stresses

In indicial notation, the equilibrium equations can be written concisely in terms of the stress tensor as

$$\sigma_{ji,j} + b_i = 0 \quad \in \Omega \tag{4.1}$$

where σ_{ij} is the stress tensor that has six components $\sigma_{ij}(i, j = x, y, z)$, each **denoting** the stress component on the plane of i and in the direction of j. The stress tensor is symmetric, $\sigma_{ij} = \sigma_{ji}$. The term $\sigma_{ji,j}$ is short for $\frac{\partial \sigma_{ij}}{\partial x_j}$, and $b_i(i = x, y, z)$ is the ith body force component. Explicitly, Eq. (4.1) can be written as

$$\begin{aligned}
&\frac{\partial \sigma_{xx}}{\partial x} + \frac{\partial \sigma_{yx}}{\partial y} + \frac{\partial \sigma_{zx}}{\partial z} + b_y = 0\\
&\frac{\partial \sigma_{xy}}{\partial x} + \frac{\partial \sigma_{yy}}{\partial y} + \frac{\partial \sigma_{zy}}{\partial z} + b_y = 0\\
&\frac{\partial \sigma_{xz}}{\partial x} + \frac{\partial \sigma_{yz}}{\partial y} + \frac{\partial \sigma_{zz}}{\partial z} + b_z = 0
\end{aligned} \tag{4.2}$$

Equations (4.1) and (4.2) form a set of three partial differential equations (PDEs) of first order in terms of stresses. It is known as the equilibrium equations for 3D solids subjected to external body forces, which can be functions of the coordinates. The unknowns in these PDEs are the stresses, and there are **six** of them.

4.1.2 *Constitutive equations*

Assuming the strain components ε_{ij} are known, the stress components σ_{ij} can be computed using (see Chapter 5 of Ref. [5])

$$\sigma_{ij} = 2G\varepsilon_{ij} + \lambda\delta_{ij}e \tag{4.3}$$

where G is the shear modulus, λ is the Lamé constant for the material, and δ_{ij} is the Kronecker delta. The term e represents the **divergence of the displacement vector**, also known as the **volume strain** or the **trace of the strain tensor**. It is given by

$$e = \frac{\partial u_x}{\partial x} + \frac{\partial u_y}{\partial y} + \frac{\partial u_z}{\partial z} = \varepsilon_{xx} + \varepsilon_{yy} + \varepsilon_{zz} = \varepsilon_{kk} \tag{4.4}$$

where $u_i(x, y, z)$ $(i = x, y, z)$ are the displacement components and are often written as $u(x, y, z)$, $v(x, y, z)$, and $w(x, y, z)$.

Expanding Eq. (4.3) explicitly, we obtain

$$
\begin{aligned}
\sigma_{xx} &= \frac{E}{(1+\nu)(1-2\nu)}\left[(1-\nu)\varepsilon_{xx} + \nu(\varepsilon_{yy}+\varepsilon_{zz})\right] \\
\sigma_{yy} &= \frac{E}{(1+\nu)(1-2\nu)}\left[(1-\nu)\varepsilon_{yy} + \nu(\varepsilon_{xx}+\varepsilon_{zz})\right] \\
\sigma_{zz} &= \frac{E}{(1+\nu)(1-2\nu)}\left[(1-\nu)\varepsilon_{zz} + \nu(\varepsilon_{xx}+\varepsilon_{yy})\right] \\
\sigma_{yz} &= \frac{E}{(1+\nu)}\varepsilon_{yz}; \quad \sigma_{xz} = \frac{E}{(1+\nu)}\varepsilon_{xz}; \quad \sigma_{xy} = \frac{E}{(1+\nu)}\varepsilon_{xy}
\end{aligned}
\tag{4.5}
$$

where E is Young's modulus and ν is Poisson's ratio, and they are related to the shear modulus by $G = \frac{E}{2(1+\nu)}$ and to the Lamé constant by $\lambda = \frac{E\nu}{(1-2\nu)(1+\nu)}$.

The constitutive equations can also be written in the reverse form as (see Section 5.13 of Ref. [5])

$$
\varepsilon_{ij} = \frac{1}{2G}\sigma_{ij} - \frac{\lambda\delta_{ij}}{2G(2G+3\lambda)}\sigma_{kk} = \frac{1}{2G}\sigma_{ij} - \frac{\nu}{E}\delta_{ij}\sigma_{kk} \tag{4.6}
$$

where $\sigma_{kk} = \sigma_{xx} + \sigma_{yy} + \sigma_{zz}$ is the trace of the stress tensor. The following snippet further simplifies Eq. (4.6):

```
E, ν, σxx, σyy, σzz = symbols("E, ν, σ_xx, σ_yy, σ_zz ")

λ = E*ν/((1+ν)*(1-2*ν))
G = E/(2*(1+ν))
un_E = λ/(2*G*(2*G+3*λ)).simplify()
display(Math(f"\small \\frac{{λ}}{{2G(2G+3λ)}} = {latex(un_E)}"))

εxx =  (σxx/(2*G) - un_E*(σxx+σyy+σzz)).simplify()
εyy =  (σyy/(2*G) - un_E*(σxx+σyy+σzz)).simplify()
εzz =  (σzz/(2*G) - un_E*(σxx+σyy+σzz)).simplify()
display(Math(f" \\text{{$(ε_{{xx}},ε_{{yy}},ε_{{zz}})$}}=\
                                    \\small {latex((εxx, εyy, εzz))}"))
```

$$
\frac{\lambda}{2G(2G+3\lambda)} = -\frac{\nu\,(2\nu-1)}{E\,(1-2\nu)}
$$

$$
(\varepsilon_{xx}, \varepsilon_{yy}, \varepsilon_{zz}) = \left(\frac{-\nu\sigma_{yy} - \nu\sigma_{zz} + \sigma_{xx}}{E}, \frac{-\nu\sigma_{xx} - \nu\sigma_{zz} + \sigma_{yy}}{E}, \frac{-\nu\sigma_{xx} - \nu\sigma_{yy} + \sigma_{zz}}{E}\right)
$$

Therefore, Eq. (4.6) can be rewritten explicitly as

$$
\begin{aligned}
\varepsilon_{xx} &= \frac{1}{E}[\sigma_{xx} - \nu(\sigma_{yy} + \sigma_{zz})] \\
\varepsilon_{yy} &= \frac{1}{E}[\sigma_{yy} - \nu(\sigma_{xx} + \sigma_{zz})] \\
\varepsilon_{zz} &= \frac{1}{E}[\sigma_{zz} - \nu(\sigma_{xx} + \sigma_{yy})] \\
\varepsilon_{yz} &= \frac{1}{2G}\sigma_{yz} = \frac{1+\nu}{E}\sigma_{yz} \\
\varepsilon_{xz} &= \frac{1}{2G}\sigma_{xz} = \frac{1+\nu}{E}\sigma_{xz} \\
\varepsilon_{xy} &= \frac{1}{2G}\sigma_{xy} = \frac{1+\nu}{E}\sigma_{xy}
\end{aligned} \tag{4.7}
$$

This equation shows clearly the roles of Young's modulus and Poisson's ratio. The proof of the relationship between the shear modulus, Young's modulus, and Poisson's ratio can be found in Section 5.15.10 of Ref. [5].

4.1.3 *Strain–displacement relations*

The strains in Eq. (4.3) are obtained from the symmetric part of the displacement gradient. They can be expressed simply as [5]

$$
\varepsilon_{ij} = \frac{1}{2}\left(\frac{\partial u_i}{\partial x_j} + \frac{\partial u_j}{\partial x_i}\right) \tag{4.8}
$$

The explicit strain–displacement relations in matrix form are given as follows:

$$
\begin{bmatrix} \varepsilon_{xx} & \varepsilon_{xy} & \varepsilon_{xz} \\ \varepsilon_{yx} & \varepsilon_{yy} & \varepsilon_{yz} \\ \varepsilon_{zx} & \varepsilon_{zy} & \varepsilon_{zz} \end{bmatrix} = \frac{1}{2}\begin{bmatrix} 2\frac{\partial u_x}{\partial x} & \frac{\partial u_x}{\partial y} + \frac{\partial u_y}{\partial x} & \frac{\partial u_x}{\partial z} + \frac{\partial u_z}{\partial x} \\ \frac{\partial u_x}{\partial y} + \frac{\partial u_y}{\partial x} & 2\frac{\partial u_y}{\partial y} & \frac{\partial u_y}{\partial z} + \frac{\partial u_z}{\partial y} \\ \frac{\partial u_x}{\partial z} + \frac{\partial u_z}{\partial x} & \frac{\partial u_y}{\partial z} + \frac{\partial u_z}{\partial y} & 2\frac{\partial u_z}{\partial z} \end{bmatrix} \tag{4.9}
$$

As clearly shown, the strain tensor is also symmetric.

4.1.4 Compatibility equations

For 3D problems, there are three displacement components, but we have six equations for strain–displacement relations, as given in Eq. (4.9). Therefore, these strain components must also be compatible. These compatibility equations can be derived in a similar procedure as that used for 2D cases, but there are a total of six of them [2, 5, 6]:

$$\frac{\partial^2 \varepsilon_{kl}}{\partial x_i \partial x_j} + \frac{\partial^2 \varepsilon_{ij}}{\partial x_k \partial x_l} = \frac{\partial^2 \varepsilon_{ik}}{\partial x_j \partial x_l} + \frac{\partial^2 \varepsilon_{jl}}{\partial x_i \partial x_k} \tag{4.10}$$

or simply

$$\varepsilon_{kl,ij} + \varepsilon_{ij,kl} = \varepsilon_{ik,jl} + \varepsilon_{jl,ik} \tag{4.11}$$

or explicitly

$$\begin{aligned}
&\frac{\partial^2 \varepsilon_{xx}}{\partial y^2} + \frac{\partial^2 \varepsilon_{yy}}{\partial x^2} = 2\frac{\partial^2 \varepsilon_{xy}}{\partial x \partial y} \\
&\frac{\partial^2 \varepsilon_{xx}}{\partial z^2} + \frac{\partial^2 \varepsilon_{zz}}{\partial x^2} = 2\frac{\partial^2 \varepsilon_{xz}}{\partial x \partial z} \\
&\frac{\partial^2 \varepsilon_{yy}}{\partial z^2} + \frac{\partial^2 \varepsilon_{zz}}{\partial y^2} = 2\frac{\partial^2 \varepsilon_{yz}}{\partial y \partial z} \\
&\frac{\partial^2 \varepsilon_{xy}}{\partial z^2} + \frac{\partial^2 \varepsilon_{zz}}{\partial x \partial y} = \frac{\partial^2 \varepsilon_{yz}}{\partial x \partial z} + \frac{\partial^2 \varepsilon_{xz}}{\partial y \partial z} \\
&\frac{\partial^2 \varepsilon_{xz}}{\partial y^2} + \frac{\partial^2 \varepsilon_{yy}}{\partial x \partial z} = \frac{\partial^2 \varepsilon_{xy}}{\partial y \partial z} + \frac{\partial^2 \varepsilon_{yz}}{\partial x \partial y} \\
&\frac{\partial^2 \varepsilon_{yz}}{\partial x^2} + \frac{\partial^2 \varepsilon_{xx}}{\partial y \partial z} = \frac{\partial^2 \varepsilon_{xz}}{\partial x \partial y} + \frac{\partial^2 \varepsilon_{xy}}{\partial x \partial z}
\end{aligned} \tag{4.12}$$

These compatibility equations must be satisfied to ensure the existence of a displacement solution. Since these equations involve second-order differentials of strains, if the strain field is linear, these equations are automatically satisfied.

Using the constitutive equations, these compatibility equations can be written in stresses [2]:

$$\nabla^2 \sigma_{ij} + \frac{1}{1+\nu}\frac{\partial^2 \sigma_{kk}}{\partial x_i \partial x_j} = -\frac{\nu}{1-\nu}\frac{\partial b_k}{\partial x_k}\delta_{ij} - \left(\frac{\partial b_i}{\partial x_j} + \frac{\partial b_j}{\partial x_i}\right) \tag{4.13}$$

where σ_{kk} is the trace of the stress tensor. Equation (4.13) can be written explicitly as

$$\begin{aligned}
\nabla^2\sigma_{xx} + \frac{1}{1+\nu}\frac{\partial^2\sigma_{kk}}{\partial x^2} &= -\frac{\nu}{1-\nu}\left(\frac{\partial b_x}{\partial x} + \frac{\partial b_y}{\partial y} + \frac{\partial b_z}{\partial z}\right) - 2\frac{\partial b_x}{\partial x} \\
\nabla^2\sigma_{yy} + \frac{1}{1+\nu}\frac{\partial^2\sigma_{kk}}{\partial y^2} &= -\frac{\nu}{1-\nu}\left(\frac{\partial b_x}{\partial x} + \frac{\partial b_y}{\partial y} + \frac{\partial b_z}{\partial z}\right) - 2\frac{\partial b_y}{\partial y} \\
\nabla^2\sigma_{zz} + \frac{1}{1+\nu}\frac{\partial^2\sigma_{kk}}{\partial z^2} &= -\frac{\nu}{1-\nu}\left(\frac{\partial b_x}{\partial x} + \frac{\partial b_y}{\partial y} + \frac{\partial b_z}{\partial z}\right) - 2\frac{\partial b_z}{\partial z} \\
\nabla^2\sigma_{xy} + \frac{1}{1+\nu}\frac{\partial^2\sigma_{kk}}{\partial x\partial y} &= -\left(\frac{\partial b_x}{\partial y} + \frac{\partial b_y}{\partial x}\right) \\
\nabla^2\sigma_{xz} + \frac{1}{1+\nu}\frac{\partial^2\sigma_{kk}}{\partial x\partial z} &= -\left(\frac{\partial b_z}{\partial x} + \frac{\partial b_x}{\partial z}\right) \\
\nabla^2\sigma_{yz} + \frac{1}{1+\nu}\frac{\partial^2\sigma_{kk}}{\partial z\partial y} &= -\left(\frac{\partial b_z}{\partial y} + \frac{\partial b_y}{\partial z}\right)
\end{aligned} \tag{4.14}$$

4.1.5 *Equilibrium in displacements*

Substituting Eq. (4.8) into (4.3) and then into (4.13), we obtain the equilibrium equation in terms of displacements:

$$Gu_{i,jj} + (G+\lambda)u_{j,ji} + b_i = \rho\ddot{u}_i \tag{4.15}$$

where the two dots stand for second-order differentiation with respect to time and $\rho\ddot{u}_i$ is the acceleration (per unit volume). It is needed only if the problem is dynamic. If we write Eq. (4.15) explicitly, it takes the following form:

$$\begin{aligned}
G\left[\left(\frac{\partial^2 u}{\partial x^2} + \frac{\partial^2 u}{\partial y^2} + \frac{\partial^2 u}{\partial z^2}\right) + \frac{1+\nu}{1-\nu}\left(\frac{\partial^2 u}{\partial x^2} + \frac{\partial^2 v}{\partial x\partial y} + \frac{\partial^2 w}{\partial x\partial z}\right)\right] + b_x &= \rho\ddot{u} \\
G\left[\left(\frac{\partial^2 v}{\partial x^2} + \frac{\partial^2 v}{\partial y^2} + \frac{\partial^2 v}{\partial z^2}\right) + \frac{1+\nu}{1-\nu}\left(\frac{\partial^2 u}{\partial x\partial y} + \frac{\partial^2 v}{\partial y^2} + \frac{\partial^2 w}{\partial y\partial z}\right)\right] + b_y &= \rho\ddot{v} \\
G\left[\left(\frac{\partial^2 w}{\partial x^2} + \frac{\partial^2 w}{\partial y^2} + \frac{\partial^2 w}{\partial z^2}\right) + \frac{1+\nu}{1-\nu}\left(\frac{\partial^2 u}{\partial x\partial z} + \frac{\partial^2 v}{\partial y\partial z} + \frac{\partial^2 w}{\partial z^2}\right)\right] + b_z &= \rho\ddot{w}
\end{aligned} \tag{4.16}$$

This set of equation can also be written as

$$
\begin{aligned}
G\nabla^2 u + (G+\lambda)\frac{\partial e}{\partial x} + b_x &= \rho\ddot{u} \\
G\nabla^2 v + (G+\lambda)\frac{\partial e}{\partial y} + b_y &= \rho\ddot{v} \\
G\nabla^2 w + (G+\lambda)\frac{\partial e}{\partial z} + b_z &= \rho\ddot{w}
\end{aligned} \tag{4.17}
$$

where ∇^2 is the Laplace operator. In 3D, it is

$$
\nabla^2 = \frac{\partial^2}{\partial x_i \partial x_i} = \frac{\partial^2}{\partial x^2} + \frac{\partial^2}{\partial y^2} + \frac{\partial^2}{\partial z^2} \tag{4.18}
$$

Equation (4.17) can be written as a single equation in indicial notation:

$$
G\left(\nabla^2 u_i + \frac{1}{1-2\nu}\frac{\partial e}{\partial x_i}\right) + b_i = \rho\ddot{u}_i \quad \text{for } i = 1, 2, 3 \tag{4.19}
$$

Or in the vector form:

$$
\mathbf{grad}\ \text{div}\ \mathbf{u} + (1-2\nu)\nabla^2\mathbf{u} + \frac{1-2\nu}{G}\mathbf{b} = \frac{\rho(1-2\nu)}{G}\ddot{\mathbf{u}} \tag{4.20}
$$

where $\mathbf{u} = [u_x\ u_y\ u_z]^\top$ is the vector of these displacement components and $\mathbf{b} = [b_x\ b_y\ b_z]^\top$ is the vector of the body force components. **Grad** is the gradient operator, and div is the divergence of a vector, whose detailed definition can be found in Ref. [7]. Equations (4.15)–(4.20) are sets of PDEs of second order in terms of displacements.

As seen in the explicit form, Eq. (4.16) is truly a set of "monster" equations. Finding an analytic solution in terms of displacement is difficult. We will do so for some problems in this chapter.

4.1.6 *Stress–displacement relations*

Substituting Eq. (4.9) into Eq. (4.3), the stresses can be written in terms of displacements:

$$
\begin{aligned}
&\sigma_{xx} = 2G\frac{\partial u_x}{\partial x} + \lambda e, \quad \sigma_{yy} = 2G\frac{\partial u_y}{\partial y} + \lambda e, \quad \sigma_{zz} = 2G\frac{\partial u_z}{\partial z} + \lambda e \\
&\sigma_{zy} = 2G\left(\frac{\partial u_z}{\partial y} + \frac{\partial u_y}{\partial z}\right) \\
&\sigma_{zx} = 2G\left(\frac{\partial u_z}{\partial x} + \frac{\partial u_x}{\partial z}\right) \\
&\sigma_{xy} = 2G\left(\frac{\partial u_x}{\partial y} + \frac{\partial u_y}{\partial x}\right)
\end{aligned} \tag{4.21}
$$

4.1.7 *Boundary conditions*

The equilibrium equations (in either stresses or displacements) govern the mechanical behavior of the solid material inside the domain. In actual engineering problems concerning the mechanics of solid materials, there are always conditions given on the boundary of the domain. In general, there are two major types of boundary conditions for solids: displacement boundary conditions (DBCs) and stress boundary conditions (SBCs). Therefore, the boundary Γ of the problem domain Ω is usually divided into two parts. One is the boundary Γ_u, where the displacements are prescribed, and another is Γ_t, where the traction is prescribed.

Often, enforcing SBCs at every point on a boundary can be difficult. Thus, we may use integral equilibrium conditions over a boundary surface using stresses. Since integrals of stresses result in forces, these types of boundary conditions are termed force boundary conditions (FBCs), similar to the 2D problems studied in Chapter 2.

4.1.7.1 *Displacement boundary conditions*

The displacement BCs can be expressed as follows:

$$u_i = u_{\Gamma i} \quad \text{on } \Gamma_u \in \Gamma \tag{4.22}$$

where $i = 1, 2, 3$, and $u_{\Gamma i}$ is the specified displacement u_i on boundary Γ_u, which can be zero but is not necessarily so.

In some cases, we may specify the derivatives of a displacement component on parts of the boundary:

$$\frac{\partial u_i}{\partial x_j} = \frac{\partial u_{\Gamma i}}{\partial x_j} \quad \text{on } \Gamma_u \in \Gamma \tag{4.23}$$

where $j = 1, 2, 3$, and $\frac{\partial u_{\Gamma i}}{\partial x_j}$ is the specified derivative of displacement u_i with respect to x_j on boundary Γ_u, which can be zero but is not necessarily so.

4.1.7.2 *Stress boundary conditions*

The stress BCs can be expressed as follows:

$$\sigma_{ij} n_j = t_{\Gamma j} \quad \text{on } \Gamma_t \in \Gamma \tag{4.24}$$

where n_j is the jth component of the outward unit normal at a point on boundary Γ_t, which is often called the direction cosines. And $t_{\Gamma j}$ is the specified traction t_j on boundary Γ_t, which can be zero but is not necessarily so.

Equation (4.24) can be written explicitly as

$$\begin{aligned}\sigma_{xx}n_1 + \sigma_{yx}n_2 + \sigma_{zx}n_3 &= t_{\Gamma 1}\\ \sigma_{xy}n_1 + \sigma_{yy}n_2 + \sigma_{zy}n_3 &= t_{\Gamma 2}\\ \sigma_{xz}n_1 + \sigma_{yz}n_2 + \sigma_{zz}n_3 &= t_{\Gamma 3}\end{aligned} \tag{4.25}$$

Our task in this chapter is to determine the solutions for stresses and displacements in 3D solids subjected to various types of forces. We choose to use stress methods, and the derived solutions must satisfy the equilibrium equations, stress boundary conditions (SBCs), force boundary conditions (FBCs), displacement boundary conditions (DBCs), and the compatibility equations. The fundamental concepts and procedures are largely similar to those used for 2D problems studied in Chapter 2.

4.2 Beams under pure bending

Consider a 3D beam with a uniform cross-section. It is subjected to a pure bending moment, as shown in Fig. 4.1. Assume the body force is zero. We aim to obtain the solutions for stresses, strains, and displacements in the beam within the Cartesian coordinate system.

For convenience, we set the origin of the coordinates at the centroid of the cross-section and the middle of the beam's length, with the x-axis aligned with the axis that passes through the centroids of the beam's cross-sections. The bending moment about the z-axis is denoted as M_z. We also require that the x–y and x–z planes are the principal planes, implying that the product moment of the cross-sectional area is zero: $\int_A yz\, dA = 0$ (see Chapter 10 of Ref. [5]).

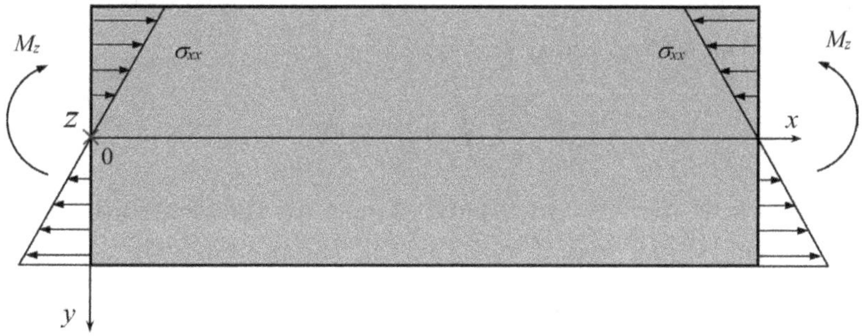

Figure 4.1. Schematic drawing of a 3D beam with a uniform cross-section of arbitrary shape with a smooth surface, subjected to pure bending.

4.2.1 *Stresses in 3D beams under pure bending*

In the stress method, we assume the stress components in some form and then ensure these assumed stresses satisfy the equilibrium equations, stress boundary conditions, and strain compatibility equations. Based on the thin beam theory (see Chapter 8 of Ref. [5]), the stresses for this problem can be assumed as follows:

$$\begin{aligned}\sigma_{xx} &= \frac{M_z y}{I} \\ \sigma_{yy} &= \sigma_{zz} = \sigma_{xz} = \sigma_{yz} = \sigma_{xy} = 0\end{aligned} \tag{4.26}$$

where $I = \int_A y^2 dA$ is the second moment of area of the cross-section, which is known for a given beam. This stress state is simple, with only one nonzero normal stress. It is clear that this stress set satisfies the equilibrium Eq. (4.2).

On the beam's circumferential surface, the surface normal component is $n_1 = 0$, and the externally applied stress is $t_{\Gamma j} = 0$ for all $j = 1, 2, 3$. Therefore, the boundary conditions in Eq. (4.25) on the circumferential surface are satisfied.

On the beam's right end surface, $n_1 = 1$, $n_2 = n_3 = 0$, and $t_{\Gamma 2} = t_{\Gamma 3} = 0$. Hence, the second and third equations of Eq. (4.25) are satisfied. The first equation in Eq. (4.25) can only be satisfied in an integral sense.

$$\begin{aligned}\int_A \sigma_{xx} dA &= \int_A t_{\Gamma 1} dA = 0 \qquad (a)\\ \int_A \sigma_{xx} z dA &= \int_A t_{\Gamma 1} z dA = 0 \qquad (b)\\ \int_A \sigma_{xx} y dA &= \int_A t_{\Gamma 1} y dA = M_z \qquad (c)\end{aligned} \tag{4.27}$$

These are the force boundary conditions (FBCs). Using the stresses given in Eq. (4.26), Eq. (4.27a) requires

$$\frac{M_z}{I} \int_A y dA = 0 \tag{4.28}$$

Since the origin of the coordinates is set at the centroid of the cross-section, we have $\int_A y dA = 0$. Thus, Eq. (4.27a) is satisfied.

Equation (4.27b) demands

$$\frac{M_z}{I} \int_A yz dA = 0 \tag{4.29}$$

Since the x–y and x–z planes are the principal planes, $\int_A yzdA = 0$, ensuring that Eq. (4.27b) is satisfied.

Equation (4.27c) requires

$$\frac{M_z}{I}\int_A y^2 dA = M_z \tag{4.30}$$

This equation is also satisfied. A similar analysis can be performed for the left end surface of the beam. This confirms that all the stress boundary conditions are satisfied. Therefore, the stress solution given in Eq. (4.26) is correct.

The actual distribution of σ_{xx} at the beam's left or right end may deviate from the linear distribution in Eq. (4.26). In such cases, the stress solution remains valid only at locations sufficiently far from the ends, according to Saint-Venant's principle.

This problem is straightforward, and the stresses are easily determined.

4.2.2 *Strains in 3D beams under pure bending*

Using the constitutive equation (4.7) and the stresses given in Eq. (4.26), the strains can be written as

$$\begin{aligned} &\varepsilon_{xx} = \frac{M_z y}{EI}, \quad \varepsilon_{yy} = -\frac{\nu M_z y}{EI}, \quad \varepsilon_{zz} = -\frac{\nu M_z y}{EI} \\ &\varepsilon_{xz} = \varepsilon_{yz} = \varepsilon_{xy} = 0 \end{aligned} \tag{4.31}$$

This strain field is linear and, therefore, satisfies the compatibility conditions given in Eq. (4.12). Consequently, the strain solution in Eq. (4.31) is valid, allowing us to determine the corresponding displacement field from these strains.

4.2.3 *Displacements in 3D beams under pure bending*

Using these strains and the strain–displacement relations in Eq. (4.9), we obtain the following six partial differential equations (PDEs):

$$\begin{aligned} &\frac{\partial u}{\partial x} = \frac{M_z y}{EI}, \quad \frac{\partial v}{\partial y} = -\frac{\nu M_z y}{EI}, \quad \frac{\partial w}{\partial z} = -\frac{\nu M_z y}{EI} \\ &\frac{\partial u}{\partial y} + \frac{\partial v}{\partial x} = 0, \quad \frac{\partial u}{\partial z} + \frac{\partial w}{\partial x} = 0, \quad \frac{\partial v}{\partial z} + \frac{\partial w}{\partial y} = 0 \end{aligned} \tag{4.32}$$

We can obtain the general expressions for the three displacement components by integrating the first three equations in Eq. (4.32) in a similar manner as for the 2D problem studied in Chapter 2. This integration introduces three unknown integral functions.

The key difference in the 3D case is that these unknown functions depend on two variables — either (y, z), (z, x) or (x, y). Consequently, determining these functions requires more careful operations. The crucial point is that when an indefinite integral is performed with respect to x, for example, it introduces an unknown function that depends only on (y, z).

The snippet is as follows:

```
# Define symbolic variables and functions:
x, y, z = sp.symbols('x, y, z', real=True)
EI, ν = sp.symbols('EI, ν', positive=True)     # properties of the beam
Mz = sp.symbols('M_z', real=True)                       # applied moment

f1  = sp.Function('f1')(y, z)         # integral function, f1 of y and z
f2  = sp.Function('f2')(z, x)         # integral function, f2 of z and x
f3  = sp.Function('f3')(x, y)         # integral function, f3 of x and y

εxx, εyy, εzz = symbols('ε_xx, ε_yy, ε_zz')       # variables for strains
εzx, εyz, εxy = symbols('ε_zx, ε_yz, ε_xy ')

εxx = Mz/EI*y; εyy = -ν*Mz/EI*y; εzz = -ν*Mz/EI*y # strains in the beam

ug = sp.integrate(εxx, x) +  f1        # general displacement u with f1
vg = sp.integrate(εyy, y) +  f2        # general displacement v with f2
wg = sp.integrate(εzz, z) +  f3        # general displacement w with f3
display(Math(f" \\text{{ General displacement solution $u_{{g}}$}}=\
                                                      {latex(ug)}"))
display(Math(f" \\text{{ General displacement solution $v_{{g}}$}}=\
                                                      {latex(vg)}"))
display(Math(f" \\text{{ General displacement solution $w_{{g}}$}}=\
                                                      {latex(wg)}"))
```

General displacement solution $u_g = f_1(y, z) + \dfrac{M_z xy}{EI}$

General displacement solution $v_g = f_2(z, x) - \dfrac{M_z y^2 \nu}{2EI}$

General displacement solution $w_g = f_3(x, y) - \dfrac{M_z yz\nu}{EI}$

4.2.4 Determination of unknown functions

To determine these three integral functions, f_1, f_2, and f_3, we shall use the last three equations in the strain–displacement relations, Eq. (4.32), which give

```
εyz = vg.diff(z) + wg.diff(y)
εzx = ug.diff(z) + wg.diff(x)
εxy = ug.diff(y) + vg.diff(x)
display(Math(f" \\text{{ $ε_{{yz}}$}}= {latex(εyz)}=0"))
display(Math(f" \\text{{ $ε_{{zx}}$}}= {latex(εzx)}=0"))
display(Math(f" \\text{{ $ε_{{xy}}$}}= {latex(εxy)}=0"))
```

$$\varepsilon_{yz} = \frac{\partial}{\partial z} f_2(z,x) + \frac{\partial}{\partial y} f_3(x,y) - \frac{M_z z\nu}{EI} = 0$$

$$\varepsilon_{zx} = \frac{\partial}{\partial z} f_1(y,z) + \frac{\partial}{\partial x} f_3(x,y) = 0$$

$$\varepsilon_{xy} = \frac{\partial}{\partial y} f_1(y,z) + \frac{\partial}{\partial x} f_2(z,x) + \frac{M_z x}{EI} = 0$$

We have three equations with three unknown functions, f_1, f_2, and f_3, which can be used to find these three unknown functions. The strategy to solve this system of equations is similar to what we do in solving algebraic equation sets by isolating the unknown variables. Since these equations are PDEs, the solution process is more involved and requires differential operations to isolate these unknown functions one by one. In addition, indefinite integrals will also be used, which introduce arbitrary unknown constants. The snippet to get this done is as follows.

Perform differentiations, $\frac{\partial}{\partial z}\varepsilon_{zx}$ and $\frac{\partial}{\partial y}\varepsilon_{xy}$, which should isolate $f_1(y,z)$ in two PDEs:

```
display(Math(f" \\frac{{∂}}{{∂z}}\\text{{$ε_{{zx}}$}} = \
                                        {latex(εzx.diff(z))}=0"))
display(Math(f" \\frac{{∂}}{{∂y}}\\text{{$ε_{{xy}}$}} = \
                                        {latex(εxy.diff(y))}=0"))
```

$$\frac{\partial}{\partial z}\varepsilon_{zx} = \frac{\partial^2}{\partial z^2} f_1(y,z) = 0$$

$$\frac{\partial}{\partial y}\varepsilon_{xy} = \frac{\partial^2}{\partial y^2} f_1(y,z) = 0$$

To satisfy both of the foregoing equations, $f_1(y,z)$ must have the following general form:

$$f_1(y,z) = a_1 + b_1 y + c_1 z + d_1 yz \tag{4.33}$$

where a_1, b_1, c_1, and d_1 are arbitrary constants.

Next, perform differentiations, $\frac{\partial}{\partial y}\varepsilon_{yz}$ and $\frac{\partial}{\partial x}\varepsilon_{zx}$, which isolates $f_3(x,y)$ in two PDEs:

```
display(Math(f" \\frac{{∂}}{{∂y}}\\text{{$ε_{{yz}}$}} = \
                                        {latex(εyz.diff(y))}=0"))
display(Math(f" \\frac{{∂}}{{∂x}}\\text{{$ε_{{zx}}$}} = \
                                        {latex(εzx.diff(x))}=0"))
```

$$\frac{\partial}{\partial y}\varepsilon_{yz} = \frac{\partial^2}{\partial y^2} f_3(x,y) = 0$$

$$\frac{\partial}{\partial x}\varepsilon_{zx} = \frac{\partial^2}{\partial x^2} f_3(x,y) = 0$$

Thus, f_3 must have the following general form:

$$f_3(x,y) = a_3 + b_3 x + c_3 y + d_3 xy \tag{4.34}$$

Substitute $f_3(x,y)$ into the expression for ε_{yz} and $f_1(y,z)$ into the expression for ε_{xy}. Then, carry out the two integrals to find two expressions for $f_2(x,y)$ only. This process will introduce additional integral functions that are one-dimensional (1D). The snippet is as follows:

```
a1, b1, c1, d1 = symbols('a1, b1, c1, d1')                    # constants
a3, b3, c3, d3 = symbols('a3, b3, c3, d3')                    # constants

f1_  = a1 + b1*y + c1*z + d1*y*z
f3_  = a3 + b3*x + c3*y + d3*x*y

h  = sp.Function('h')(x)                             # integral function of x
εyz_h = Mz*z*ν/EI - f3_.diff(y)                      # shear strain with h(x)
f2_h = (sp.integrate(εyz_h, z) + h).expand()           # f2(x, z) with h(x)
display(Math(f" \\text{{$f_2(x, z)$}}={latex(f2_h)}"))

s  = sp.Function('s')(z)                             # integral function of z
εxy_s =- Mz*x/EI - f1_.diff(y)                       # shear strain with s(z)
f2_s = (sp.integrate(εxy_s, x) + s).expand()           # f2(x, z) with s(z)
display(Math(f" \\text{{$f_2(x, z)$}}={latex(f2_s)}"))
```

$$f_2(x,z) = -c_3 z - d_3 xz + h(x) + \frac{M_z z^2 \nu}{2EI}$$

$$f_2(x,z) = -b_1 x - d_1 xz + s(z) - \frac{M_z x^2}{2EI}$$

These two expressions must be equal. This can only happen if we set $d_1 = d_3$ and define the integral function $h(x)$ as

$$h(x) = -b_1 x - \frac{M_z x^2}{2EI} \tag{4.35}$$

Hence, $f_2(x,y)$ can be written in the general form:

$$f_2(x,y) = a_2 + b_2 z + c_2 x + d_3 xz - \frac{M_z x^2}{2EI} + \frac{M_z z^2 \nu}{2EI} \tag{4.36}$$

The following snippet defines $f_2(x,y)$ and then updates the expressions for these three displacements and shear stresses using Eq. (4.32):

```
a2, b2, c2 = symbols('a2, b2, c2')                              # constants

f2_ = a2 + b2*z + c2*x + d3*z*x + Mz*ν*z**2/EI/2 - Mz*x**2/EI/2
ug = sp.integrate(εxx, x) +  f1_                        # general displacements
vg = sp.integrate(εyy, y) +  f2_
wg = sp.integrate(εzz, z) +  f3_

εyz = (vg.diff(z)+wg.diff(y)).collect(x).collect(y).collect(z)
εzx = (ug.diff(z)+wg.diff(x)).collect(x).collect(y).collect(z)
εxy = (ug.diff(y)+vg.diff(x)).collect(x).collect(y).collect(z)
display(Math(f" \\text{{ $ε_{{yz}}$}}= {latex(εyz)}=0"))
display(Math(f" \\text{{ $ε_{{zx}}$}}= {latex(εzx)}=0"))
display(Math(f" \\text{{ $ε_{{xy}}$}}= {latex(εxy)}=0"))
```

$\varepsilon_{yz} = b_2 + c_3 + 2d_3 x = 0$

$\varepsilon_{zx} = b_3 + c_1 + y\left(d_1 + d_3\right) = 0$

$\varepsilon_{xy} = b_1 + c_2 + z\left(d_1 + d_3\right) = 0$

All three equations must be zero for any x, y, and z. Thus, we must set $d_1 = 0$ and $d_3 = 0$ to meet this condition.

Now, we use these three expressions again for the shear stresses to determine the relationships among the constants a_i, b_i, and c_i:

```
dic_d = {d1:0, d3:0}
ug = ug.subs(dic_d); vg = vg.subs(dic_d); wg = wg.subs(dic_d)

abc = [a1, b1, c1,  a2, b2, c2,  a3, b3, c3 ]
eqn1 = vg.diff(z)+wg.diff(y)
eqn2 = ug.diff(z)+wg.diff(x)
eqn3 = ug.diff(y)+vg.diff(x)

abc_sln = sp.solve([eqn1, eqn2, eqn3], abc)
abc_sln
```

$\{b_1 : -c_2,\ b_2 : -c_3,\ c_1 : -b_3\}$

Substituting these relations back into the displacement expressions, we obtain the general expression for the displacements as follows:

```
ug = ug.subs(abc_sln).expand()
vg = vg.subs(abc_sln).expand()
wg = wg.subs(abc_sln).expand()
display(Math(f" \\text{{ General displacement solution $u_{{g}}$}}=\
                                        {latex(ug)}"))
display(Math(f" \\text{{ General displacement solution $v_{{g}}$}}=\
                                        {latex(vg)}"))
display(Math(f" \\text{{ General displacement solution $w_{{g}}$}}=\
                                        {latex(wg)}"))
```

General displacement solution, $u_g = a_1 - b_3 z - c_2 y + \frac{M_z x y}{EI}$

General displacement solution, $v_g = a_2 + c_2 x - c_3 z - \frac{M_z x^2}{2EI} - \frac{M_z y^2 \nu}{2EI} + \frac{M_z z^2 \nu}{2EI}$

General displacement solution, $w_g = a_3 + b_3 x + c_3 y - \frac{M_z y z \nu}{EI}$

4.2.5 *Displacement boundary conditions, final solution*

There are six unknown constants remaining in these expressions. The beam has six degrees of freedom (DOFs) corresponding to rigid body movements that do not induce strain or stress: three translational, represented by a_1, a_2, and a_3, and three rotational, represented by b_3, c_2, and c_3. Assuming the beam is constrained at $(0, 0, 0)$ with zero displacements and zero displacement gradients at the origin, we have

$$\begin{aligned} & u = v = w = 0 \\ & \frac{\partial u}{\partial z} = \frac{\partial v}{\partial x} = \frac{\partial w}{\partial y} = 0 \end{aligned} \tag{4.37}$$

These six conditions can be used to determine the six constants:

```
# Use DBCs:
dic0 = {x:0, y:0, z:0}
abc_sln = sp.solve([ug.subs(dic0), vg.subs(dic0), wg.subs(dic0),
          ug.diff(z).subs(dic0),   vg.diff(x).subs(dic0),
          wg.diff(y).subs(dic0)],  [a1, a2, a3, b3, c2, c3 ])
display(Math(f"\\text{{Constants found using DBCs}}={latex(abc_sln)}"))

# Update the displacements:
u = ug.subs(abc_sln).expand()
v = vg.subs(abc_sln).expand().collect(Mz/(2*EI))
w = wg.subs(abc_sln).expand()
display(Math(f"\\text{{Final displacement solution $u$}}={latex(u)}"))
display(Math(f"\\text{{Final displacement solution $v$}}={latex(v)}"))
display(Math(f"\\text{{Final displacement solution $w$}}={latex(w)}"))
display(Math(f"\\text{{Deflection of the centroid axis, $v(x,0,0)$}}=\
                                         {latex(v.subs({y:0, z:0}))}"))
```

Constants found using DBCs $= \{a_1 : 0,\ a_2 : 0,\ a_3 : 0,\ b_3 : 0,\ c_2 : 0,\ c_3 : 0\}$

Final displacement solution, $u = \dfrac{M_z xy}{EI}$

Final displacement solution, $v = \dfrac{M_z\left(-x^2 - y^2\nu + z^2\nu\right)}{2EI}$

Final displacement solution, $w = -\dfrac{M_z yz\nu}{EI}$

Deflection of the centroid axis, $v(x,0,0) = -\dfrac{M_z x^2}{2EI}$

The first and last expressions were derived using the thin beam theory [5]. The other two can only be derived from the current 3D elasticity formulation.

4.2.6 *Curvatures of the beam*

The curvature of the deformed beam can be determined from the second derivatives of the deflection v:

$$
\begin{aligned}
\frac{1}{R_x} &= -\frac{\partial^2 v}{\partial x^2} = \frac{M_z}{2EI} \\
\frac{1}{R_y} &= -\frac{\partial^2 v}{\partial y^2} = \frac{\nu M_z}{2EI} \\
\frac{1}{R_z} &= -\frac{\partial^2 v}{\partial z^2} = -\frac{\nu M_z}{2EI}
\end{aligned}
\tag{4.38}
$$

where R_x, R_y, and R_z are the radii of curvature along the x-, y-, and z-axes of the bent beam. These curvatures are all constants. The first expression in Eq. (4.38) is derived using the thin beam theory [5]. The other two can only be obtained from the current 3D elasticity formulation.

4.3 General formulation for 3D beam under an end force

4.3.1 *Setting of the problem*

Consider a 3D beam of length L with a uniform cross-section A of arbitrary shape and a smooth circumferential surface. It is subjected to a force at the right end in the x-direction, as shown in Fig. 4.2. Assume the body force is zero. Our goal is to determine the stresses, strains, and displacements in the beam within the Cartesian coordinate system.

For convenience, we set the origin of the coordinate system at the centroid of the cross-section and the middle of the beam's length, with the x-axis aligned with the axial axis passing through the centroids of the beam. We also require that the x–y and x–z planes are the principal planes, implying that $\int_A yz dA = 0$.

4.3.2 *Stresses in 3D beam under end force*

As usual, in the stress method, we shall assume the stress components in some form and then ensure these assumed stresses satisfy the equilibrium equations and stress boundary conditions. Based on the setup above,

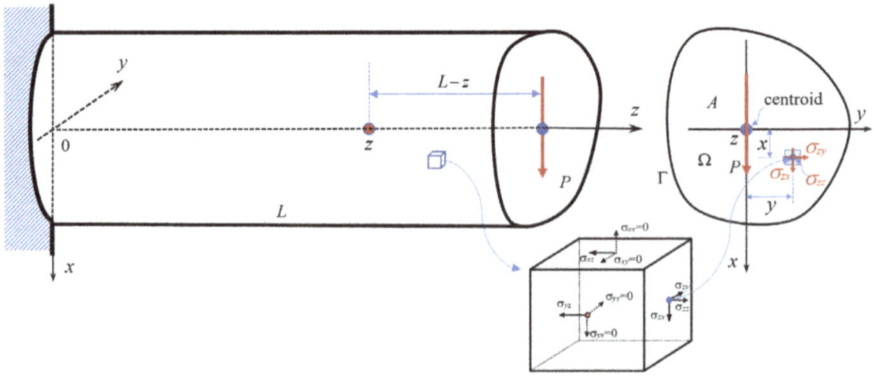

Figure 4.2. Schematic drawing of a 3D beam with a uniform cross-section of arbitrary shape and a smooth circumferential surface, subjected to a force at the right end.

the stresses can be assumed following the thin beam theory, as discussed in Chapter 8 of Ref. [5].

$$\begin{aligned}
&\sigma_{zz} = P(q_0 + q_1 x + q_2 y)(L - z), \quad \text{on beam cross-section} \\
&\sigma_{zx} = \text{nonzero unknown, on beam cross-section} \\
&\sigma_{zy} = \text{nonzero unknown, on beam cross-section} \\
&\sigma_{xx} = \sigma_{yy} = \sigma_{xy} = 0
\end{aligned} \tag{4.39}$$

where $q_i (i = 0, 1, 2)$ are unknown constants to be determined. The differences between this problem and the previous pure bending problem are as follows:

1. The external force is not a moment, and the moment on a cross-section of the beam changes linearly along the axial direction. Hence, the normal stress on the beam's cross-section includes a term of $(L - z)$.
2. The shear stresses σ_{zx} and σ_{zy} will be nonzero and unknown. We thus have three nonzero stress components.

These differences significantly increase the difficulty of finding the stress and displacement solutions. The following derives the solution, using the equilibrium equations and stress boundary conditions.

4.3.3 *Determine the normal stress via equilibrium*

First, on any cross-section of the 3D beam, the stresses must satisfy the following integral equilibrium equations or FBCs:

$$\begin{aligned}
&\sum F_z = 0 : \iint_A \sigma_{zz} dxdy = 0 \\
&\sum M_x = 0 : \iint_A \sigma_{zz} y dxdy = 0 \\
&\sum M_y = 0 : \iint_A \sigma_{zz} x dxdy = -P(L - z) \\
&\sum F_x = 0 : \iint_A \sigma_{zx} dxdy = P \\
&\sum F_y = 0 : \iint_A \sigma_{zy} dxdy = 0 \\
&\sum M_z = 0 : \iint_A (\sigma_{zy} x - \sigma_{zx} y) dxdy = 0
\end{aligned} \tag{4.40}$$

These are force boundary conditions on a cross-section at arbitrary z. Substituting the normal stress expression σ_{zz} given in Eq. (4.39) into the first three equations of Eq. (4.40), we obtain

$$\begin{aligned}
&\iint_A P(q_0 + q_1 x + q_2 y) dxdy = 0 \\
&\iint_A P(q_0 + q_1 x + q_2 y) y \, dxdy = 0 \\
&\iint_A P(q_0 + q_1 x + q_2 y) x \, dxdy = -P
\end{aligned} \tag{4.41}$$

which gives

$$\begin{aligned}
&q_1 \underbrace{\iint_A x dxdy}_{Q_y} + q_2 \underbrace{\iint_A y dxdy}_{Q_x} + q_0 \underbrace{\iint_A dxdy}_{A} = 0 \\
&q_1 \underbrace{\iint_A xy dxdy}_{I_{xy}} + q_2 \underbrace{\iint_A y^2 dxdy}_{I_{xx}} + q_0 \underbrace{\iint_A y dxdy}_{Q_x} = 0 \\
&q_1 \underbrace{\iint_A x^2 dxdy}_{I_{yy}} + q_2 \underbrace{\iint_A xy dxdy}_{I_{xy}} + q_0 \underbrace{\iint_A x dxdy}_{Q_y} = -1
\end{aligned} \tag{4.42}$$

Or

$$\begin{aligned}
q_1 Q_y + q_2 Q_x + q_0 A &= 0 \\
q_1 I_{xy} + q_2 I_{xx} + q_0 Q_x &= 0 \\
q_1 I_{yy} + q_2 I_{xy} + q_0 Q_y &= -1
\end{aligned} \tag{4.43}$$

where I_{xx}, I_{yy}, I_{xy} are the second moments and Q_x, Q_y are the first moments of area of the cross-section of the beam. All these are known for a given beam with a uniform cross-section (see Chapter 10 of Ref. [5]). Equation (4.43) can be solved for the unknown constants q_0, q_1, and q_2:

$$q_1 = \frac{-AI_{xx} + Q_x^2}{D}; \quad q_2 = \frac{AI_{xy} - Q_x Q_y}{D}; \quad q_0 = \frac{I_{xx} Q_y - I_{xy} Q_x}{D} \tag{4.44}$$

where the term in the denominator is

$$D = AI_{xx} I_{yy} - AI_{xy}^2 - I_{xx} Q_y^2 + 2I_{xy} Q_x Q_y - I_{yy} Q_x^2 \tag{4.45}$$

Therefore, q_i, where $i = 0, 1, 2$, are all constants computed from the areal properties of the beam's cross-section. If the cross-section of the beam is symmetric, we can then conveniently choose the x- and y-axes as the axes of symmetry with the centroid at the origin. The snippet to find q_0, q_1, and q_2 for a beam with an arbitrary cross-section is as follows:

```
# Define symbolic variables and functions:
x, y, z = sp.symbols('x, y, z', real=True)
q0, q1, q2, D = sp.symbols('q0, q1, q2, D', real=True)
A = sp.symbols('A', positive=True)         # Length and area of the beam
P = sp.symbols('P', real=True)                              # applied moment
Ixx, Iyy, Ixy = symbols('I_{xx}, I_{yy}, I_{xy}') # 2nd moments of area
Qx, Qy = symbols('Q_x, Q_y')                        # 1st moments of area

eqn1 = q1* Qy + q2*Qx  + q0*A
eqn2 = q1*Ixy + q2*Ixx + q0*Qx
eqn3 = q1*Iyy + q2*Ixy + q0*Qy + 1
abc_sln = sp.solve([eqn1, eqn2, eqn3], [q0, q1, q2])

D_ = abc_sln[q1].as_numer_denom()[1]
display(Math(f" \\text{{Denominator of coefficients, $D$}}{latex(D_)}"))
abc_sln = {k: v.subs(D_,D) for k, v in abc_sln.items()}
display(Math(f" \\text{{Coefficients found}}=\\small {latex(abc_sln)}"))
```

Denominator of coefficients, $D = AI_{xx}I_{yy} - AI_{xy}^2 - I_{xx}Q_y^2 + 2I_{xy}Q_xQ_y - I_{yy}Q_x^2$

Coefficients found

$$= \left\{ q_0 : \frac{I_{xx}Q_y - I_{xy}Q_x}{D}, \ q_1 : \frac{-AI_{xx} + Q_x^2}{D}, \ q_2 : \frac{AI_{xy} - Q_xQ_y}{D} \right\}$$

With q_i, where $i = 0, 1, 2$ given by Eq. (4.44), the **normal stress** σ_{zz} is now fully determined and is given by Eq. (4.39), which applies to any 3D beam with an arbitrary shaped cross-section.

4.3.4 *Shear stresses in stream functions*

Next, let us find the **shear stresses** σ_{zx} and σ_{zy}. Using the equilibrium Eq. (4.2) with no body forces and the first equation in Eq. (4.39), we have

$$\begin{aligned} &\frac{\partial \sigma_{zx}}{\partial z} = 0; \quad \frac{\partial \sigma_{zy}}{\partial z} = 0 \\ &\frac{\partial \sigma_{xz}}{\partial x} + \frac{\partial \sigma_{yz}}{\partial y} = P(q_1x + q_2y + q_0) \end{aligned} \tag{4.46}$$

This means that the shear stresses σ_{zx} and σ_{zy} are independent of z, and hence are only functions of (x, y). One may notice the similarity and differences between this set of equations given in Eq. (4.46) and Eq. (9.59) in Ref. [5], which can help to understand the techniques: converting the problem into one of finding a stream function governed by a Poisson equation. The details are presented in the following.

4.3.5 *Derivation of stream functions*

The last equation in Eq. (4.46) can be rewritten in the form of

$$\frac{\partial}{\partial x}\underbrace{\left[\frac{2}{P}\sigma_{zx}-(q_1x^2+q_0x)\right]}_{V_y}=\frac{\partial}{\partial y}\underbrace{\left[-\frac{2}{P}\sigma_{zy}+(q_2y^2+q_0y)\right]}_{V_x} \tag{4.47}$$

This means that the vector field $[V_x \quad V_y]^\top$ is a **gradient field**, and there must exist a potential function ϕ for this vector field, such that [7]

$$\begin{aligned} V_x &= \frac{\partial\phi}{\partial x} = -\frac{2}{P}\sigma_{zy}+(q_2y^2+q_0y) \\ V_y &= \frac{\partial\phi}{\partial y} = \frac{2}{P}\sigma_{zx}-(q_1x^2+q_0x) \end{aligned} \tag{4.48}$$

If such a potential function can be found, the shear stresses can then be found as

$$\begin{aligned} \sigma_{zx} &= \frac{P}{2}\frac{\partial\phi}{\partial y}+\frac{P}{2}(q_1x^2+q_0x) \\ \sigma_{zy} &= -\frac{P}{2}\frac{\partial\phi}{\partial x}+\frac{P}{2}(q_2y^2+q_0y) \end{aligned} \tag{4.49}$$

It is clear that a shear stress consists of two parts: the first part relates to the unknown function ϕ, and the other is a known function, the last terms in Eq. (4.49). Since $[\sigma_{zx} \; \sigma_{zy}]^\top$ can be viewed as a shear stress vector on a cross-section of the beam, the function ϕ is in fact the **stream function** [7] of the first part of the shear stress vector.

4.3.6 *Control equations for stream functions*

Since these shear stresses are independent of z, the stream function ϕ can only be a function of x and y: $\phi(x, y)$. The governing equation for it can be derived using the compatibility equations. Using the last two equations in Eq. (4.14) (without body forces) and $I_1 = \sigma_{zz} + \sigma_{xx} + \sigma_{yy} = P(q_0 + q_1 x + q_2 y)(L - z)$, we found

$$\frac{\partial}{\partial y}(\nabla^2\phi) = -\frac{2\nu q_1}{(1+\nu)}$$
$$\frac{\partial}{\partial x}(\nabla^2\phi) = \frac{2\nu q_2}{(1+\nu)} \tag{4.50}$$

Integrating once for both equations gives

$$\nabla^2\phi = -\frac{2\nu}{(1+\nu)}q_1 y + C_y(x)$$
$$\nabla^2\phi = \frac{2\nu}{(1+\nu)}q_2 x + C_x(y) \tag{4.51}$$

where C_y is an arbitrary function of only x (constant to y), and C_x is an arbitrary function of only y (constant to x). Since the foregoing two equations must be equal, we set

$$C_y(x) = \frac{2\nu}{(1+\nu)}q_2 x - C_0$$
$$C_x(y) = -\frac{2\nu}{(1+\nu)}q_1 y - C_0 \tag{4.52}$$

where C_0 is an arbitrary constant. We thus obtain

$$\nabla^2\phi = \frac{2\nu}{(1+\nu)}(q_2 x - q_1 y) - C_0, \quad \text{in } \Omega \tag{4.53}$$

This means the stream function of the shear stress vector on the cross-section satisfies a nonhomogeneous Laplace equation, known as the Poisson equation.

The boundary conditions for it are based on the stress condition on the boundary curve Γ of the beam's cross-section. Using the third equation in Eq. (4.25), and $n_3 = 0$ on Γ, we obtain

$$\sigma_{zx}n_1 + \sigma_{zy}n_2 = 0 \tag{4.54}$$

At any point on Γ, we also have

$$n_1 = \frac{dy}{ds}; n_1 = -\frac{dx}{ds} \tag{4.55}$$

Equation (4.54) becomes

$$\sigma_{zx}\frac{dy}{ds} - \sigma_{zy}\frac{dx}{ds} = 0 \tag{4.56}$$

Substituting Eq. (4.49) into Eq. (4.56) gives

$$\frac{\partial \phi}{\partial s} = -(q_1 x^2 + q_0 x)\frac{dy}{ds} + (q_2 y^2 + q_0 y)\frac{dx}{ds}, \quad \text{on } \Gamma \tag{4.57}$$

The control equations for the stream function is the Poisson equation (4.53) in Ω with boundary condition (4.57) on Γ. We now need to find such a stream function.

4.3.7 *Splitting the stream function*

One approach to finding such a ϕ is to split it into two functions:

$$\phi = \phi_B + C_0 \phi_T \tag{4.58}$$

The problem defined in (4.53) and (4.57) can be split into a torsional problem described by ϕ_T:

$$\begin{aligned} \nabla^2 \phi_T &= -1 \quad &\in \Omega \\ \frac{\partial \phi_T}{\partial s} &= 0 \quad &\in \Gamma \end{aligned} \tag{4.59}$$

and a bending problem described by ϕ_B:

$$\begin{aligned} \nabla^2 \phi_B &= \frac{2\nu}{(1+\nu)}(q_2 x - q_1 y) \quad &\in \Omega \\ \frac{\partial \phi_B}{\partial s} &= -(q_1 x^2 + q_0 x)\frac{dy}{ds} + (q_2 y^2 + q_0 y)\frac{dx}{ds} \quad &\in \Gamma \end{aligned} \tag{4.60}$$

This splitting reveals that this 3D cantilever beam problem is divided into two problems:

1. One is a torsional problem for the beam that is represented by the function ϕ_T, which is, in fact, the Prandtl stress function or a stream function of the torsional shear stress vector. Such a torsional problem can be solved by techniques presented in Chapter 9 in Ref. [5]. We shall not repeat that here.
2. Another is the bending problem for the beam, specifically for the shear stresses on its cross-section, which is represented by the function ϕ_B, which needs to be determined here.

The integral constant C_0 in Eq. (4.58) must relate to the torsional problem. It is therefore no longer our concern in solving the bending effects of the cantilever beam.

Our task now becomes solving Eq. (4.60) for ϕ_B. Once ϕ_B is found, these shear stresses can then be calculated using Eq. (4.49). To simplify the problem, we consider only beams with symmetric cross-sections, which is often the case in engineering applications. In such cases, the x- and y-axes become the principal axes, and we have (Chapter 10 of Ref. [5])

$$Q_x = 0; \quad Q_y = 0; \quad I_{xy} = 0 \tag{4.61}$$

which leads to

$$q_1 = \frac{-1}{I_{yy}}; \quad q_2 = 0; \quad q_0 = 0 \tag{4.62}$$

The normal stress becomes

$$\sigma_{zz} = \frac{-P}{I_{yy}} x(L-z) \tag{4.63}$$

The expressions for the stresses are simplified accordingly, and Eq. (4.60) becomes

$$\begin{aligned} \nabla^2 \phi_B &= \frac{2\nu}{(1+\nu)I_{yy}} y \quad && \in \Omega \\ \frac{\partial \phi_B}{\partial s} &= \frac{x^2}{I_{yy}} \frac{dy}{ds} && \in \Gamma \end{aligned} \tag{4.64}$$

Also, the integral constant C_0 in Eq. (4.58) can be set to zero because we can purposely let the force P pass through the shear center [5], ensuring that P does not introduce any torque to the beam cross-section. This results in a pure bending problem.

This Poisson equation still has a nonzero boundary condition. Thus, Timoshenko suggested [2] further splitting ϕ_B as

$$\phi_B = \frac{1}{P} \left[\psi(x,y) + F(y)\right] \tag{4.65}$$

where $\psi(x,y)$ remains a function of (x,y), while $F(y)$ depends only on y. This transformation introduces no loss of generality. Substituting Eq. (4.65) into Eq. (4.64) yields

$$\begin{aligned} \nabla^2 \psi &= \frac{2\nu P}{(1+\nu)I_{yy}} y - \frac{df}{dy} \quad && \in \Omega \\ \frac{\partial \psi}{\partial s} &= \left(\frac{P}{I_{yy}} x^2 - f\right) \frac{dy}{ds} && \in \Gamma \end{aligned} \tag{4.66}$$

where $f = \frac{dF}{dy}$, which is a function of only y.

If we can choose a $f(y)$ in Eq. (4.66) such that

$$\left(\frac{P}{I_{yy}}x^2 - f\right)\frac{dy}{ds} = 0 \tag{4.67}$$

we can then have the following homogeneous boundary condition for ψ:

$$\frac{\partial \psi}{\partial s} = 0 \quad \in \Gamma \tag{4.68}$$

Once ψ is found, the shear stresses can be computed using

$$\begin{aligned} \sigma_{zx} &= \frac{1}{2}\frac{\partial \psi}{\partial y} - \frac{Px^2}{2I_{yy}} + \frac{f}{2} \\ \sigma_{zy} &= -\frac{1}{2}\frac{\partial \psi}{\partial x} \end{aligned} \tag{4.69}$$

It is evident that ψ serves as the stream function for a portion of the shear stress vector, which needs to be determined. The remaining contribution comes from the function $-\frac{Px^2}{2I_{yy}} + \frac{f}{2}$, which affects only σ_{zx}, where $f(y)$ must be determined for a given beam cross-section shape.

We are now ready to analyze beams with specific cross-section geometries.

4.4 Cantilevers with an elliptic cross-section

4.4.1 *Problem setup*

Consider a 3D cantilever beam with a uniform elliptic cross-section, as illustrated in Fig. 4.3. The beam is subjected to a downward force at its right end and is free from body forces. Our goal is to determine the stress distribution in the beam using a Cartesian coordinate system.

The boundary equation of the elliptic cross-section is given by

$$\frac{x^2}{r_a^2} + \frac{y^2}{r_b^2} = 1 \tag{4.70}$$

where r_a and r_b are the major and minor semi-axes of the ellipse. In this case, the function $f(y)$, which satisfies Eq. (4.67), is given by

$$f(y) = -\frac{Pr_a^2}{I_{yy}r_b^2}(y^2 - r_b^2) \tag{4.71}$$

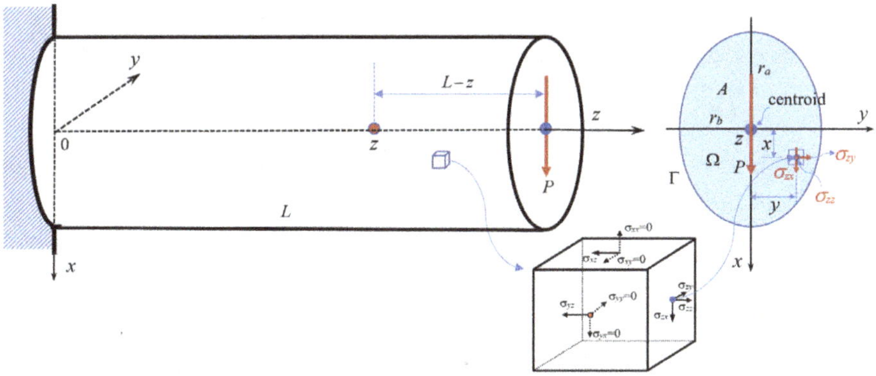

Figure 4.3. Schematic of a 3D beam with a uniform elliptic cross-section subjected to a force at the right end. The three nonzero unknown stress components on the z-plane are shown.

4.4.2 *Stream function for elliptic beam*

Substituting Eq. (4.71) into Eq. (4.66) gives

$$\begin{aligned} \nabla^2\psi &= \frac{2Py}{I_{yy}}\left(\frac{r_a^2}{r_b^2}+\frac{\nu}{1+\nu}\right) \quad \in \Omega \\ \psi &= 0 \in \Gamma \end{aligned} \tag{4.72}$$

Here, we modify the boundary condition from $\frac{\partial\psi}{\partial s}=0$ to $\psi=0$. This adjustment is justified as follows:

1. The condition $\frac{\partial\psi}{\partial s}=0$ implies that ψ remains constant at all points on Γ.
2. Since the elliptical cross-section forms a simply-connected domain, this constant merely shifts ψ by a uniform value throughout Ω.
3. When computing shear stresses using Eq. (4.69), any constant in ψ has no effect and can be omitted.

With this modification, the function satisfying Eq. (4.72) is easily found as

$$\psi = q\left(\frac{x^2}{r_a^2}+\frac{y^2}{r_b^2}-1\right)y \tag{4.73}$$

where q is a parameter that depends only on the applied force and cross-section properties of the problem.

$$q = \frac{P}{I_{yy}}\left(\frac{r_a^2 r_b^2}{3r_a^2+r_b^2}\right)\left(\frac{r_a^2}{r_b^2}+\frac{\nu}{(1+\nu)}\right) \tag{4.74}$$

We observe that the function ψ is an odd function in y, ensuring a symmetric solution for σ_{zx} when applying the first equation of Eq. (4.69). Similarly, ψ is an even function in x, which leads to an antisymmetric solution for σ_{zy} when using the second equation of Eq. (4.69). This behavior aligns with the symmetry of the elliptic cross-section about both the x- and y-axes, as well as the fact that the applied force P acts along the x-axis.

4.4.3 *Stresses in an elliptic beam*

The shear stress expressions can now be derived using Eq. (4.69). The results are presented in the following section.

```
# Define symbolic variables and functions:
x, y, z = sp.symbols('x, y, z', real=True)
ν = sp.symbols('ν', positive=True)                          # Poisson's ratio
ra, rb, A, L = sp.symbols('r_a, r_b, A, L', positive=True)  # geometry
P = sp.symbols('P', positive=True)         # applied force at the right end
Iyy = symbols('I_{yy}', positive=True)              # 2nd moments of area
Iyy_ = A*ra**2/4                                         # Iyy for ellipse
A_ = pi*ra*rb

f_ellipse = x**2/ra**2 + y**2/rb**2 - 1
f_y = -P*ra**2/(Iyy*rb**2)*(y**2-rb**2)
q = P/Iyy*(ra**2*rb**2/(3*ra**2+rb**2))*(ra**2/rb**2+ν/(1+ν))

ψ = q*f_ellipse*y
d2ψ = sp.nsimplify((ψ.diff(x,2) + ψ.diff(y,2))).simplify()
display(Math(f" \\text{{ $∇^2ψ$}}={latex(d2ψ)}"))

σzx = sp.nsimplify(1/2*(ψ.diff(y))).simplify() \
    + sp.nsimplify(1/2*f_y - P*x**2/Iyy/2).simplify()
σzy = sp.nsimplify(-1/2*ψ.diff(x)).simplify()

σzz = P*(q0+q1*x+q2*y)*(L-z)
σzz_sym = σzz.subs({q1:-1/Iyy, q2:0, q0:0})
display(Math(f" \\text{{ Shear stress $σ_{{zx}}$}}={latex(σzx)}"))
display(Math(f" \\text{{ Shear stress $σ_{{zy}}$}}={latex(σzy)}"))
display(Math(f" \\text{{ Normal stress $σ_{{zz}}$}}={latex(σzz_sym)}"))
```

$$\nabla^2\psi = \frac{2Py\left(r_a^2\left(\nu+1\right)+r_b^2\nu\right)}{I_{yy}r_b^2\left(\nu+1\right)}$$

Shear stress,

$$\sigma_{zx} = \frac{P\left(r_a^2\left(r_b^2-y^2\right)-r_b^2x^2\right)}{2I_{yy}r_b^2} + \frac{P\left(r_a^2\left(\nu+1\right)+r_b^2\nu\right)\left(-r_a^2r_b^2+3r_a^2y^2+r_b^2x^2\right)}{2I_{yy}r_b^2\cdot\left(3r_a^2+r_b^2\right)\left(\nu+1\right)}$$

Shear stress, $\sigma_{zy} = -\dfrac{Pxy\left(r_a^2\left(\nu+1\right)+r_b^2\nu\right)}{I_{yy}\left(3r_a^2+r_b^2\right)\left(\nu+1\right)}$

Normal stress, $\sigma_{zz} = -\dfrac{Px\left(L-z\right)}{I_{yy}}$

4.4.4 *Checking of FBCs*

In the derivation above, we used the first three equations in Eq. (4.40) and left the last three equations unaddressed. We will now check if the last three equations are also satisfied by the stress solution obtained.

First, the fifth equation in Eq. (4.40) is clearly satisfied due to the antisymmetry of σ_{zy}. Its integral over the symmetric area of an ellipse will always yield zero.

Second, the sixth equation in Eq. (4.40) is also satisfied for similar symmetry reasons. Both $\sigma_{zy}x$ and $\sigma_{zx}y$ are antisymmetric with respect to y, implying the integral over the symmetric area of the ellipse will result in zero.

Lastly, the integrand in the fourth equation of Eq. (4.40) is an even function in both x and y, so its integral can produce a nonzero value. Let's perform the integration to determine this value.

```
Int = sp.integrate
yb = rb*sqrt(1-x**2/ra**2)
fourth_eq = Int(Int(σzx, (y, -yb, yb)), (x,-ra, ra)).subs(Iyy, Iyy_)\
                                             .subs(A, A_).simplify()
display(Math(f"\\text{{The 4th Eq. of FBCs}}={latex(fourth_eq)}"))
```

The fourth Eq. of FBCs $= P$

This means that all the FBCs are satisfied. We conclude that the solution obtained is correct. It also agrees with the formulas provided in Ref. [2].

4.4.5 *Strains in an elliptic beam*

Using the constitutive equation (4.7) and the normal stresses, the strains can be computed using

```
E, G, L = sp.symbols('E, G, L', positive=True)        # beam parameters
#σzz, σzx, σzy, q1 = symbols('σ_{zz}, σ_{zx}, σ_{zy}, q1')
dicG = {G:E/(2*(1+ν))}

σzz = -P/Iyy*x*(L-z)        # P*q1*x*(L-z) #symmetric, q1=-1/Iyy,q2=q0=0
σxx = -ν*σzz; σyy = -ν*σzz

εxx = σxx/E; εyy = σyy/E; εzz = σzz/E                             # strains
γzx = 2*(1+ν)*σzx/E                                           # γzx = σzx/G
γzy = 2*(1+ν)*σzy/E                                           # γzy = σzy/G
γxy = 0*x
εzx = γzx/2; εzy = γzy/2; εxy = γxy/2

display(Math(f"\\text{{$ε_{{xx}}$}}={latex(εxx)}; \\text{{ $ε_{{yy}}$}}\
                   = {latex(εyy)}; \\text{{ $ε_{{zz}}$}}= {latex(εzz)}"))
display(Math(f"\\text{{$γ_{{zy}}$}}={latex(γzy.simplify())}"))
display(Math(f"\\text{{$γ_{{zx}}$}}={latex(γzx.simplify())}"))
display(Math(f"\\text{{$γ_{{xy}}$}}={latex(γxy)}"))
```

$$\varepsilon_{xx} = \frac{Px\nu\left(L-z\right)}{EI_{yy}}; \varepsilon_{yy} = \frac{Px\nu\left(L-z\right)}{EI_{yy}}; \ \varepsilon_{zz} = -\frac{Px\left(L-z\right)}{EI_{yy}}$$

$$\gamma_{zy} = -\frac{2Pxy\left(r_a^2\left(\nu+1\right)+r_b^2\nu\right)}{EI_{yy}\left(3r_a^2+r_b^2\right)}$$

$$\gamma_{zx} = \frac{P\left(2r_a^4\nu+2r_a^4+r_a^2r_b^2-2r_a^2x^2\nu-2r_a^2x^2+2r_a^2y^2\nu-r_a^2y^2-r_b^2x^2\right)}{EI_{yy}\left(3r_a^2+r_b^2\right)}$$

$$\gamma_{xy} = 0$$

We can confirm that these strains satisfy these six compatibility equations (4.11):

```
εxx.diff(y,2) + εyy.diff(x,2) - γxy.diff(x).diff(y),\
εxx.diff(z,2) + εzz.diff(x,2) - γzx.diff(x).diff(z),\
εyy.diff(z,2) + εzz.diff(y,2) - γzy.diff(y).diff(z),\
εxy.diff(z,2) + εzz.diff(x).diff(y) \
                              - εzy.diff(x).diff(z)- εzx.diff(y).diff(z),\
(εzx.diff(y,2) + εyy.diff(x).diff(z) \
                   - εxy.diff(y).diff(z)- εzy.diff(x).diff(y)).simplify(),\
(εzy.diff(x,2) + εxx.diff(y).diff(z) \
                     - εzx.diff(x).diff(y)- εxy.diff(x).diff(z)).simplify()
```

(0, 0, 0, 0, 0, 0)

This means that this set of stresses and strains is correct, and a displacement field exists corresponding to this set of strains.

The maximum shear stress σ_{zx} is found at $x = y = 0$:

```
display(Math(f" \\text{{ Maximum $σ_{{zx}}$}}=\
                                    {latex(σzx.subs({x:0,y:0}))}"))
```

$$\text{Maximum, } \sigma_{zx} = \frac{Pr_a^2}{2I_{yy}} - \frac{Pr_a^2\left(r_a^2(\nu+1) + r_b^2\nu\right)}{2I_{yy}\left(3r_a^2 + r_b^2\right)(\nu+1)}$$

It can be seen that the normal stress is the same as that obtained using the thin beam theory. However, the two shear stresses differ. When $r_b \ll r_a$, which corresponds to a very tall beam, the shear stresses become

```
σzxoo = sp.limit(σzx.subs(y,0), rb, 0).simplify().collect(ra).collect(x)
display(Math(f" \\text{{ When $r_b→0: σ_{{zx}}$}}={latex(σzxoo)}"))
display(Math(f" \\text{{ When $r_b→0, x=0: σ_{{zx}}$}}=\
            {latex(σzxoo.subs(Iyy,Iyy_).subs({x:0,rb:0}).simplify())}"))
σzyoo = sp.limit(σzy, rb, 0).simplify()
display(Math(f" \\text{{ When $r_b→0: σ_{{zy}}$}}={latex(σzyoo)}"))
```

$$\text{When } r_b \to 0 : \sigma_{zx} = \frac{P\left(r_a^2 - x^2\right)}{3I_{yy}}$$

$$\text{When } r_b \to 0, x = 0 : \sigma_{zx} = \frac{4P}{3A}$$

$$\text{When } r_b \to 0 : \sigma_{zy} = -\frac{Pxy}{3I_{yy}}$$

The result for σ_{zx} is the same as that from the thin beam theory. However, σ_{zy} cannot be obtained using the thin beam theory.

When $r_a \ll r_b$, we have a very wide beam, and the shear stresses become

```
σzxoo = sp.limit(σzxoo, rb, oo).simplify()
display(Math(f" \\text{{ When $r_b→∞: σ_{{zx}}$}}={latex(σzxoo)}"))

σzyoo = sp.limit(σzy, rb, oo).simplify()
display(Math(f" \\text{{ When $r_b→∞: σ_{{zy}}$}}={latex(σzyoo)}"))
```

$$\text{When } r_b \to \infty : \sigma_{zx} = \frac{P\left(r_a^2 - x^2\right)}{3I_{yy}}$$

$$\text{When } r_b \to \infty : \sigma_{zy} = -\frac{Pxy\nu}{I_{yy}(\nu+1)}$$

The Poisson effects are observed in σ_{zy} because the beam is constrained in the y-direction.

4.4.6 *Cantilevers with circular cross-section*

The shear stresses for a cantilever with a circular cross-section subjected to a force at the right end of the beam can be found easily by setting $r_a = r_b$. The snippet is as follows:

```
A_= sp.pi*ra**2                          # used for expressions in radius only:
σzx_circ = σzx.subs(rb,ra).simplify()    #.subs({Iyy:Iyy_}).subs({A:A_})
display(Math(f" \\text{{ Shear stress $σ_{{zx}}$}}={latex(σzx_circ)}"))
display(Math(f" \\text{{ Maximum $σ_{{zx,x=0,y=0}}$}}=\
                   {latex(σzx_circ.subs({x:0,y:0}).simplify())}"))
σzy_circ = σzy.subs(rb,ra).simplify()    #.subs({Iyy:Iyy_}).subs({A:A_})
display(Math(f" \\text{{ Shear stress $σ_{{zy}}$}}={latex(σzy_circ)}"))
```

Shear stress

$$\sigma_{zx} = \frac{P\left(-4\left(\nu+1\right)\left(-r_a^2+x^2+y^2\right)+\left(2\nu+1\right)\left(-r_a^2+x^2+3y^2\right)\right)}{8I_{yy}\left(\nu+1\right)}$$

Maximum $\sigma_{zx,x=0,y=0} = \dfrac{Pr_a^2\cdot\left(2\nu+3\right)}{8I_{yy}\left(\nu+1\right)}$

Shear stress $\sigma_{zy} = -\dfrac{Pxy\left(2\nu+1\right)}{4I_{yy}\left(\nu+1\right)}$

The normal stress has no changes.

4.5 Cantilevers with rectangular cross-section

4.5.1 *Setting of the problem*

Assume the 3D cantilever has a uniform rectangular cross-section of $2b \times 2h$, as shown in Fig. 4.4. It is subjected to a downward force at the right end of the beam and is free from body forces. The domain of the cross-section Ω is a rectangle bounded by two pairs of parallel straight edges Γ. We would like to obtain the solutions for stresses in the beam in the Cartesian coordinate system.

The expression for the geometry of the boundary Γ of the beam cross-section can be written as

$$(x^2 - h^2)(y^2 - b^2) = 0 \tag{4.75}$$

where h and b are the half-height and half-base-width of the beam cross-section. We purposely let P be along the x-axis and pass through the centroid of the cross-section at the right end, so that there is no twist deformation

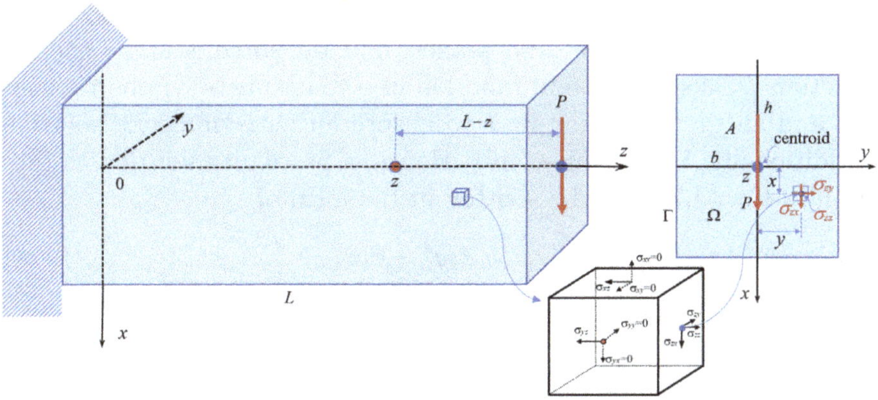

Figure 4.4. Schematic drawing of a 3D beam with a uniform rectangular cross-section subjected to a force at the right end. Three nonzero unknown stress components on the z-plane are shown and need to be determined.

in the cantilever. If P is not passing through the centroid, the torque (with respect to the z-axis) generated by P can be treated separately as a torsional problem, and its solution can be found in Chapter 9 of Ref. [5]. Thus, our focus will only be on the bending effects resulting from P.

Since the rectangular cross-section is symmetric, our derivation starts from Eq. (4.66). Hence, we need to find a function $f(y)$ that satisfies Eq. (4.67). However, the rectangular cross-section boundary consists of line segments, making it difficult to find a single-piece function $f(y)$. Therefore, we will proceed in stages.

4.5.2 *Determination of stream functions*

First, consider the boundary $y = \pm b$, where $\frac{dy}{ds} = 0$. We simply set $f(y) = \frac{Ph^2}{I_{yy}}$, which is a constant. Equation (4.66) becomes

$$\begin{aligned} \nabla^2 \psi &= \frac{2\nu P}{(1+\nu) I_{yy}} y \quad \in \Omega \\ \frac{\partial \psi}{\partial s} &= 0 \quad \text{at } y = \pm b \quad \text{on } \Gamma \end{aligned} \tag{4.76}$$

Using Eq. (4.69), the shear stresses can be given as

$$\begin{aligned} \sigma_{zx} &= \frac{1}{2}\frac{\partial \psi}{\partial y} + \frac{P(h^2 - x^2)}{2 I_{yy}} \\ \sigma_{zy} &= -\frac{1}{2}\frac{\partial \psi}{\partial x} \end{aligned} \tag{4.77}$$

where ψ satisfies Eq. (4.76). Since the rectangular cross-section is symmetric with respect to both the x- and y-axes, and the force is along the x-axis, the function ψ must be an odd function in y to ensure a symmetric solution for σ_{zx} and an even function in x to ensure an antisymmetric solution for σ_{zy}. Additionally, $\nabla^2\psi$ is linear in y. Hence, a particular solution to the first equation in Eq. (4.76) can be written in the form of

$$\psi = c_3 y^3 + c_1 y \tag{4.78}$$

Substituting Eq. (4.78) into Eq. (4.76) finds

$$c_3 = \frac{\nu P}{3(1+\nu)I_{yy}} \tag{4.79}$$

Now, we can write the function ψ as

$$\psi = \Theta + \frac{\nu P}{3(1+\nu)I_{yy}} y^3 + c_1 y \tag{4.80}$$

where Θ is an arbitrary unknown function of (x, y).

Substituting Eq. (4.80) into Eq. (4.76), we obtain

$$\begin{aligned} &\nabla^2\Theta = 0 \quad \in \Omega \\ &\Theta = -\frac{\nu P}{3(1+\nu)I_{yy}} y^3 - c_1 y \quad \in \Gamma \end{aligned} \tag{4.81}$$

We have now a standard Laplace equation for Θ. To ensure $\Theta = 0$ on the boundary $y = \pm b$, we set c_1 as follows:

$$c_1 = -\frac{\nu P b^2}{3(1+\nu)I_{yy}} \tag{4.82}$$

Equation (4.80) becomes

$$\psi = \Theta + \frac{\nu P}{3(1+\nu)I_{yy}} (y^3 - b^2 y) \tag{4.83}$$

It is clear that ψ and Θ differ by a known function. If Θ can be found, we can then determine ψ. The function Θ is easier to find because it satisfies the following equation:

$$\begin{aligned} &\nabla^2\Theta = 0 \quad \in \Omega \\ &\Theta = \begin{cases} 0 & \text{for } y = \pm b \\ \dfrac{\nu P}{3(1+\nu)I_{yy}} (b^2 y - y^3) & \text{for } x = \pm h \end{cases} \end{aligned} \tag{4.84}$$

Such a Θ ensures that $\psi = 0$ on all edges of the rectangular cross-section. Our task now becomes finding a Θ that satisfies Eq. (4.84).

Since Θ satisfies the Laplace equation, it must be a harmonic function, which can be solved using the method of **separation of variables**, as it can be written in the following product of two unknown functions:

$$\Theta(x, y) = g_1(x)g_2(y) \tag{4.85}$$

where the first function is a function of only x and the second is a function of only y. Substituting Eq. (4.85) into the first equation in Eq. (4.84), we obtain

$$\nabla^2\Theta = g_2(y)g_1''(x) + g_1(x)g_2''(y) = 0 \tag{4.86}$$

which can be rewritten as

$$\frac{g_1''(x)}{g_1(x)} = -\frac{g_2''(y)}{g_2(y)} = \lambda^2 \tag{4.87}$$

The first term is a function of only x and the second is a function of only y. Thus, λ can only be an unknown constant independent of both x and y. Equation (4.87) gives two separate equations:

$$g_1''(x) - \lambda^2 g_1(x) = 0 \tag{4.88}$$

$$g_2''(y) + \lambda^2 g_2(y) = 0 \tag{4.89}$$

Each of these equations is a 1D differential equation (DE). The solutions to them are, respectively,

$$g_1(x) = A\cosh(\lambda x) + B\sinh(\lambda x) \tag{4.90}$$

$$g_2(y) = C\cos(\lambda y) + D\sin(\lambda y) \tag{4.91}$$

The solutions contain four constants. Since we know that ψ is odd in y and even in x, Θ should have the same properties. Therefore, $g_1(x)$ must be an even function, and $g_2(y)$ must be odd. Thus, $B = C = 0$. Substituting Eq. (4.90) and Eq. (4.91) into Eq. (4.85), we obtain

$$\Theta(x, y) = A\cosh(\lambda x)\sin(\lambda y) \tag{4.92}$$

where A (absorbing C) is a to-be-determined constant, in addition to the constant λ. This is done using the boundary conditions given in Eq. (4.84). The above derivation can be confirmed using the following snippet:

```
x, y, λ, A, B, C, D = symbols("x, y, λ, A, B, C, D")
g1 = A*sp.cosh(λ*x) + B*sp.sinh(λ*x )
g2 = C*sp.cos(λ*y)  + D*sp.sin(λ*y )
Θ = g1*g2
print((g1.diff(x,2)-λ**2*g1).simplify(),
      (g2.diff(y,2)+λ**2*g2).simplify(),
      (Θ.diff(x,2)+Θ.diff(y,2)).simplify())                # Laplace of Θ
```

```
0 0 0
```

Next, to satisfy the boundary condition at $y = \pm b$, we set

$$\lambda = \frac{n\pi}{b}, \quad n = 1, 2, 3, \ldots \tag{4.93}$$

Equation (4.92) becomes

$$\Theta(x, y) = \sum_{n=1}^{\infty} A_n \cosh \frac{n\pi x}{b} \sin \frac{n\pi y}{b} \tag{4.94}$$

Using the boundary condition at $x = \pm h$, we have

$$\sum_{n=1}^{\infty} A_n \cosh \frac{n\pi h}{b} \sin \frac{n\pi y}{b} = -\frac{\nu P}{3(1+\nu)I_{yy}}(y^3 - b^2 y) \tag{4.95}$$

To solve the foregoing equation for A_n, we multiply $\sin\left(\frac{m\pi y}{b}\right)$ on both sides and integrate along y from $-b$ to b, which gives

$$\begin{aligned}
&\left(A_n \cosh \frac{n\pi b}{2h}\right) \underbrace{\int_{-b}^{b} \sin \frac{n\pi y}{b} \sin \frac{m\pi y}{b} dy}_{b} \\
&= \frac{-\nu P}{3(1+\nu)I_{yy}} \int_{-b}^{b} (y^3 - b^2 y) \sin \frac{m\pi y}{b} dy \\
&= -\frac{4\nu P}{(1+\nu)I_{yy}} \frac{(-1)^n p b^4}{n^3 \pi^3}
\end{aligned} \tag{4.96}$$

Here, we used the integral orthogonal property of $\sin\left(\frac{n\pi y}{b}\right)$ and $\sin\left(\frac{m\pi y}{b}\right)$: the integral on the left side is zero if $n \neq m$, and it is b when $n = m$. These integrals in the foregoing equation can be done using the following snippet:

```
x, y, b = symbols("x, y, b", real=True)
m, n = symbols('m, n', integer=True, positive=True)

# The integral = on the left-hand side
I_left = integrate(sp.sin(m*sp.pi*y/(b))*sp.sin(n*sp.pi*y/(b)),(y,-b,b))

# The integral = on the right-hand side
I_right = integrate((y**3-b**2*y)*sp.sin(n*sp.pi*y/(b)), (y, -b, b))

display(Math(f" \\text{{Integral on the left }}={latex(I_left) }; \;\;\
                    \\text{{Integral on the right}}={latex(I_right)}"))
```

$$\text{Integral on the left } = \begin{cases} 0 & \text{for } m \neq n \\ b & \text{otherwise} \end{cases};$$

$$\text{Integral on the right} = \frac{12\left(-1\right)^{n} b^{4}}{\pi^{3} n^{3}}$$

Hence, A_n is found as

$$A_n = -\frac{4\nu P b^3(-1)^n}{(1+\nu) I_{yy} n^3 \pi^3 \cosh\frac{n\pi h}{b}} \tag{4.97}$$

4.5.3 *Final solution of the stream function* ψ

Function Θ is found as

$$\Theta = -\frac{4\nu P b^3}{(1+\nu) I_{yy} \pi^3} \sum_{n=1}^{\infty} \frac{(-1)^n \cosh\frac{n\pi x}{b} \sin\frac{n\pi y}{b}}{n^3 \cosh\frac{n\pi h}{b}} \tag{4.98}$$

The stream function ψ can be found using Eq. (4.83):

$$\psi = \frac{\nu P}{3(1+\nu) I_{yy}} \left((y^3 - b^2 y) - \frac{12 b^3}{\pi^3} \sum_{n=1}^{\infty} \psi_n \right) \tag{4.99}$$

where ψ_n is the nth base term in the infinite series:

$$\psi_n = \frac{(-1)^n \cosh\frac{n\pi x}{b} \sin\frac{n\pi y}{b}}{n^3 \cosh\frac{n\pi h}{b}} \tag{4.100}$$

Note that in the denominator of ψ_n, there is an n^3 term, which means that it decays cubically with n. Most importantly, the cosh function has the following Taylor expansion:

$$\cosh x = 1 + \frac{x^2}{2!} + \frac{x^4}{4!} + \cdots \tag{4.101}$$

Therefore, ψ_n in Eq. (4.99) goes to zero extremely fast with increasing n because of the factorials that decay faster than exponentials. Therefore, the series converges very quickly. This means that in actual computation, we need to take only a very few terms. Additionally because ψ_n is a harmonic function, which is infinitely smooth, differentiable, and integrable with respect to both x and y, such differentiation or integration will not affect the convergence property of the series. This property is important for our later operations, which require performing differentiations and integrations on this series. Due to this excellent convergence property, we can swap the orders of the summation with the integration or differentiation when needed.

Let us code these functions in Python:

```
x, y, z = symbols("x, y, z ", real=True)
h, b, P, ν, Iyy = symbols("h, b, P, ν, I_{yy}", positive=True)
n = symbols('n', integer=True, positive=True)
Sum = sp.summation; Int = sp.integrate          # re-define the operators

ψn = sp.Function('ψ_n')(x, y)          # symbol of the base function ψ_n
Σψn = sp.Function('Σψ_n')(x, y)       # symbol of Sum of the base function

# define these functions (denoted with _):
ψn_ = (-1)**n*sp.cosh(n*sp.pi/b*x)*sp.sin(n*sp.pi/b*y)\
                                        /(n**3*sp.cosh(n*sp.pi/b*h))
display(Math(f"\\text{{Series term $ψ_{{n}}$}}={latex(ψn_)}"))
Σψn_ = Sum(ψn_, (n, 1, oo))
display(Math(f"\\text{{Summation of terms $Σψ_{{n}}$}}={latex(Σψn_)}"))

C = ν*P/(3*(1+ν)*Iyy)
ψ_final = C*(y**3-b**2*y - 12*b**3/sp.pi**3*Σψn)
display(Math(f" \\text{{Final stream function $ψ(x,y)$}}=\
                                                  {latex(ψ_final)}"))
```

Series term, $\psi_n = \dfrac{(-1)^n \sin\left(\frac{\pi n y}{b}\right)\cosh\left(\frac{\pi n x}{b}\right)}{n^3\cosh\left(\frac{\pi h n}{b}\right)}$

Summation of terms, $\sum \psi_n = \sum_{n=1}^{\infty} \dfrac{(-1)^n \sin\left(\frac{\pi n y}{b}\right)\cosh\left(\frac{\pi n x}{b}\right)}{n^3\cosh\left(\frac{\pi h n}{b}\right)}$

Final stream function, $\psi(x,y) = \dfrac{P\nu\left(-\frac{12b^3\sum\psi_n(x,y)}{\pi^3} - b^2 y + y^3\right)}{I_{yy}\left(3\nu+3\right)}$

The following are examples of differentiation and integration of the series term. It is important to note that the outcome of different operations on ψ_n can produce very similar results. For example:

$$\int \frac{\partial \psi_n}{\partial x}\, dy = -\int \frac{\partial \psi_n}{\partial y}\, dx \tag{4.102}$$

This equality arises due to the periodicity of the series term functions. This kind of property will be frequently used in the later derivations. Readers may take a moment to examine these operations and notations used in the code:

```
ψndx = sp.Function('ψ_ndx')(x,y)                    # define as a function
ψndx_= ψn_.diff(x)                # with _: the expression of the function

display(Math(f" \\frac{{∂ψ_n}}{{∂x}}={latex(ψndx_)}"))
display(Math(f" \\ ∇^2ψ_n ={latex(ψn_.diff(x,2)+ψn_.diff(y,2))}"))
display(Math(f"∇^2zψ_n={latex((z*ψn_).diff(x,2)+(z*ψn_).diff(y,2))}"))

ψndxIy = sp.Function('ψ_ndxIy')(x,y)
ψndxIy_ = sp.integrate(ψn_.diff(x), y)

ψndyIx = sp.Function('ψ_ndyIx')(x,y)
ψndyIx_= sp.integrate(ψn_.diff(y), x)

ψndxIy_, ψndyIx_  # display the expressions, both differ only by a sign
```

$$\frac{\partial \psi_n}{\partial x} = \frac{(-1)^n \pi \sin\left(\frac{\pi n y}{b}\right) \sinh\left(\frac{\pi n x}{b}\right)}{b n^2 \cosh\left(\frac{\pi h n}{b}\right)}$$

$$\nabla^2 \psi_n = 0$$

$$\nabla^2 z\psi_n = 0$$

$$\left(-\frac{(-1)^n \cos\left(\frac{\pi n y}{b}\right) \sinh\left(\frac{\pi n x}{b}\right)}{n^3 \cosh\left(\frac{\pi h n}{b}\right)},\ \frac{(-1)^n \cos\left(\frac{\pi n y}{b}\right) \sinh\left(\frac{\pi n x}{b}\right)}{n^3 \cosh\left(\frac{\pi h n}{b}\right)} \right)$$

4.5.4 *Final shear stress solution*

With ψ found, using Eq. (4.77) and $f(y) = \frac{Ph^2}{I_{yy}}$, the final expressions for these shear stresses can be found as follows, along with the normal stress obtained earlier:

$$\sigma_{zz} = -\frac{Px\,(L-z)}{I_{yy}}$$

$$\sigma_{zx} = \frac{P(h^2-x^2)}{2I_{yy}} + \frac{\nu P}{6(1+\nu)I_{yy}}\left(3y^2 - b^2 - \frac{12b^2}{\pi^2}\sum_{n=1}^{\infty}\frac{(-1)^n\cosh\frac{n\pi x}{b}\cos\frac{n\pi y}{b}}{n^2\cosh\frac{n\pi h}{b}}\right)$$

$$\sigma_{zy} = \frac{2\nu P b^2}{\pi^2(1+\nu)I_{yy}}\sum_{n=1}^{\infty}\frac{(-1)^n\sinh\frac{n\pi x}{b}\sin\frac{n\pi y}{b}}{n^2\cosh\frac{n\pi h}{b}} \tag{4.103}$$

This is the final solution of shear stresses on the cross-section of a 3D beam with a uniform rectangular cross-section subjected to a vertical force at the right end passing through the centroid. The σ_{zy} component is resulting from the Poisson's ratio. It is zero if the material has a zero Poisson's ratio.

The snippet to obtain the Python expressions for these shear stresses is given in the following. The normal stress is the same as that for a beam with an elliptic cross-section obtained earlier:

```
σzx_final =( ψ_final.diff(y)/2).simplify() + P/2/Iyy*(h**2-x**2)
σzy_final =(-ψ_final.diff(x)/2).simplify()
```

4.5.5 *Check on stress boundary conditions*

On the surface of the cantilever, it is obvious from Eq. (4.103) that $\sigma_{zy} = 0$ on $x = \pm h$. However, the shear stress $\sigma_{zx} = 0$ on $y = \pm b$ is not so obvious. Its first term is clearly zero. For the second term to be zero, the term in the big parentheses in Eq. (4.103) must be zero. This depends on the converged result of the following infinite series:

$$\sum_{n=1}^{\infty}\frac{(-1)^n\cos\frac{n\pi y}{b}}{n^2} \tag{4.104}$$

However, SymPy could not produce the closed form for this series. We suspect that this infinite series converges to the following simple polynomial:

$$\sum_{n=1}^{\infty}\frac{(-1)^n\cos\frac{n\pi y}{b}}{n^2} = \frac{\pi^2}{12}\left(3\frac{y^2}{b^2} - 1\right) \tag{4.105}$$

Let us verify this equality numerically. The snippet for this numerical "proof" is as follows:

```
# Define the series:
analytic_result = Sum((-1)**n*cos(pi*n*y/b)/(n**2), (n, 1, oo))
#display(analytic_result)  # failed to evaluate, returns the expression
N = symbols('N', integer=True, positive=True)          # use finite terms
series_N = Sum((-1)**n*cos(pi*n*y/b)/(n**2), (n, 1, N))

# Compute a numerical value of the series with finite terms:
y_val = random.random()                        # random number between 0 and 1
b_val = y_val + random.random()/2  # be sure b_val >= y_val
dic_yb = {b:b_val, N:2000}
np_series_N = lambdify(y, series_N.subs(dic_yb), "numpy")     # to numpy

numer_result = np_series_N(y_val) - np.pi**2/12*(3*(y_val/b_val)**2-1)
print('Is it close to zero?', numer_result )  # should be close to zero
```

```
Is it close to zero? -1.6634228612588675e-07
```

It is seen that the stress boundary conditions are satisfied. In the foregoing code cell, we used random numbers. Hence, each time the above code is executed, it produces a different value, but all close to zero. Readers may give it a try. This is not a rigorous proof, but should be a convincing validation.

Equation (4.105) will be used in the derivations later and in the next chapter.

4.5.6 *Checking of FBCs*

In obtaining the solution above, we used the first three equations in Eq. (4.40) and left alone the last three equations in it. We shall now check whether the last three equations are also satisfied by the stress solution obtained.

The fifth and sixth equations in Eq. (4.40) are clearly satisfied, for the similar symmetry reasons mentioned earlier for the elliptic beam.

The integrand in the fourth equation in Eq. (4.40) is σ_{zx}, which is an even function in both x and y. Its integral can be some value. We need to perform the integration and find it. First, let us integrate the term with the series and the term with $(3y^2 - b^2)$:

```
term_coshcos = (-1)**n*sp.cosh(n*sp.pi/b*x)*sp.cos(n*sp.pi/b*y)\
                                             /(n**2*sp.cosh(n*sp.pi/b*h))
#display(Math(f"\\text{{Series term}}={term_coshcos}"))
Int = sp.integrate
int_coshcos = Int(Int(term_coshcos, (y, -b, b)), (x,-h, h)).simplify()
display(Math(f"\\text{{Integral of the series}}={latex(int_coshcos)}"))

term_yb = 3*y**2 - b**2
int_yb = Int(Int(term_yb, (y, -b, b)), (x,-h, h)).simplify()
display(Math(f"\\text{{Integral of $(3y^2 - b^2)$}}={latex(int_yb)}"))
```

Integral of the series $= 0$

Integral of $(3y^2 - b^2) = 0$

Both are zero. This implies that this correction term does not affect the force in the axial direction. Let us integrate the first term in Eq. (4.103):

```
Iyy_ = (2*b)*(2*h)**3/12
term_1st = P/2/Iyy*(h**2-x**2)
int_1st = Int(Int(term_1st, (y, -b, b)), (x,-h,h)).subs(Iyy, Iyy_)
display(Math(f"\\text{{Integral of the 1st term}}={latex(int_1st)}"))
```

Integral of the 1st term $= P$

This means that all the FBCs are satisfied. We conclude that the solution obtained is correct.

4.5.7 *Shear stresses at the neutral axis*

4.5.7.1 *General formula*

For rectangular cross-sections, the neutral axis (where the normal stress is zero) is at $x = 0$. The shear stresses there become

$$\sigma_{zx} = \frac{Ph^2}{2I_{yy}}\left\{1 \underbrace{-\frac{\nu}{3(1+\nu)}\left[\left(\frac{b}{h}\right)^2 - 3\left(\frac{y}{h}\right)^2 + \frac{12}{\pi^2}\left(\frac{b}{h}\right)^2\sum_{n=1}^{\infty}\frac{(-1)^n\cos\frac{n\pi y}{b}}{n^2\cosh\frac{n\pi h}{b}}\right]}_{\alpha(y),\ \text{correction term}}\right\}$$

$$\sigma_{zy} = 0 \tag{4.106}$$

For tall beams, the maximum shear stress is registered along $x = 0$.

The first term $\frac{Ph^2}{2I_{yy}}$ is the same as that from the thin beam theory. The additional term is the correction from the rigorous theory for 3D beams. It represents the Poisson effect. If Poisson's ratio is zero, this term disappears.

When $h >> b$, the correction approaches zero quadratically with respect to b/h, and the solution approaches that from the thin beam theory.

4.5.7.2 *At the center of the cross-section*

At $x = y = 0$, the shear stress σ_{zx} is found as

$$\sigma_{zx} = \frac{Ph^2}{2I_{yy}} \left\{ 1 \underbrace{-\frac{\nu}{3(1+\nu)} \left(\frac{b}{h}\right)^2 \left[1 + \frac{12}{\pi^2} \sum_{n=1}^{\infty} \frac{(-1)^n}{n^2 \cosh \frac{n\pi h}{b}}\right]}_{\alpha,\ \text{correction term}} \right\} \tag{4.107}$$

These formulas can be derived using the following snippet. Readers may uncomment these lines to display the formulas:

```
#display(Math(f" \\text{{$x=0$}}: {latex(σzx_final.subs({x:0}))}"))
#display(Math(f" \\text{{$x=0$}}: {latex(σzy_final.subs({x:0}))}"))
σzx_finalx0y = σzx_final.subs({x:0,y:0})
display(Math(f" \\text{{$x=y=0$}}: {latex(σzx_finalx0y)}"))
α_correction = -ν*σzx_finalx0y.args[1].args[6]/3/pi**2/(1+ν)/h**2
display(Math(f" \\text{{The correction term}}: {latex(α_correction)}"))

ν_ = 1/3; N = 100                                  # Examine special cases:
α_correction = -ν/3/(ν + 1)*(b/h)**2*(1+12/pi**2\
                  *Sum((-1)**n/(n**2*cosh(n*pi*h/b)),(n,1,N)))
print(α_correction.subs({ν:ν_, b:h    }).evalf())
print(α_correction.subs({ν:ν_, b:h/2}).evalf())
print(α_correction.subs({ν:ν_, b:h*2}).evalf())
```

$$x = y = 0\text{:}\ \frac{Ph^2}{2I_{yy}} - \frac{P\nu\left(12b^3 \frac{d}{dy} \sum \psi_n(0,y)\Big|_{y=0} + \pi^3 b^2\right)}{6\pi^3 I_{yy}(\nu+1)}$$

$$\text{The correction term:}\ -\frac{\nu\left(12b^3 \frac{d}{dy} \sum \psi_n(0,y)\Big|_{y=0} + \pi^3 b^2\right)}{3\pi^2 h^2(\nu+1)}$$

```
-0.0746855166864324
-0.0207387720403617
-0.179827690893943
```

As shown, for materials with Poisson's ratio of 1/3, we found:

- when $h/b = 1$, the correction is about -7.5%.
- when $h/b = 2$ (very tall beam), the correction is about -2.07%.
- when $h/b = 0.5$ (very wide beam), the correction is about -18%.

The thin beam theory is an overestimation (on the conservative side).

4.5.7.3 *At the vertical edges of the cross-section*

At $x = 0$ and $y = b$, which is at the right edge of the cross-section, the shear stress σ_{zx} is given by

$$\sigma_{zx} = \frac{Ph^2}{2I_{yy}} \left\{ 1 - \underbrace{\frac{\nu}{3(1+\nu)} \left(\frac{b}{h}\right)^2 \left[\frac{1}{4} - \frac{12}{\pi^2} \sum_{n=1}^{\infty} \frac{1}{n^2 \cosh \frac{n\pi h}{b}} \right]}_{\alpha,\ \text{correction term}} \right\} \tag{4.108}$$

These formulas can be derived using the following snippet. Readers may uncomment these lines to display the formulas:

```
σzx_finalx0yb = σzx_final.subs({x:0,y:b})
display(Math(f" \\text{{$x=y=b$}}: {latex(σzx_finalx0yb)}"))
α_correction = -ν*σzx_finalx0yb.args[1].args[6]/3/pi**2/(1+ν)/h**2
display(Math(f" \\text{{The correction term}}: {latex(α_correction)}"))
#display(ψn_.diff(y).subs({x:0,y:b}))

ν_ = 1/3; N = 100                                    # Examine special cases:
α_correction = -ν/3/(ν + 1)*(b/h)**2*(-2 + 12/pi**2\
                                *Sum(1/(n**2*cosh(n*pi*h/b)),(n,1,N)))
print(α_correction.subs({ν:ν_, b:h    }).evalf())
print(α_correction.subs({ν:ν_, b:h/2}).evalf())
print(α_correction.subs({ν:ν_, b:h*2}).evalf())
```

$$x = y = b:\ \frac{Ph^2}{2I_{yy}} - \frac{P\nu\left(12b^3 \frac{d}{db} \sum \psi_n(0,b) - 2\pi^3 b^2\right)}{6\pi^3 I_{yy}(\nu+1)}$$

$$\text{The correction term:}\ -\frac{\nu\left(12b^3 \frac{d}{db} \sum \psi_n(0,b) - 2\pi^3 b^2\right)}{3\pi^2 h^2(\nu+1)}$$

```
0.157829550762841
0.0415720170382042
0.495486792419625
```

As shown, for materials with Poisson's ratio of 1/3, we found at $x = 0$ and $y = b$:

- when $h/b = 1$, the correction is about 15.8%.
- when $h/b = 2$ (very tall beam), the correction is about 4.14%.
- when $h/b = 0.5$ (very wide beam), the correction is about 49.5%!

In other words, the largest shear stress is not at the centroid of the cross-section, but at the two side edges. This is particularly true if the beam is wide! This fact can only be revealed using the current 3D theory.

The thin beam theory is a severe underestimation (on the severely unsafe side). Readers may take a look at the shear flows on the cross-section given at the end of this section.

4.5.8 *Strains in the beam*

Using the constitutive equation (4.7) and the normal and shear stresses, the strains can be obtained using

$$\begin{aligned}
\varepsilon_{zz} &= \frac{\sigma_{zz}}{E} = -\frac{Px}{EI_{yy}}(L-z) \\
\varepsilon_{xx} &= \frac{\sigma_{xx}}{E} = \frac{\nu\sigma_{zz}}{E} = -\frac{\nu Px}{EI_{yy}}(L-z) \\
\varepsilon_{yy} &= \frac{\sigma_{yy}}{E} = \frac{\nu\sigma_{zz}}{E} = -\frac{\nu Px}{EI_{yy}}(L-z) \\
\gamma_{xy} &= 0; \quad \gamma_{zx} = \frac{\sigma_{zx}}{G}; \quad \gamma_{zy} = \frac{\sigma_{zy}}{G}
\end{aligned} \tag{4.109}$$

```
E, G, L = sp.symbols('E, G, L', positive=True)          # beam parameters

σzz_final = -P/Iyy*x*(L-z) # P*q1*x*(L-z) #symmetric, q1=-1/Iyy,q2=q0=0
σxx_final = -ν*σzz_final;      σyy_final = -ν*σzz_final

εxx_final = σxx_final/E                                          # strains
εyy_final = σyy_final/E
εzz_final = σzz_final/E

γzx_final = 2*(1+ν)*σzx_final/E                       # γzx = σzx_final/G
γzy_final = 2*(1+ν)*σzy_final/E                       # γzy = σzy_final/G
γxy_final = 0*x
εzx_final = γzx_final/2
εzy_final = γzy_final/2
εxy_final = γxy_final/2
display(Math(f" \\text{{ $γ_{{zy}}$}}= {latex(γzy_final.simplify())}"))
display(Math(f" \\text{{ $γ_{{zx}}$}}= {latex(γzx_final)}"))
display(Math(f" \\text{{ $γ_{{xy}}$}}= {latex(γxy_final)}"))
```

$$\gamma_{zy} = \frac{4Pb^3\nu\frac{\partial}{\partial x}\sum\psi_n(x,y)}{\pi^3 EI_{yy}}$$

$$\gamma_{zx} = \frac{(2\nu+2)\left(-\frac{P\nu\left(12b^3\frac{\partial}{\partial y}\sum\psi_n(x,y)+\pi^3\left(b^2-3y^2\right)\right)}{6\pi^3 I_{yy}(\nu+1)} + \frac{P\left(h^2-x^2\right)}{2I_{yy}}\right)}{E}$$

$$\gamma_{xy} = 0$$

We confirm that these strains satisfy these six compatibility equations (4.11):

```
# six compatibility equations:
εxx_final.diff(y,2) + εyy_final.diff(x,2) - γxy_final.diff(x).diff(y),\
εxx_final.diff(z,2) + εzz_final.diff(x,2) - γzx_final.diff(x).diff(z),\
εyy_final.diff(z,2) + εzz_final.diff(y,2) - γzy_final.diff(y).diff(z),\
εxy_final.diff(z,2) + εzz_final.diff(x).diff(y) \
                    - εzy_final.diff(x).diff(z)- εzx_final.diff(y).diff(z),\
(εzx_final.diff(y,2) + εyy_final.diff(x).diff(z) \
    - εxy_final.diff(y).diff(z)- εzy_final.diff(x).diff(y)).simplify(),\
(εzy_final.diff(x,2) + εxx_final.diff(y).diff(z) \
    - εzx_final.diff(x).diff(y)- εxy_final.diff(x).diff(z)).simplify()
```

$$\left(0,\ 0,\ 0,\ 0,\ \frac{2Pb^3\nu\left(-\frac{\partial^3}{\partial y^3}\sum\psi_n(x,y)-\frac{\partial^3}{\partial y\partial x^2}\sum\psi_n(x,y)\right)}{\pi^3 EI_{yy}},\right.$$
$$\left.\frac{2Pb^3\nu\left(\frac{\partial^3}{\partial x^3}\sum\psi_n(x,y)+\frac{\partial^3}{\partial y^2\partial x}\sum\psi_n(x,y)\right)}{\pi^3 EI_{yy}}\right)$$

It is easy to confirm that the last two terms are also zero:

```
(-Σψn_.diff(y,3)-Σψn_.diff(x,2).diff(y)).simplify(),\
(-Σψn_.diff(x,3)-Σψn_.diff(y,2).diff(x)).simplify()
```

$(0,\ 0)$

This means that the set of stresses and strains is correct. We can now find the displacements for this given set of compatible strain fields.

4.5.9 *Displacements in a 3D beam*

Using these strains and the strain–displacement relations (4.9), we have the following six partial differential equations (PDEs) that relate the displacements to these strains obtained:

$$\begin{gathered}\frac{\partial u}{\partial x}=\varepsilon_{xx},\quad \frac{\partial v}{\partial y}=\varepsilon_{yy},\quad \frac{\partial w}{\partial z}=\varepsilon_{zz}\\ \frac{\partial u}{\partial y}+\frac{\partial v}{\partial x}=\gamma_{xy}=0,\quad \frac{\partial u}{\partial z}+\frac{\partial w}{\partial x}=\gamma_{zx},\quad \frac{\partial v}{\partial z}+\frac{\partial w}{\partial y}=\gamma_{zy}\end{gathered}\tag{4.110}$$

First, we can obtain the general expressions for the three displacement components by integrating using the first three equations in Eq. (4.32), which will introduce three unknown integral functions. The snippet is as follows:

```
# Computing the displacements:

f1 = sp.Function('f1')(y,z)                                # integral functions
f2 = sp.Function('f2')(z,x)
f3 = sp.Function('f3')(x,y)

ug = sp.integrate(εxx_final, x).collect(P*ν) + f1
display(Math(f"\\text{{General displacement $u_g(x,y,z)$}}={latex(ug)}"))
vg = sp.integrate(εyy_final, y).collect(P*ν) + f2
display(Math(f"\\text{{General displacement $v_g(x,y,z)$}}={latex(vg)}"))
wg = sp.integrate(εzz_final, z).collect(P*x*z/E/Iyy) + f3
display(Math(f"\\text{{General displacement $w_g(x,y,z)$}}={latex(wg)}"))
```

$$\text{General displacement } u_g(x,y,z) = f_1(y,z) + \frac{Px^2\nu\,(L-z)}{2EI_{yy}}$$

$$\text{General displacement } v_g(x,y,z) = f_2(z,x) + \frac{Pxy\nu\,(L-z)}{EI_{yy}}$$

$$\text{General displacement } w_g(x,y,z) = f_3(x,y) + \frac{Pxz\left(-L+\frac{z}{2}\right)}{EI_{yy}}$$

Next, to determine these three integral functions, $f_1(y,z)$, $f_2(z,x)$, and $f_3(x,y)$, we shall use the last three equations in Eq. (4.110), which give

```
γzy_ = vg.diff(z)+wg.diff(y)-γzy_final
γzx_ = ug.diff(z)+wg.diff(x)-γzx_final
γxy_ = (ug.diff(y)+vg.diff(x)-γxy_final).expand()                  # γxy=0
display(Math(f"\\text{{ $γ_{{zy}}$_}}=\\scriptsize {latex(γzy_)}=0"))
display(Math(f"\\text{{ $γ_{{zx}}$_}}=\\scriptsize {latex(γzx_)}=0"))
display(Math(f"\\text{{ $γ_{{xy}}$_}}=\\scriptsize {latex(γxy_)}=0"))
```

$$\gamma_{zy-} = \frac{\partial}{\partial z}f_2(z,x) + \frac{\partial}{\partial y}f_3(x,y) - \frac{2Pb^3\nu\,(2\nu+2)\,\frac{\partial}{\partial x}\sum\psi_n(x,y)}{\pi^3 EI_{yy}\,(\nu+1)} - \frac{Pxy\nu}{EI_{yy}} = 0$$

$$\gamma_{zx-} = \frac{\partial}{\partial z}f_1(y,z) + \frac{\partial}{\partial x}f_3(x,y) - \frac{(2\nu+2)\left(-\frac{P\nu\left(12b^3\frac{\partial}{\partial y}\sum\psi_n(x,y)+\pi^3\left(b^2-3y^2\right)\right)}{6\pi^3 I_{yy}(\nu+1)} + \frac{P\left(h^2-x^2\right)}{2I_{yy}}\right)}{E} - \frac{Px^2\nu}{2EI_{yy}} + \frac{Pz\left(-L+\frac{z}{2}\right)}{EI_{yy}} = 0$$

$$\gamma_{xy-} = \frac{\partial}{\partial y}f_1(y,z) + \frac{\partial}{\partial x}f_2(z,x) + \frac{LPy\nu}{EI_{yy}} - \frac{Pyz\nu}{EI_{yy}} = 0$$

4.5.10 *Determination of unknown functions*

Performing differentiations, $\frac{\partial}{\partial z}\gamma_{zx}$ and $\frac{\partial}{\partial y}\gamma_{xy}$, gives two PDEs for $f_1(y, z)$:

```
γzx_dz = γzx_.diff(z).simplify().expand()
γxy_dy = γxy_.diff(y).expand()
display(Math(f" {latex(γzx_dz)}=0"))
display(Math(f" {latex(γxy_dy)}=0"))
```

$$\frac{\partial^2}{\partial z^2} f_1(y, z) - \frac{LP}{EI_{yy}} + \frac{Pz}{EI_{yy}} = 0$$

$$\frac{\partial^2}{\partial y^2} f_1(y, z) + \frac{LP\nu}{EI_{yy}} - \frac{Pz\nu}{EI_{yy}} = 0$$

Now, integrating both equations twice shall give two expressions for the function f_1, which introduces four integral functions:

```
h0  = sp.Function('h0')(y)
h1  = sp.Function('h1')(y)
s0  = sp.Function('s0')(z)
s1  = sp.Function('s1')(z)
f1_h = (sp.integrate(sp.integrate(γzx_dz, z) + h1, z) + h0).expand()
f1_s = (sp.integrate(sp.integrate(γxy_dy, y) + s1, y) + s0).expand()

display(f1_h.args)              # to locate these terms in the expression
display(f1_s.args)
display(Math(f" {latex(f1_h)}=0"))
display(Math(f" {latex(f1_s)}=0"))
```

$$\left(zh_1(y),\ \frac{Pz^3}{6EI_{yy}},\ -\frac{LPz^2}{2EI_{yy}},\ f_1(y, z),\ h_0(y)\right)$$

$$\left(ys_1(z),\ \frac{LPy^2\nu}{2EI_{yy}},\ -\frac{Py^2z\nu}{2EI_{yy}},\ f_1(y, z),\ s_0(z)\right)$$

$$zh_1(y) + f_1(y, z) + h_0(y) - \frac{LPz^2}{2EI_{yy}} + \frac{Pz^3}{6EI_{yy}} = 0$$

$$ys_1(z) + f_1(y, z) + s_0(z) + \frac{LPy^2\nu}{2EI_{yy}} - \frac{Py^2z\nu}{2EI_{yy}} = 0$$

To satisfy both the foregoing equations, $f_1(y, z)$ must have the following general form:

$$f_1(y, z) = a_1 + b_1y + c_1z + d_1yz - \frac{LPy^2\nu}{2EI_{yy}} + \frac{LPz^2}{2EI_{yy}} + \frac{Py^2z\nu}{2EI_{yy}} - \frac{Pz^3}{6EI_{yy}} \tag{4.111}$$

where a_1, b_1, c_1, and d_1 are arbitrary constants. This can be validated using the following snippet:

```
a1, b1, c1, d1 = symbols('a1, b1, c1, d1')                          # constants
f1_ = a1 + b1*y + c1*z + d1*y*z - f1_h.args[1] - f1_h.args[2] \
                              - f1_s.args[1] - f1_s.args[2]
f1_.diff(y,2).expand(), f1_.diff(z,2).expand()
```

$$\left(-\frac{LP\nu}{EI_{yy}} + \frac{Pz\nu}{EI_{yy}},\ \frac{LP}{EI_{yy}} - \frac{Pz}{EI_{yy}}\right)$$

These are the same as those obtained by computing $\frac{\partial}{\partial z}\gamma_{zx}$ and $\frac{\partial}{\partial y}\gamma_{xy}$. This means that $f_1(y, z)$ given in Eq. (4.111) is correct.

Similarly, performing differentiations, $\frac{\partial}{\partial y}\gamma_{zy}$ and $\frac{\partial}{\partial x}\gamma_{zx}$, should give two PDEs for $f_3(y, z)$:

```
γzy_dy = γzy_.diff(y).simplify().expand()
γzx_dx = γzx_.diff(x).simplify().expand()
display(γzy_dy.args)          # to locate these terms in the expression
display(γzx_dx.args)
display(Math(f" {latex(γzy_dy)}=0"))
display(Math(f" {latex(γzx_dx)}=0"))
```

$$\left(-\frac{Px\nu}{EI_{yy}},\ -\frac{4Pb^3\nu\frac{\partial^2}{\partial y\partial x}\sum\psi_n(x,y)}{\pi^3 EI_{yy}},\ \frac{\partial^2}{\partial y^2}f_3(x,y)\right)$$

$$\left(\frac{2Px}{EI_{yy}},\ \frac{Px\nu}{EI_{yy}},\ \frac{4Pb^3\nu\frac{\partial^2}{\partial y\partial x}\sum\psi_n(x,y)}{\pi^3 EI_{yy}},\ \frac{\partial^2}{\partial x^2}f_3(x,y)\right)$$

$$\frac{\partial^2}{\partial y^2}f_3(x,y) - \frac{4Pb^3\nu\frac{\partial^2}{\partial y\partial x}\sum\psi_n(x,y)}{\pi^3 EI_{yy}} - \frac{Px\nu}{EI_{yy}} = 0$$

$$\frac{\partial^2}{\partial x^2}f_3(x,y) + \frac{4Pb^3\nu\frac{\partial^2}{\partial y\partial x}\sum\psi_n(x,y)}{\pi^3 EI_{yy}} + \frac{Px\nu}{EI_{yy}} + \frac{2Px}{EI_{yy}} = 0$$

Now, integrating the foregoing two equations twice will give two expressions for the function f_3. To ease this integration process, the terms involving the series $\Sigma\psi_n(x, y)$ are isolated and integrated separately. We note that

$$\int\int \frac{\partial^2}{\partial y\partial x}\sum\psi_n(x,y)dydy = -\int\int \frac{\partial^2}{\partial x\partial y}\sum\psi_n(x,y)dxdx \tag{4.112}$$

or

$$\underbrace{\int \frac{\partial}{\partial x} \sum \psi_n(x,y)dy}_{\psi_{ndxIy}} = -\int \frac{\partial}{\partial y} \sum \psi_n(x,y)dx \tag{4.113}$$

which can be confirmed using the following snippet:

```
ψndxIy_  = Int(ψn_.diff(x),y)
ψndyIx_  = Int(ψn_.diff(y),x)
ψndxIy_, ψndyIx_
```

$$\left(-\frac{(-1)^n \cos\left(\frac{\pi n y}{b}\right) \sinh\left(\frac{\pi n x}{b}\right)}{n^3 \cosh\left(\frac{\pi h n}{b}\right)}, \frac{(-1)^n \cos\left(\frac{\pi n y}{b}\right) \sinh\left(\frac{\pi n x}{b}\right)}{n^3 \cosh\left(\frac{\pi h n}{b}\right)}\right)$$

```
# Define the integrated isolated term:
ΣψndxIy = sp.Function('Σψ_{ndxIy}')(x,y)
term_ψndxIy = 4*P*b**3*ν*ΣψndxIy/pi**3/E/Iyy
term_ψndxIy_= 4*P*b**3*ν*Sum(ψndxIy_,(n,1,oo))/pi**3/E/Iyy
display(Math(f" \\text{{The integrated isolated term: $Σψ_{{ndxIy}}$}}=\
                                                    {latex(term_ψndxIy_)}"))
```

The integrated isolated term:

$$\sum \psi_{ndxIy} = \frac{4Pb^3\nu \sum_{n=1}^{\infty} -\frac{(-1)^n \cos\left(\frac{\pi n y}{b}\right) \sinh\left(\frac{\pi n x}{b}\right)}{n^3 \cosh\left(\frac{\pi h n}{b}\right)}}{\pi^3 E I_{yy}}$$

We need to add this back to the expression for $f_3(x,y)$ later.

The integrations for the remaining terms are carried out as follows, which again introduce four integral functions:

```
h0   = sp.Function('h0')(x)                                # integral functions
h1   = sp.Function('h1')(x)
s0   = sp.Function('s0')(y)
s1   = sp.Function('s1')(y)
f3_h = (Int(Int(γzy_dy.args[0]+γzy_dy.args[2], y)+h1, y) + h0).expand()
f3_s = (Int(Int(γzx_dx.args[0]+γzx_dx.args[1]+γzx_dx.args[3], x)\
                                                   +s1, x) + s0).expand()
display(f3_h.args)                  # to locate these terms in the expression
display(f3_s.args)
display(Math(f" {latex(f3_h)}=0"))
display(Math(f" {latex(f3_s)}=0"))
```

$$\left(yh_1(x),\ -\frac{Pxy^2\nu}{2EI_{yy}},\ f_3(x,y),\ h_0(x)\right)$$

$$\left(xs_1(y),\ \frac{Px^3}{3EI_{yy}},\ \frac{Px^3\nu}{6EI_{yy}},\ f_3(x,y),\ s_0(y)\right)$$

$$yh_1(x) + f_3(x,y) + h_0(x) - \frac{Pxy^2\nu}{2EI_{yy}} = 0$$

$$xs_1(y) + f_3(x,y) + s_0(y) + \frac{Px^3\nu}{6EI_{yy}} + \frac{Px^3}{3EI_{yy}} = 0$$

$f_3(x,y)$, satisfying both the previous equations (including the terms involving $\sum \psi_n(x,y)$), must have the following form:

$$\begin{aligned} f_3(x,y) = a_3 + b_3x + c_3y + d_3xy - \frac{4Pb^3\nu \sum \psi_{ndxIy}}{\pi^3 EI_{yy}} \\ - \frac{Pax^3\nu}{6E} - \frac{Paxy^2\nu}{2E} - \frac{Px^3\nu}{3EI_{yy}} - \frac{Px^3}{3EI_{yy}} \end{aligned} \tag{4.114}$$

Here, we have added the integrated isolated term back in. The following snippet defines $f_3(x,y)$ and validates it:

```
# Define f3(x,y):
a3, b3, c3, d3 = symbols('a3, b3, c3, d3')                    # constants

f3_ = a3 + b3*x + c3*y + d3*x*y + term_ψndxIy_ \
                    - f3_h.args[1] -f3_s.args[1]-f3_s.args[2]
# Validation:
display(Math(f"\\frac{{∂^2f_3}}{{∂y^2}}={latex(f3_.diff(y,2).expand())}"))
display(Math(f"\\frac{{∂^2f_3}}{{∂x^2}}={latex(f3_.diff(x,2).expand())}"))
```

$$\frac{\partial^2 f_3}{\partial y^2} = \frac{4Pb\nu \sum_{n=1}^{\infty} \frac{(-1)^n \cos\left(\frac{\pi n y}{b}\right)\sinh\left(\frac{\pi n x}{b}\right)}{n\cosh\left(\frac{\pi h n}{b}\right)}}{\pi EI_{yy}} + \frac{Px\nu}{EI_{yy}}$$

$$\frac{\partial^2 f_3}{\partial x^2} = -\frac{4Pb\nu \sum_{n=1}^{\infty} \frac{(-1)^n \cos\left(\frac{\pi n y}{b}\right)\sinh\left(\frac{\pi n x}{b}\right)}{n\cosh\left(\frac{\pi h n}{b}\right)}}{\pi EI_{yy}} - \frac{Px\nu}{EI_{yy}} - \frac{2Px}{EI_{yy}}$$

The results are the same as those obtained earlier using $\frac{\partial}{\partial y}\gamma_{zy}$ and $\frac{\partial}{\partial x}\gamma_{zx}$.

Substitute $f_3(x,y)$ into the expression for γ_{zy} and $f_1(x,y)$ into the expression for γ_{xy}. Then, carry out the two integrals, and we find two expressions for $f_2(x,y)$. This process will introduce new integral functions. The snippet is as follows:

```
display(γzy_.args)
display(γxy_.args)
display(Math(f" \\text{{ $γ_{{zy}}$_}}= {latex(γzy_)}=0"))
display(Math(f" \\text{{ $γ_{{xy}}$_}}= {latex(γxy_)}=0"))
```

$$\left(-\frac{Pxy\nu}{EI_{yy}},\ -\frac{2Pb^3\nu\left(2\nu+2\right)\frac{\partial}{\partial x}\sum\psi_n(x,y)}{\pi^3 EI_{yy}\left(\nu+1\right)},\ \frac{\partial}{\partial z}f_2(z,x),\ \frac{\partial}{\partial y}f_3(x,y)\right)$$

$$\left(\frac{LPy\nu}{EI_{yy}},\ -\frac{Pyz\nu}{EI_{yy}},\ \frac{\partial}{\partial y}f_1(y,z),\ \frac{\partial}{\partial x}f_2(z,x)\right)$$

$$\gamma_{zy}=\frac{\partial}{\partial z}f_2(z,x)+\frac{\partial}{\partial y}f_3(x,y)-\frac{2Pb^3\nu\left(2\nu+2\right)\frac{\partial}{\partial x}\sum\psi_n(x,y)}{\pi^3 EI_{yy}\left(\nu+1\right)}-\frac{Pxy\nu}{EI_{yy}}=0$$

$$\gamma_{xy-}=\frac{\partial}{\partial y}f_1(y,z)+\frac{\partial}{\partial x}f_2(z,x)+\frac{LPy\nu}{EI_{yy}}-\frac{Pyz\nu}{EI_{yy}}=0$$

```
h0  = sp.Function('h0')(x)                          # integral function of x
# f2(z,x) with h(x):
f2_h = (sp.integrate(-γzy_.args[0]-γzy_.args[1]-f3_.diff(y), z) + h0)
display(Math(f" \\text{{$f_2(x, z)$}}={latex(f2_h.simplify())}"))

s0  = sp.Function('s0')(z)                          # integral function of z
# f2(x, z) with s(z)
f2_s = (sp.integrate(-γxy_.args[0]-γxy_.args[1] - f1_.diff(y), x) + s0)
display(Math(f" \\text{{$f_2(x, z)$}}={latex(f2_s.expand())}"))
```

$$f_2(x,z)=-c_3z-d_3xz+h_0(x)+\frac{4Pb^3z\nu\frac{\partial}{\partial x}\sum\psi_n(x,y)}{\pi^3 EI_{yy}}-\frac{4Pb^2z\nu\sum_{n=1}^{\infty}\frac{(-1)^n\sin\left(\frac{\pi ny}{b}\right)\sinh\left(\frac{\pi nx}{b}\right)}{n^2\cosh\left(\frac{\pi hn}{b}\right)}}{\pi^2 EI_{yy}}$$

$$f_2(x,z)=-b_1x-d_1xz+s_0(z)$$

These two terms involving ψ_n cancel each other out. This is because

```
display(Math(f"\\frac{{∂}}{{∂x}}Σψ_n(x,y)={latex(ψn_.diff(x))}"))
```

$$\frac{\partial}{\partial x}\sum\psi_n(x,y)=\frac{(-1)^n\pi\sin\left(\frac{\pi ny}{b}\right)\sinh\left(\frac{\pi nx}{b}\right)}{bn^2\cosh\left(\frac{\pi hn}{b}\right)}$$

These two expressions for $f_2(x,z)$ must be equal. This can only happen if we set $d_1 = d_3$ and set the integral function $h(x) = -b_1x$.

Hence, $f_2(x, z)$ can be written as

$$f_2(x, z) = a_2 - b_1 x - c_3 z - d_1 xz \tag{4.115}$$

The following snippet defines $f_2(x, y)$. Since we have obtained all the integral functions f_1, f_2, and f_3, we can now update the expressions for the displacements:

```
a2 = symbols('a2 ')                                          # constant

f2_ = a2 -b1*x - c3*z - d1*x*z
ug =  sp.integrate(εxx_final, x) +  f1_          # general displacements
vg =  sp.integrate(εyy_final, y) +  f2_
wg = (sp.integrate(εzz_final, z) +  f3_).subs(d3,d1)
display(Math(f" \\text{{ $u_g$}}= \\scriptsize {latex(ug)}"))
display(Math(f" \\text{{ $v_g$}}= \\scriptsize {latex(vg)}"))
display(Math(f" \\text{{ $w_g$}}= \\scriptsize {latex(wg)}"))
```

$$u_g = a_1 + b_1 y + c_1 z + d_1 yz - \frac{LPy^2\nu}{2EI_{yy}} + \frac{LPz^2}{2EI_{yy}} + \frac{Py^2 z\nu}{2EI_{yy}} - \frac{Pz^3}{6EI_{yy}} + \frac{x^2 (LP\nu - Pz\nu)}{2EI_{yy}}$$

$$v_g = a_2 - b_1 x - c_3 z - d_1 xz + \frac{Pxy\nu (L - z)}{EI_{yy}}$$

$$w_g = a_3 + b_3 x + c_3 y + d_1 xy - \frac{LPxz}{EI_{yy}} + \frac{4Pb^3\nu \sum_{n=1}^{\infty} - \frac{(-1)^n \cos\left(\frac{\pi n y}{b}\right) \sinh\left(\frac{\pi n x}{b}\right)}{n^3 \cosh\left(\frac{\pi h n}{b}\right)}}{\pi^3 EI_{yy}} - \frac{Px^3\nu}{6EI_{yy}} - \frac{Px^3}{3EI_{yy}} + \frac{Pxy^2\nu}{2EI_{yy}} + \frac{Pxz^2}{2EI_{yy}}$$

The expressions for these three shear strains can be updated using the last three equations in Eq. (4.110):

```
γzy = (vg.diff(z)+wg.diff(y)).collect(x).collect(y).collect(z)
γzx = (ug.diff(z)+wg.diff(x)).collect(x).collect(y).collect(z)
γxy = (ug.diff(y)+vg.diff(x)).collect(x).collect(y).collect(z)
display(Math(f" \\text{{ $γ_{{zy}}$}}= {latex(γzy)}"))
display(Math(f" \\text{{ $γ_{{zx}}$}}= {latex(γzx.simplify())}"))
display(Math(f" \\text{{ $γ_{{xy}}$}}= {latex(γxy.simplify())}"))
```

$$\gamma_{zy} = \frac{4Pb^3\nu \sum_{n=1}^{\infty} \frac{(-1)^n \pi \sin\left(\frac{\pi n y}{b}\right) \sinh\left(\frac{\pi n x}{b}\right)}{bn^2 \cosh\left(\frac{\pi h n}{b}\right)}}{\pi^3 EI_{yy}}$$

$$\gamma_{zx} = b_3 + c_1 + 2d_1 y - \frac{4Pb^2\nu \sum_{n=1}^{\infty} \frac{(-1)^n \cos\left(\frac{\pi n y}{b}\right) \cosh\left(\frac{\pi n x}{b}\right)}{n^2 \cosh\left(\frac{\pi h n}{b}\right)}}{\pi^2 E I_{yy}}$$

$$- \frac{Px^2\nu}{EI_{yy}} - \frac{Px^2}{EI_{yy}} + \frac{Py^2\nu}{EI_{yy}}$$

$$\gamma_{xy} = 0$$

The strains γ_{zy} and γ_{xy} are the same as the final strains we obtained earlier. However, γ_{zx} is different. It contains three terms with unknown constants, which were introduced during the integration process for computing the displacements. Since the strain must match the final strain, we can use this condition to determine the relationships between these three constants:

```
b3_c1 = (γzx_final - γzx).subs({x:0,y:0}).simplify().expand()
display(b3_c1)
```

$$-b_3 - c_1 - \frac{4Pb^3\nu \left.\frac{d}{dy} \sum \psi_n(0,y)\right|_{y=0}}{\pi^3 E I_{yy}} + \frac{4Pb^2\nu \sum_{n=1}^{\infty} \frac{(-1)^n}{n^2 \cosh\left(\frac{\pi h n}{b}\right)}}{\pi^2 E I_{yy}}$$

$$- \frac{Pb^2\nu}{3EI_{yy}} + \frac{Ph^2\nu}{EI_{yy}} + \frac{Ph^2}{EI_{yy}}$$

These two terms involving ψ_n cancel each other out because

```
ψn_.diff(y).subs({x:0,y:0})
```

$$\frac{(-1)^n \pi}{bn^2 \cosh\left(\frac{\pi h n}{b}\right)}$$

We found b_3 by excluding these two terms. These terms can be viewed using

```
b3_c1.args                                              # to view these terms
```

$$\left(-b_3,\ -c_1,\ \frac{Ph^2}{EI_{yy}},\ \frac{Ph^2\nu}{EI_{yy}},\ -\frac{Pb^2\nu}{3EI_{yy}},\ -\frac{4Pb^3\nu \left.\frac{d}{dy} \sum \psi_n(0,y)\right|_{y=0}}{\pi^3 E I_{yy}},\right.$$

$$\left.\frac{4Pb^2\nu \sum_{n=1}^{\infty} \frac{(-1)^n}{n^2 \cosh\left(\frac{\pi h n}{b}\right)}}{\pi^2 E I_{yy}}\right)$$

The expression for b_3 is formed using the following snippet, and it is then used to compute the shear strain γ_{zx}:

```
b3_ = b3_c1.args[1] + b3_c1.args[2] + b3_c1.args[3] + b3_c1.args[4]
display(Math(f" \\text{{$b_{{3}}$}}={latex(b3_)}"))
γzx = γzx.subs({b3:b3_})
display(Math(f" \\text{{$γ_{{zx}}$}}=\\scriptsize {latex(γzx)}"))
```

$$b_3 = -c_1 - \frac{Pb^2\nu}{3EI_{yy}} + \frac{Ph^2\nu}{EI_{yy}} + \frac{Ph^2}{EI_{yy}}$$

$$\gamma_{zx} = 2d_1 y + x^2\left(-\frac{P\nu}{EI_{yy}} - \frac{P}{EI_{yy}}\right) + \frac{4Pb^3\nu\sum_{n=1}^{\infty} - \frac{(-1)^n \pi \cos\left(\frac{\pi n y}{b}\right)\cosh\left(\frac{\pi n x}{b}\right)}{bn^2\cosh\left(\frac{\pi h n}{b}\right)}}{\pi^3 EI_{yy}}$$
$$- \frac{Pb^2\nu}{3EI_{yy}} + \frac{Ph^2\nu}{EI_{yy}} + \frac{Ph^2}{EI_{yy}} + \frac{Py^2\nu}{EI_{yy}}$$

Strain γ_{zx} still has a term, $2d_1y$, which is not present in its final solution. Therefore, d_1 must be set to zero.

```
γzx = γzx.subs({d1:0})
(γzx_final - γzx).simplify()                               # Check again
```

$$-\frac{4Pb^2\nu\left(b\frac{\partial}{\partial y}\sum\psi_n(x,y) - \pi\sum_{n=1}^{\infty}\frac{(-1)^n\cos\left(\frac{\pi n y}{b}\right)\cosh\left(\frac{\pi n x}{b}\right)}{n^2\cosh\left(\frac{\pi h n}{b}\right)}\right)}{\pi^3 EI_{yy}}$$

This is zero, because:

```
ψn_.diff(y)
```

$$\frac{(-1)^n\pi\cos\left(\frac{\pi n y}{b}\right)\cosh\left(\frac{\pi n x}{b}\right)}{bn^2\cosh\left(\frac{\pi h n}{b}\right)}$$

Now, all the strains obtained from the general expressions of the displacement are correct.

4.5.11 *Final displacement solution*

Next, we need to determine the relations of the constants a_i, b_i, and c_i in the displacement expressions. Substituting these relations for these constants back into the displacement expressions, we obtain the general form for the displacements as follows:

```
ug = ug.subs({d1:0, b3:b3_}).expand()
vg = vg.subs({d1:0, b3:b3_}).expand()
wg = wg.subs({d1:0, b3:b3_}).expand().collect(E*Iyy)
display(Math(f" \\text{{ Displacement $u_{{g}}$}}=\\small {latex(ug)}"))
display(Math(f" \\text{{ Displacement $v_{{g}}$}}=\\small {latex(vg)}"))
wg_ = r"\small{\text{Displacement $w_{{g}}$=}" + latex(wg)\
                    .replace('+ P h^{2} x \nu', '\\\\ + P h^{2} x \nu') + "}"
display(Math(wg_))
```

Displacement, $u_g = a_1 + b_1 y + c_1 z + \frac{LPx^2\nu}{2EI_{yy}} - \frac{LPy^2\nu}{2EI_{yy}} + \frac{LPz^2}{2EI_{yy}} - \frac{Px^2z\nu}{2EI_{yy}} + \frac{Py^2z\nu}{2EI_{yy}} - \frac{Pz^3}{6EI_{yy}}$

Displacement, $v_g = a_2 - b_1 x - c_3 z + \frac{LPxy\nu}{EI_{yy}} - \frac{Pxyz\nu}{EI_{yy}}$

Displacement,

$$w_g = a_3 - c_1 x + c_3 y + \frac{\begin{array}{c} -LPxz + \frac{4Pb^3\nu \sum_{n=1}^{\infty} - \frac{(-1)^n \cos\left(\frac{\pi n y}{b}\right)\sinh\left(\frac{\pi n x}{b}\right)}{n^3 \cosh\left(\frac{\pi h n}{b}\right)}}{\pi^3} - \frac{Pb^2x\nu}{3} \\ + Ph^2x\nu + Ph^2x - \frac{Px^3\nu}{6} - \frac{Px^3}{3} + \frac{Pxy^2\nu}{2} + \frac{Pxz^2}{2} \end{array}}{EI_{yy}}$$

There are six unknown constants remaining in these expressions. These represent the rigid body movement of the 3D beam. It has six degrees of freedom (DOFs): three translational, represented by a_1, a_2, and a_3, and three rotational, represented by b_1, c_1, and c_3. Assuming the beam does not move at $(0, 0, 0)$ and the slopes of the displacements are also zero at the origin, we have the following DBCs:

$$\begin{aligned} &u = v = w = 0 \\ &\frac{\partial u}{\partial z} = \frac{\partial v}{\partial x} = \frac{\partial w}{\partial y} = 0 \end{aligned} \tag{4.116}$$

These six conditions can be used to determine these six constants:

```
# enforcing DBCs:
dic0 = {x:0, y:0, z:0}
abc_sln = sp.solve([ug.subs(dic0), vg.subs(dic0), wg.subs(dic0),
          ug.diff(z).subs(dic0),   vg.diff(x).subs(dic0),
          wg.diff(y).subs(dic0)],  [a1, a2, a3, b1, c1, c3 ])
abc_sln
```

$\{a_1 : 0,\ a_2 : 0,\ a_3 : 0,\ b_1 : 0,\ c_1 : 0,\ c_3 : 0\}$

The final solution for the displacements are found using:

```
abc_sln[a3] = 0; abc_sln[c3] = 0
u = ug.subs(abc_sln).simplify().collect(3*L).collect(ν)
v = vg.subs(abc_sln).expand().simplify()
w = wg.subs(abc_sln).expand().collect(P/E/Iyy).collect(h**2)

#display(Math(f"\\text{{Final displacement solution $u$}}={latex(u)}"))
#display(Math(f"\\text{{Final displacement solution $v$}}={latex(v)}"))
#display(Math(f"\\text{{Final displacement solution $w$}}="))
#display(w)
```

The final displacement results are rearranged as follows:

$$
\begin{aligned}
u(x,y,z) &= \frac{Pz^2}{6EI_{yy}}(3L-z) + \frac{P\nu}{2EI_{yy}}(x^2-y^2)(L-z) \\
v(x,y,z) &= \frac{P\nu xy(L-z)}{EI_{yy}} \\
w(x,y,z) &= \frac{P}{EI_{yy}}\left(-\frac{4b^3\nu}{\pi^3}\sum_{n=1}^{\infty}\frac{(-1)^n\cos\left(\frac{\pi ny}{b}\right)\sinh\left(\frac{\pi nx}{b}\right)}{n^3\cosh\left(\frac{\pi hn}{b}\right)}\right) \\
&\quad + \frac{P}{EI_{yy}}\left(-Lxz - \frac{b^2x\nu}{3} + h^2x(\nu+1) - \frac{x^3}{6}(\nu+2) + \frac{x}{2}(y^2\nu+z^2)\right)
\end{aligned}
\tag{4.117}
$$

The first term in the expression for the beam deflection u was found using the thin beam theory [5]. The other term in u and the expressions for v and w can only be obtained from our 3D theory. The displacement in the beam axial direction w consists of two terms: one in the normal polynomial form, and another in a harmonic term in a fast converging series. The polynomial terms are of third order, which is similar to the 2D cantilever studied in the previous chapter. The 3D rigorous theory results in a quite complicated displacement in the axial direction. The plane-remaining-plane assumption is not correct based on our 3D theory.

4.5.12 *Displacements along the support end*

The displacement distributions along the support end can be computed by setting $z = 0$:

```
u_z0 = u.subs(z,0)
v_z0 = v.subs(z,0)
w_z0 = w.subs(z,0).collect(x**3).collect(x)
display(Math(f"\\text{{Displacement $u$ at root}}={latex(u_z0)}"))
display(Math(f"\\text{{$v$ at root}}={latex(v_z0)}"))
display(Math(f"\\text{{$w$ at root}}=\\small {latex(w_z0)}"))
```

$$\text{Displacement } u \text{ at root} = \frac{EI_{yy}\left(a_1 + b_1 y\right) + \nu\left(\frac{LPx^2}{2} - \frac{LPy^2}{2}\right)}{EI_{yy}}$$

$$v \text{ at root} = \frac{EI_{yy}\left(a_2 - b_1 x\right) + LPxy\nu}{EI_{yy}}$$

$$w \text{ at root} = -c_1 x + \frac{P\left(\frac{4b^3\nu\sum_{n=1}^{\infty} - \frac{(-1)^n \cos\left(\frac{\pi n y}{b}\right)\sinh\left(\frac{\pi n x}{b}\right)}{n^3\cosh\left(\frac{\pi h n}{b}\right)}}{\pi^3} + x^3\left(-\frac{\nu}{6} - \frac{1}{3}\right) + x\left(-\frac{b^2\nu}{3} + h^2(\nu+1) + \frac{y^2\nu}{2}\right)\right)}{EI_{yy}}$$

4.5.13 *Curvatures of the 3D beam*

As we did earlier, the curvatures of the bent beam can be computed using the second derivatives of the deflection u:

```
display(Math(f"\\frac{{∂^2u}}{{∂z^2}}= {latex(u.diff(z,2))};\;\;\; \
                    \\frac{{∂^2u}}{{∂y^2}}= {latex(u.diff(y,2))};\;\;\; \
                    \\frac{{∂^2u}}{{∂x^2}}= {latex(u.diff(x,2))}"))
```

$$\frac{\partial^2 u}{\partial z^2} = \frac{P(L-z)}{EI_{yy}}; \quad \frac{\partial^2 u}{\partial y^2} = -\frac{P\nu(L-z)}{EI_{yy}}; \quad \frac{\partial^2 u}{\partial x^2} = \frac{P\nu(L-z)}{EI_{yy}}$$

The first expression is the same as the curvature from the thin beam theory. The other two can only be obtained from 3D elasticity. They result from Poisson's ratio.

We have finally found a full set of solutions for 3D beams with rectangular cross-sections subjected to an end force, including stresses, strains, and displacements. This set of exact solutions can be useful as a benchmark for evaluating numerical methods.

4.5.14 *Plot of a stream function*

We write the following code to compute the stream function on the rectangular cross-section of a beam, using Eq. (4.99):

```
def sf_Psi(b, h, nu, P, x, y, n_end = 21):
    '''Compute the steam function Psi for shear stresses on a
       rectangular cross-section of a cantilever with end load P.
       It is effectively a series solution to a 2D Laplace PDE with
       boundary conditions.
    b:        half-height of the domain
    h:        half-width  of the domain
    P:        constant - force P at the end of the cantilever
    n_end:    terms used in the series of the solution
    x and y: point at which the solution to be found.'''

    I = (2*b)*(2*h)**3/12
    C = nu*P/(3*(1+nu)*I)
    sum1 = 0.0
    for n in range(1, n_end):
        sum1 += (-1)**n/(n**3*np.cosh(n*np.pi*h/b))*\
                 np.cosh(n*np.pi*x/b)*np.sin(n*np.pi*y/b)

    Psi = C*(y**3-b**2*y)-12*b**3/sp.pi**3*sum1
    return C*sum1
```

```
# Plot of a Steam Function:

plt.rcParams['figure.dpi'] = 500
h, b, L, z = 2., 1., 1., 1/2                          # define geometry values
P, E, ν, n_end = 100, 1., 1/3, 21                          # define parameters

xl, xr, dx = -h, h, 0.09
yl, yr, dy = -b, b, 0.09

X = np.arange(xl, xr+dx, dx)
Y = np.arange(yl, yr+dy, dy)
X, Y = np.meshgrid(X, Y)
xn, yn = X.shape
ψ = X*0
W = X*0
```

Figure 4.5. A schematic view of the stream function of the shear stress vector on a rectangular cross-section.

```
for xk in range(xn):
    for yk in range(yn):
        ψ[xk, yk] = sf_Psi(b, h, ν, P, X[xk,yk], Y[xk,yk], n_end=n_end)

fig = plt.figure()      # Create a 3D plot
plt.ioff()
#plt.ion()                         # uncomment this to display the plots
ax = fig.add_subplot(211, projection='3d')
surf = ax.plot_surface(X,Y,ψ-0.7,rstride=1,cstride=1,cmap='viridis')
ax.set_xlim(-4, 2)
ax.set_ylim(-3.5, 1.)
ax.set_zlim(-5, 2.5)
ax.zaxis.line.set_visible(False)
ax.yaxis.pane.fill = False  # Remove background pane for z-axis
ax.grid(False);
#plt.show()                        # uncomment this to display the plots
```

Figure 4.5 plots the stream function of the shear stress vector on the cross-section of a cantilever beam subjected to a force at the right end.

4.6 Review questions

1. Consider a 3D solid with an arbitrary geometry domain (which may have holes) bounded by smooth boundaries or piecewise smooth convex boundaries. The solid is free from body force and is pressurized on all these

boundaries with p. Show that the stress state

$$\begin{aligned} \sigma_{xx} &= \sigma_{yy} = \sigma_{zz} = -p \\ \sigma_{yz} &= \sigma_{xz} = \sigma_{yz} = 0 \end{aligned} \tag{4.118}$$

satisfies all the necessary equations and conditions and hence is the exact stress solution for this problem.

2. Consider a 3D beam with an elliptic cross-section. Derive formulas for the general form of displacements using the displacement boundary conditions given in Eq. (4.116).

4.7 Concluding remarks

This chapter discusses in detail the stress method for solving 3D beams subjected to different loads. For the challenging problem of beams under an end load, the following remarks may be made:

1. The general problem is split into two parts: shear stresses caused by torsion and shear stresses caused by bending. The torsional problem was solved previously [5], allowing us to focus on obtaining solutions for the shear stresses due to bending.
2. The basic solution strategy is to use stream functions to construct part of the shear stresses.
3. To find the displacement solutions, techniques for handling 2D unknown functions are presented.
4. A complete set of closed-form exact formulas for the displacements and stresses in a 3D beam with a rectangular cross-section is obtained in this chapter for the first time, to the best of the author's knowledge. These formulas partially agree with those given in Chapter 7 of Ref. [2].
5. The exact solutions for both stresses and displacements obtained are useful as a benchmark for evaluating numerical methods. Exact solutions for 3D problems are rare in the existing literature. We thus believe the exact solutions presented in this chapter are valuable additions.

The next chapter presents displacement methods for 3D problems.

References

[1] Z.L. Xu, *Elasticity*, Vol. 1 & 2, People's Publisher, China, 1979.
[2] A.P. Boresi, K. Chong and J.D. Lee, *Elasticity in Engineering Mechanics*, John Wiley & Sons, New York, 2010.
[3] A.I. Lure, *Theory of Elasticity. Foundations of Engineering Mechanics*, Springer, Berlin, 2005.

[4] S. Timoshenko and J.N. Goodier, *Theory of Elasticity*, McGraw-Hill, New York, 1970. Available online http://books.google.com/books?id=yFISAAAAIAAJ& dq=theory+of+elasticity&ei=ICiKSsr3G4jwkQSbxMyPCg.

[5] G.R. Liu, *Mechanics of Materials: Formulations and Solutions with Python*, World Scientific, New Jersey, 2024.

[6] A.P. Boresi, R.J. Schmidt and O.M. Sidebottom, *Advanced Mechanics of Materials*, John Wiley & Sons, New York, 1985.

[7] G.R. Liu, *Vector Calculus: Formulation, Applications, and Python Codes*, World Scientific, New Jersey, 2025.

Chapter 5

Displacement Methods for 3D Solids

```
# Place cursor in this cell, and press Ctrl+Enter to import dependencies.
import sys                          # For accessing the computer system
sys.path.append('../grbin/')   # Change to the directory in your system

from commonImports import *          # Import dependencies from '../grbin/'
import grcodes as gr                    # Import the module of the author
importlib.reload(gr)                # When grcodes is modified, reload it

from continuum_mechanics.solids import sym_grad
init_printing(use_unicode=True)         # For latex-like quality printing
np.set_printoptions(precision=4,suppress=True,
                    formatter={'float_kind': '{:.4e}'.format})
```

This chapter discusses three-dimensional (3D) solid mechanics problems in Cartesian coordinates using the displacement methods.

The displacement methods treat the displacements as the primary variables. They construct the displacement solution first and then use it to derive the strains and stresses. The procedure is thus more direct because the strains and stresses can be obtained via differentiation, which is much easier compared to indefinite integration. In addition, differentiation does not introduce additional unknowns (such as integral functions and constants, as in the stress methods).

The displacement method works extremely well for 1D problems, as demonstrated in Ref. [1]. For higher-dimensional (2D and 3D) problems, however, challenges arise due to multiple unknown displacement components, compounded by the stress boundary conditions (SBCs). The governing equations for displacements form a set of partial differential equations (PDEs). The key idea is to avoid solving this PDE set directly by assuming the displacements are linear combinations of known basis functions with unknown constants [2], which are then used to find the stresses. These unknown constants can be determined step by step using various conditions. This process has been demonstrated in detail for 2D solids in Chapter 3.

The displacement method for finite 3D domains discussed in this chapter was proposed by the author. The procedure is largely the same as that for 2D solids, but we require special basis functions as an enrichment, in addition to polynomial ones. We will demonstrate how this approach applies to solving a 3D problem with finite dimensions. This idea is similar to those used in modern numerical methods, such as the finite element methods [3], smoothed finite element methods [4], and meshfree methods [5].

There are also special techniques to reduce the problem to 1D-like problems. One effective way to do this is by finding a proper special function or functions, such as displacement potential functions or displacement functions. The displacement components can then be written as the derivatives of these functions. This is somewhat similar to the stress methods that use stress functions. We will demonstrate how this can be done for 3D problems with infinite dimensions. For these techniques, we refer to the textbook [6].

The necessary equations needed to handle this are given at the beginning of the previous chapter. We will use those directly without deriving them here. Interested readers are referred to [1] for the derivation process.

5.1 3D half-space subjected to pressure

Let us begin with a simple 3D problem that can be reduced to a 1D problem. We will then solve it step by step using a displacement method similar to the one proposed in Chapter 3 for 2D problems.

5.1.1 *Setting of the problem*

Consider a three-dimensional (3D) half-space. It is subjected to a distributed force q on the entire top surface and a vertical body force b_z, as shown in Fig. 5.1. We aim to find the stress and displacement distributions in the half-space.

5.1.2 *Analyzing the problem*

Since the pressure is uniform across the entire infinite surface of the 3D half-space, and the gravity force does not change in any plane parallel to the

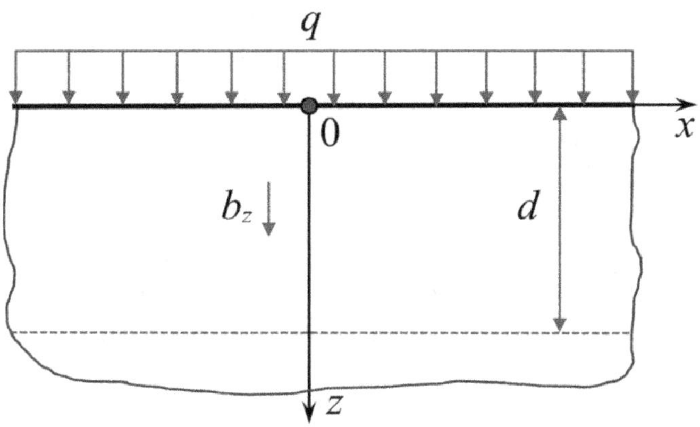

Figure 5.1. Schematic view of a cross-section of a 3D half-space, subjected to a surface traction and a body force in the vertical direction.

x–y plane, any vertical plane is a symmetric plane. Therefore, the displacements u and v will be zero throughout the entire domain.

$$u(x, y, z) = 0; \quad v(x, y, z) = 0 \tag{5.1}$$

The only nonzero displacement is w, which does not change in any x–y plane. Therefore, w is a function of z only. The first two equilibrium equations (Eq. 4.16) are automatically satisfied, and the third equation becomes

$$(2G + \lambda)\frac{\partial^2 w}{\partial z^2} + b_z(z) = 0 \tag{5.2}$$

or

$$\frac{\partial^2 w}{\partial z^2} = -\frac{b_z(z)}{2G + \lambda} \tag{5.3}$$

This 3D half-space problem is essentially 1D. It is similar to the 1D strain problem of bars discussed in Chapter 7 of Ref. [1]. The only difference is that the 1D strain problem has a constraint in one direction, whereas here, constraints exist in both the x and y directions (and hence in all directions). Since Eq. (5.3) is 1D, it can be solved analytically for any arbitrary body force b_z, provided it is integrable.

5.1.3 *A standard method via indefinite integration*

Let us use a standard method to solve this problem, assuming b_z is a constant. By integrating Eq. (5.3) twice, we obtain

$$w = -\frac{b_z}{2(2G + \lambda)}z^2 + c_0 z + c_1 \tag{5.4}$$

where c_0 and c_1 are two integral constants introduced during the two stages of integration. These constants can be determined using the stress boundary condition (SBC) on the top surface of the half-space and the displacement boundary condition (DBC). This process straightforwardly leads to the solution. The key point here is that this standard method works for a 1D PDE written in terms of displacement. When the PDE is multi-dimensional, the integration process can become complicated because these unknown constants turn into unknown functions. Also, since there are multiple displacement components, separating the variables of integration can be challenging. Thus, alternative approaches are necessary.

5.1.4 *A displacement method without indefinite integration*

In our displacement method, we do not perform indefinite integration of the PDE Eq. (5.3) written in terms of displacement. The solution process consists of the following four major steps.

5.1.4.1 *Step 1: Assume displacement*

We directly assume the displacement w has the following general form:

$$w = c_0 + c_1 z + c_2 z^2 \tag{5.5}$$

where $[1 \quad z \quad z^2]^\top$ are monomial basis functions and c_0, c_1, and c_2 are unknown constants. In other words, the displacement w is a linear combination of polynomial basis functions. The following snippet constructs the displacement:

```
z, d = sp.symbols('z, d', nonnegative=True)                      # Coordinate
G, λ, ν = sp.symbols('G, λ, ν', positive=True)        # Material constants
c0, c1, c2 = sp.symbols('c0, c1, c2', real=True)       # unknown constants
bz, q, n3 = sp.symbols('b_z, q, n_3')                                # forces

σxx, σyy, σzz = sp.symbols('σ_xx, σ_yy, σ_zz')                     # stresses
cs = sp.Matrix([c0, c1, c2])                               # unknown constants
w_bfs = sp.Matrix([1, z, z**2])          # Basis functions for displacement
display(Math(f"\\text{{Basis function for w, w_bfs}}={latex(w_bfs.T)}"))

wg = (cs.T@Matrix(w_bfs))[0]                             # assumed displacement

display(Math(f"\\text{{Assumed general displacement $w_g(z)$}}=\
                                                          {latex(wg)}"))
```

Basis function for w, w_bfs $= \begin{bmatrix} 1 & z & z^2 \end{bmatrix}$

Assumed general displacement $w_g(z) = c_0 + c_1 z + c_2 z^2$

The assumed displacement contains three unknown constants.

5.1.4.2 *Step 2: Use displacement boundary conditions*

Assume $w(z)\big|_{z=d} = 0$, which means that the point at depth $z = d$ serves as the reference point for measuring the displacement, as shown in Fig. 5.1. We then have

$$w(z)\Bigg|_{z=d} = c_2 d^2 + c_1 d + c_0 = 0 \tag{5.6}$$

This DBC gives

$$c_0 = -c_1 d - c_2 d^2 \tag{5.7}$$

We have found c_0 in terms of the other constants. We can then update the values of the constants and the displacement. The Python code is as follows:

```
sln_DBC = sp.solve([wg.subs(z,d)], cs)
sln_DBC

display(Math(f" \\text{{Conditions for fixed displacements}}=\
                                                    {latex(sln_DBC)}"))
sln_cs = {**sln_DBC}    # update the solutions for the unknown constants
w = wg.subs(sln_cs).simplify()                  # update the displacement
display(Math(f" \\text{{General displacement $w_g$}}= {latex(w)}"))
```

Conditions for fixed displacements $= \left\{c_0 : -c_1 d - c_2 d^2\right\}$

General displacement $w_g = -c_1 d + c_1 z - c_2 d^2 + c_2 z^2$

The displacement now contains two unknown constants.

5.1.4.3 *Step 3: Compute stresses using displacements*

Using Eq. (5.5), the volume strain or the divergence of the displacement vector becomes

$$e = \underbrace{\varepsilon_{xx}}_{=0} + \underbrace{\varepsilon_{yy}}_{=0} + \underbrace{\varepsilon_{zz}}_{=\partial w/\partial z} = -\frac{b_z}{2G+\lambda} z + c_0 \tag{5.8}$$

Substituting Eqs. (5.8) and (5.5) into the constitutive Eq. (4.21), we obtain

$$\begin{aligned}
\sigma_{xx} &= \lambda e + 2G \underbrace{\varepsilon_{xx}}_{=0} = -\lambda \frac{b_z}{2G+\lambda} z + \lambda c_0 = -\frac{\nu}{1-\nu} b_z z + \lambda c_0 \\
\sigma_{yy} &= \lambda e + 2G \underbrace{\varepsilon_{yy}}_{=0} = -\lambda \frac{b_z}{2G+\lambda} z + \lambda c_0 = -\frac{\nu}{1-\nu} b_z z + \lambda c_0 \\
\sigma_{zz} &= \lambda e + 2G \underbrace{\varepsilon_{zz}}_{e} = (\lambda + 2G) e = -b_z z + (\lambda + 2G) c_0 \\
\sigma_{yz} &= 0; \quad \sigma_{xz} = 0; \quad \sigma_{xy} = 0
\end{aligned} \tag{5.9}$$

It is clear that the nonzero stresses are the three normal stresses. The following snippet computes these stresses:

```
# Compute the stresses:
e = w.diff(z,1)                                   # the volume strain for 1D
σxxg = λ*e;    σyyg = λ*e                                  # general stresses
σzzg = (λ+2*G)*e
display(Math(f" σ_{{xx}} ={latex(σxxg)}; σ_{{yy}} ={latex(σyyg)}; \
                                         σ_{{zz}} ={latex(σzzg)}"))
```

$\sigma_{xx} = \lambda\left(c_1 + 2c_2 z\right); \sigma_{yy} = \lambda\left(c_1 + 2c_2 z\right); \sigma_{zz} = \left(2G + \lambda\right)\left(c_1 + 2c_2 z\right)$

We have found the stresses, but they still contain unknown constants. Next, we use the SBC on the top surface of the half-space to solve for one of the constants.

5.1.4.4 *Step 4: Use stress boundary conditions*

On the top surface of the half-space, the tractions are given as

$$t_{\Gamma 1} = 0; \quad t_{\Gamma 2} = 0; \quad t_{\Gamma 3} = q \quad \text{at } z = 0 \tag{5.10}$$

where q is the surface load. Since the direction cosines of the top surface are

$$n_1 = 0; \quad n_2 = 0; \quad n_3 = -1 \quad \text{at } z = 0 \tag{5.11}$$

the first two equations of the traction boundary conditions in Eq. (4.25) are automatically satisfied. The third condition becomes

$$\sigma_{zz} n_3 = -\sigma_{zz} = q \tag{5.12}$$

This SBC allows us to determine one more constant:

```
n3 = -1                              # q=t_{Γ3} on the top surface where n_3=-1
sln_SBC = sp.solve([(σzzg.subs({z:0})*n3 - q)], cs)
display(sln_SBC)
sln_cs = {**sln_DBC, **sln_SBC}             # update the unknown constants
sln_cs = {key: value.subs(sln_cs) for key, value in sln_cs.items()}
print(f"Current len of sln_cs = {len(sln_cs)}; Total cs = {len(cs)}")

# Update the displacement:
w = wg.subs(sln_cs).expand().simplify()
display(Math(f"\\text{{ Current displacement $w$}}=\\small {latex(w)}"))
# Update the stresses:
e = w.diff(z,1).simplify()
σxx = λ*e; σyy = λ*e                                          # stresses
σzz = (λ+2*G)*e
display(Math(f"\\text{{ Current stresses:}}"))
display(Math(f" σ_{{xx}}=\\small {latex(σxx)}; \
       σ_{{yy}}=\\small {latex(σyy)}; σ_{{zz}}=\\small {latex(σzz)}"))
```

$$\left\{c_1 : -\frac{q}{2G+\lambda}\right\}$$

```
Current len of sln_cs = 2; Total cs = 3
```

Current displacement $w = \dfrac{c_2 (2G+\lambda)\left(-d^2+z^2\right)+dq-qz}{2G+\lambda}$

Current stresses:

$\sigma_{xx} = \dfrac{\lambda (2c_2 z (2G+\lambda)-q)}{2G+\lambda}; \quad \sigma_{yy} = \dfrac{\lambda (2c_2 z (2G+\lambda)-q)}{2G+\lambda};$

$\sigma_{zz} = 2c_2 z (2G+\lambda) - q$

These updated solutions still contain one unknown constant, c_2. Next, we use the force boundary condition (FBC) on any horizontal plane at z in the half-space.

5.1.4.5 *Step 5: Use integral equilibrium conditions*

Integral equilibrium provides an FBC on any horizontal plane at z in the half-space. It can be written in an integral form as

$$\int_{-\infty}^{\infty}\int_{-\infty}^{\infty} (\sigma_{zz} + z b_z + q)\, dx\, dy = 0 \tag{5.13}$$

where b_z is the body force. Equation (5.13) implies that the total force on the free-body diagram at z must balance the total force resulting from σ_{zz} on the surface at z, as shown in Fig. 5.2.

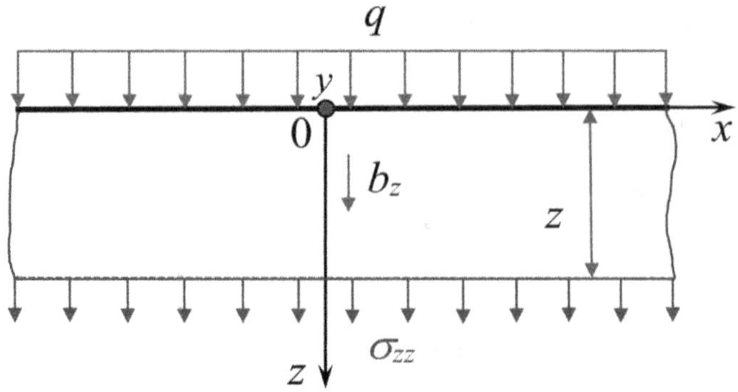

Figure 5.2. Schematic view of the free-body diagram of a 3D half-space, subjected to a surface traction and a body force in the vertical direction.

To ensure that Eq. (5.13) is satisfied, we must have

$$\sigma_{zz} = -zb_z - q \tag{5.14}$$

This FBC determines the last constant c_2. The Python snippet is as follows:

```
# Enforce the force boundary condition at free-body diagram cut:

sln_FBC = sp.solve([(σzz + z*bz + q)], cs)
display(sln_FBC)
sln_cs = {**sln_DBC, **sln_SBC, **sln_FBC}   # update unknown constants
sln_cs = {key: value.subs(sln_cs) for key, value in sln_cs.items()}
print(f"Current len of sln_cs = {len(sln_cs)}; Total cs = {len(cs)}")

# Update the displacement:
w_final = wg.subs(sln_cs).expand().simplify().expand()
display(Math(f" \\text{{ Final displacement solution $w$}}=\
                                                 {latex(w_final)}"))
# Update the stresses:
λ_ = 2*G*ν/(1-2*ν)                          # relation of elastic constants
e_final = w_final.diff(z,1).simplify()
σxx_final = (λ*e_final).subs(λ,λ_).simplify()                    # stresses
σyy_final = (λ*e_final).subs(λ,λ_).simplify()
σzz_final = ((λ+2*G)*e_final).subs(λ,λ_).simplify()
display(Math(f" \\text{{ Final stress solution:}} "))
display(Math(f" σ_{{xx}} ={latex(σxx_final)}; \
          σ_{{yy}} ={latex(σyy_final)}; σ_{{zz}} ={latex(σzz_final)}"))

# put all solution in a list for later use:
slns = [w_final.subs(λ,λ_), σxx_final, σyy_final, σzz_final]
```

$$\left\{c_2 : -\frac{b_z}{4G + 2\lambda}\right\}$$

```
Current len of sln_cs = 3; Total cs = 3
```

Final displacement solution $w = \dfrac{b_z d^2}{4G+2\lambda} - \dfrac{b_z z^2}{4G+2\lambda} + \dfrac{dq}{2G+\lambda} - \dfrac{qz}{2G+\lambda}$

Final stress solution:

$\sigma_{xx} = \dfrac{\nu\left(b_z z + q\right)}{\nu - 1}; \sigma_{yy} = \dfrac{\nu\left(b_z z + q\right)}{\nu - 1}; \sigma_{zz} = -b_z z - q$

All constants have now been determined.

5.1.5 *Final solution for displacements and stresses*

The full set of final solutions is rewritten as follows. The displacement solution is

$$w = \frac{1}{2G+\lambda}\left[q(d-z) + \frac{b_z}{2}(d-z)^2\right] \tag{5.15}$$

The vertical displacement is quadratic in z. The maximum displacement clearly occurs on the top surface of the half-space at $z = 0$, referenced to $z = d$.

$$w_{\max} = \frac{1}{2G+\lambda}\left[-qd + \frac{b_z}{2}d^2\right] \tag{5.16}$$

The stress solutions are

$$\begin{aligned} \sigma_{xx} &= \sigma_{yy} = -\frac{\nu}{1-\nu}(b_z z + q) \\ \sigma_{zz} &= -(b_z z + q) \\ \sigma_{yz} &= 0; \quad \sigma_{xz} = 0; \quad \sigma_{xy} = 0 \end{aligned} \tag{5.17}$$

We observe that all these stresses are linear functions of z, so the strains are also linear functions of z. Therefore, the compatibility equations are automatically satisfied. For positive b_z and q, the solid is compressed. Furthermore, if $b_z = 0$, both the stresses and strains remain constant.

Note also that the ratio of the following stresses is constant:

$$\frac{\sigma_{xx}}{\sigma_{zz}} = \frac{\sigma_{yy}}{\sigma_{zz}} = \frac{\nu}{1-\nu} \tag{5.18}$$

This indicates that the ratio of lateral to vertical pressure is constant and depends solely on Poisson's ratio of the material. Since $\nu < 0.5$, the lateral pressure is always smaller than the vertical pressure. Assuming $\nu = \frac{1}{3}$, we have $\frac{\sigma_{xx}}{\sigma_{zz}} = \frac{\sigma_{yy}}{\sigma_{zz}} = \frac{1}{2}$, implying that the lateral pressure is half of the vertical pressure.

5.1.6 *Plot the distribution of solutions*

All the solutions are functions of z. We write the following code to plot these solutions. As we are only interested in the distribution pattern of the solutions, we set all relevant parameters to unity:

```
# Set the parameters to unity:
dic_ = {q:1,bz:1,G:1,ν:sp.Rational(1,3),d:1}
slnz=[slns[i].subs(dic_).simplify() for i in range(len(slns))]
display(Math(f" \\text{{ Displacement function}}:{latex(slnz[0])}"))
display(Math(f" \\text{{ Stress functions}}:{latex(slnz[1:])}"))
```

Displacement function: $-\frac{z^2}{8} - \frac{z}{4} + \frac{3}{8}$

Stress functions: $\left[-\frac{z}{2} - \frac{1}{2}, -\frac{z}{2} - \frac{1}{2}, -z - 1\right]$

```
np.set_printoptions(formatter={'float': '{: 0.4e}'.format})

dE = 2.0; dz = 0.002
Z  = np.arange(0.001, dE, dz)          # coordinate z in depth-direction

# convert SymPy solution func. to numpy arrays.
n = len(slnz)
z = symbols("z")
slnx = [sp.lambdify(z, slnz[i], 'numpy') for i in range(n)]
sz   = [slnx[i](Z)  for i in range(n)]

sz_max = [np.max(abs(sz[i])) for i in range(n)]   # absolute max values
print(f"Absolute maximum values: {np.array(sz_max)}")
```

```
Absolute maximum values: [ 6.2425e-01  1.4995e+00  1.4995e+00  2.9990e+00]
```

```
# Plot the distribution curves:
y_data = [sz[i]/sz_max[i] for i in range(n)]  # List of function values
labels=["Displacement w", "$σ_{xx}$", "$σ_{yy}$", "$σ_{zz}$"]
x_data = [Z for i in range(n)]

sc = sz_max[3]                                              # stress scalar
y_data = [sz[0]/sz_max[0], sz[1]/sc, sz[2]/sc, sz[3]/sc]
gr.plot_fs(x_data,y_data,labels,'z','Solutions','imagesSM/w_s_3D_HS',\
                                                              ioff=True)
```

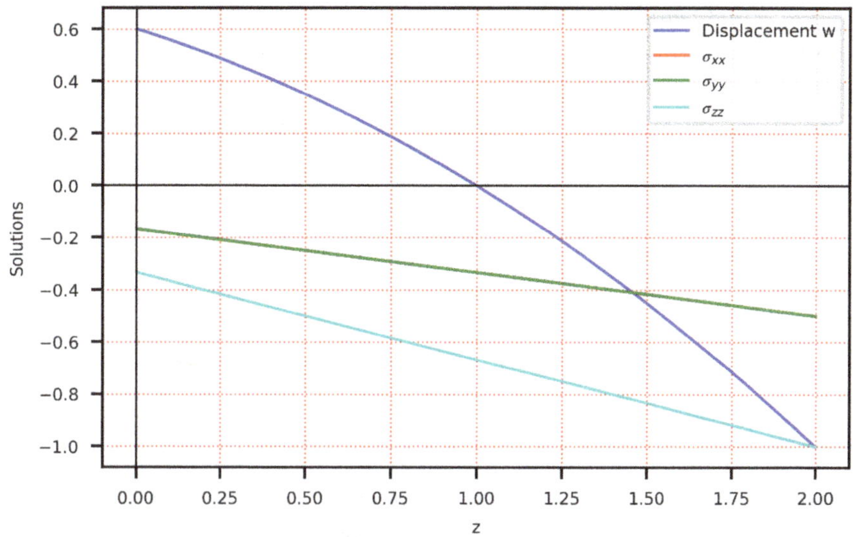

Figure 5.3. Distribution of the vertical displacement and the three normal stresses in the 3D half-space subjected to a surface pressure and a constant vertical body force.

The distributions of the vertical displacement w and the three normal stresses σ_{zz}, σ_{xx}, and σ_{yy} in the 3D half-space subjected to a surface pressure and a constant vertical body force are shown in Fig. 5.3. The displacement is normalized by its maximum value in the range. All these stresses are normalized by the maximum value of σ_{zz}. The displacement varies quadratically, and the stresses vary linearly with depth z.

5.1.7 *Nonconstant body forces*

As mentioned earlier, this 3D half-space problem is essentially 1D, allowing the body force b_z to be an arbitrary function of z. Assume the body force is linearly distributed:

$$b_z = q_0 + q_1 z \tag{5.19}$$

where q_0 and q_1 are constants.

In this case, an additional higher-order basis function is required for the assumed displacement w:

$$w = c_0 + c_1 z + c_2 z^2 + c_3 z^3 \tag{5.20}$$

where c_0, c_1, c_2, and c_3 are unknown constants.

All procedures and codes remain the same as in the constant body force case studied previously, yielding the displacement solution:

$$w = \frac{1}{2G+\lambda}\left(\frac{d^3 q_1}{6} + \frac{d^2 q_0}{2} + dq - qz - \frac{q_0 z^2}{2} - \frac{q_1 z^3}{6}\right) \tag{5.21}$$

The solution for the stresses is

$$\begin{aligned}
\sigma_{xx} &= \sigma_{yy} = \frac{-\lambda}{2G+\lambda}\left(q + q_0 z + \frac{q_1 z^2}{2}\right) \\
\sigma_{zz} &= -q - q_0 z - \frac{q_1 z^2}{2} \\
\sigma_{yz} &= 0; \quad \sigma_{xz} = 0; \quad \sigma_{xy} = 0
\end{aligned} \tag{5.22}$$

It is a good exercise for readers to find these solutions. It would be even better if readers could put all these snippets together to create a general solver for similar problems that can be cast in 1D.

This 3D half-space problem is straightforward and essentially 1D. There are several simple methods to solve it. For example, one may use the differential equation solver `sp.dsolve()` to obtain the solution with just a few lines of code. Our four-step displacement method is not necessary for this simple problem. However, for complex 3D problems, PDE solvers may fail and will not always work. The displacement method can be powerful in such cases.

Although simple, this 1D problem illustrates the principles and major steps of the displacement method, which largely follows the same approach for both 2D and 3D problems. The general procedure is outlined for 2D problems in Section 3.1 and is summarized below for 3D problems.

5.2 Displacement method procedure using basis functions

The general solution procedure in a displacement method is largely the same as that for 2D problems given in Section 3.1. The major differences are as follows:

1. For a 3D problem, there are three displacement components: u, v, and w, and hence assumptions are needed for each of these.
2. The displacement field for a 3D problem is much more complicated. Hence, in general, one needs to add special basis functions, such as

harmonic functions in infinite series form, rational harmonic functions, and logarithmic harmonic functions, in addition to the standard monomial basis functions.

The displacement methods presented in this chapter possess the features stated in Section 3.1. Since the solution construction starts from displacements, and the strains (and stresses) are computed from the displacements, the strain compatibility is naturally satisfied from the start. However, in the later process of enforcing various conditions for the stresses, the stresses may be altered, which could result in the dissatisfaction of the compatibility equations. Therefore, we should reinforce these equations in such cases.

In the following studies, we first examine a few 3D beams, which were studied in the previous chapter using stress methods and stream functions. However, we only obtained stress solutions for them. In this study, we aim to obtain a complete set of solutions, including displacements, strains, and stresses.

Since a 3D problem is much more complicated in coding, we first write the following utility codes to handle some of the standard operations, including computing stresses from given displacements, constitutive equations, checking differential equilibrium equations, and examining compatibility conditions for a given set of displacements, stresses, strains, etc. These utility codes are self-explanatory with quite detailed comments. Readers may skip reading these codes and refer to them when they are used.

```
def displacement2stress(X, G, ν, U):
    '''Compute the stresses for given displacements in 3D.
       U: array like with u, v, w.
       G: shear modulus, ν: Poisson's ratio.
       It returns 6 stress components in a list.
    '''
    x, y, z = X
    u, v, w = U
    #G = E/(2*(1+ν))                     # Use G for more concise formulas
    e = u.diff(x) + v.diff(y) + w.diff(z)   # volume strain (disp. div.)
    σxx = (2*G*(ν/(1-2*ν)*e+u.diff(x))).expand().collect(G*x).collect(z)
    σyy = (2*G*(ν/(1-2*ν)*e+v.diff(y))).expand().collect(G*x).collect(z)
    σzz = (2*G*(ν/(1-2*ν)*e+w.diff(z))).simplify()
    σyz = (G*(w.diff(y)+v.diff(z)))
    σxz = (G*(w.diff(x)+u.diff(z))).simplify()
    σxy = (G*(v.diff(x)+u.diff(y))).expand().collect(G*y).collect(z)

    return [σxx, σyy, σzz, σyz, σxz, σxy]
```

```
def stress2strain(E, ν, σ):
    '''3D constitutive equation, from stresses to strains'''
    εxx = ((σxx - ν*σyy - ν*σzz)/E).simplify()
    εyy = ((σyy - ν*σxx - ν*σzz)/E).simplify()
    εzz = ((σzz - ν*σxx - ν*σyy)/E).simplify()
    εyz = (1+ν)*σyz/E; εxz = (1+ν)*σxz/E; εxy = (1+ν)*σxy/E
    return [εxx,εyy,εzz,εyz,εxz,εxy]

def strain2stress(E, ν, ε):
    '''3D constitutive equation, from strains to stresses'''
    E_ = E/((1+ν)*(1-2*ν))
    σxx = E_*( (1-ν)*εxx + ν*(εyy + εzz) ).simplify()
    σyy = E_*( (1-ν)*εyy + ν*(εxx + εzz) ).simplify()
    σzz = E_*( (1-ν)*εxx + ν*(εxx + εyy) ).simplify()
    σyz = E*εyz/(1+ν); σxz = E*εxz/(1+ν); σxy = E*εxy/(1+ν)
    return [σxx,σyy,σzz,σyz,σxz,σxy]
```

```
def comptblty_check(E, ν, U, ε, σ, given = 'E'):
    '''Check compatibility conditions for given a displacement, stress,
       or strain fields in 3D. It returns the compatibility expressions.
    '''
    if given == "U":         # If the displacement expressions are given
        u, v, w = U
        εxx = u.diff(x); εyy=v.diff(y); εzz= w.diff(z) # compute strains
        εyz = ((w.diff(y)+v.diff(z))/2).simplify()
        εxz = ((w.diff(x)+u.diff(z))/2).simplify()
        εxy = ((v.diff(x)+u.diff(y))/2).simplify()
    elif given == "S":             # If the stress expressions are given
        σxx, σyy, σzz, σyz, σxz, σxy = σ
        εxx = (σxx - ν*σyy - ν*σzz)/E
        εyy = (σyy - ν*σxx - ν*σzz)/E
        εzz = (σzz - ν*σxx - ν*σyy)/E
        εyz = (1+ν)*σyz/E; εxz = (1+ν)*σxz/E; εxy = (1+ν)*σxy/E
    else:                          # If the strain expressions are given
        εxx, εyy, εzz, εyz, εxz, εxy = ε

    exprs = []
    exprs.append((εxx.diff(y,2)+εyy.diff(x,2) - 2*εxy.diff(x).diff(y))\
                                                          .simplify())
    exprs.append((εxx.diff(z,2)+εzz.diff(x,2) - 2*εxz.diff(x).diff(z))\
                                                          .simplify())
    exprs.append((εyy.diff(z,2)+εzz.diff(y,2) - 2*εyz.diff(y).diff(z))\
                                                          .simplify())
    exprs.append((εxy.diff(z,2) + εzz.diff(x).diff(y) \
       - εyz.diff(x).diff(z)- εxz.diff(y).diff(z)).expand().simplify())
```

```
    exprs.append((εxz.diff(y,2) + εyy.diff(x).diff(z) \
        - εxy.diff(y).diff(z)- εyz.diff(x).diff(y)).expand().simplify())
    exprs.append((εyz.diff(x,2) + εxx.diff(y).diff(z) \
        - εxz.diff(x).diff(y)- εxy.diff(x).diff(z)).expand().simplify())
    return exprs
```

```
def equilibrium_check(X, σ):
    '''Check the differential equilibrium equations for a given set
    of stresses in 3D.'''
    x, y, z = X
    σxx, σyy, σzz, σyz, σxz, σxy = σ
    eqn1 = (σxx.diff(x) + σxy.diff(y) + σxz.diff(z)).simplify()
    eqn2 = (σxy.diff(x) + σyy.diff(y) + σyz.diff(z)).simplify()
    eqn3 = (σxz.diff(x) + σyz.diff(y) + σzz.diff(z)).simplify()
    return [eqn1, eqn2, eqn3]
```

5.3 Beams with elliptic cross-section

This section studies a 3D beam with an **elliptic** cross-section. The detailed setting and the use of Cartesian coordinates for this problem are given in Section 4.3. We aim to obtain a complete set of solutions, including displacements, strains, and stresses. The general expressions for displacements utilize polynomial basis functions.

5.3.1 *Constructing general displacements*

The following code defines monomial bases, generates symbolic constants, and constructs these three displacement functions. In this process, we make use of the known symmetry properties for this problem:

1. The displacement function $u(x, y, z)$ is symmetric in x and antisymmetric in y.
2. The displacement function $v(x, y, z)$ is antisymmetric in both x and y.
3. The displacement function $w(x, y, z)$ is antisymmetric in x and symmetric in y.

These conditions are used to filter out the unnecessary basis functions. The snippet is as follows:

```
# Define necessary symbolic variables:
x, y, z = symbols("x, y, z ", real=True)
X = sp.Matrix([x, y, z])
ra, rb, L, Iyy = symbols("r_a,r_b, L, I_{yy}",positive=True) # geometry
P, E, G, ν = symbols("P, E, G, ν ",positive=True)          # parameters
Iyy_ = pi/4*ra**3*rb                                      # I for ellipse

σxx,σyy,σzz,σyz,σxz,σxy = sp.symbols("σ_xx,σ_yy,σ_zz,σ_yz,σ_xz,σ_xy")
σ = sp.Matrix([σxx, σyy, σzz, σyz, σxz, σxy])          # stress vector

εxx,εyy,εzz,εyz,εxz,εxy = sp.symbols("ε_xx,ε_yy,ε_zz,ε_yz,ε_xz,ε_xy")
ε = sp.Matrix([εxx, εyy, εzz, εyz, εxz, εxy])          # strain vector

order = 3                 # Set to use upto complete 3rd order monomials:
# Generate the complete set of the basis functions (monomials):
bfs_list = [x**i*y**j*z**k for i in range(order + 1)
                           for j in range(order + 1)
                           for k in range(order + 1) if i+j+k <= order]

bfs = sp.Matrix(bfs_list)            # form sympy vector with these bases
display(Math(f" \\text{{Complete 3rd order monomials in 3D}}:"))
display(Math(f" \\text{{bfs_all}} = {latex(bfs_list)}"))

# Select basis functions (bfs) for u(x,y,z) using symmteric conditions:
def is_even_in_x_and_y(poly):
    return poly.subs(x, -x) == poly and poly.subs(y, -y) == poly
u_bfs = [poly for poly in bfs if is_even_in_x_and_y(poly)]
display(Math(f" \\text{{u_bfs}}={latex(u_bfs)}"))          # Bases for u

# Select bfs for v(x,y,z) using symmteric conditions:
def is_odd_in_x_and_y(poly):
    return -poly.subs(x, -x) == poly and -poly.subs(y, -y) == poly
v_bfs = [poly for poly in bfs if is_odd_in_x_and_y(poly)]
display(Math(f" \\text{{v_bfs}}={latex(v_bfs)}"))          # Bases for v

# Select bfs for w(x,y,z) using symmteric conditions:
def is_odd_in_x_and_even_y(poly):
    return -poly.subs(x, -x) == poly and poly.subs(y, -y) == poly
w_bfs = [poly for poly in bfs if is_odd_in_x_and_even_y(poly)]
display(Math(f" \\text{{w_bfs}}={latex(w_bfs)}"))          # Bases for w
```

Complete third-order monomials in 3D:

$$\text{bfs\_all} = \left[1,\ z,\ z^2,\ z^3,\ y,\ yz,\ yz^2,\ y^2,\ y^2z,\ y^3,\ x,\ xz,\ xz^2,\ xy,\ xyz,\ xy^2,\ x^2,\ x^2z,\ x^2y,\ x^3\right]$$

$$\text{u\_bfs} = \left[1,\ z,\ z^2,\ z^3,\ y^2,\ y^2 z,\ x^2,\ x^2 z\right]$$

$$\text{v\_bfs} = [xy,\ xyz]$$

$$\text{w\_bfs} = \left[x,\ xz,\ xz^2,\ xy^2,\ x^3\right]$$

Based on the number of basis functions to be used, we generate the necessary number of symbolic unknown constants. This process is done separately for each of the displacement components. The general expressions for these displacement components are the dot products of the bases and their corresponding coefficients.

```
# unknown constants for displacements:
ai = sp.symbols(f'a:{len(u_bfs)}')                      # unknown constants
display(Math(f" \\text{{Unknown constants for $u$}}={latex(ai)}"))
ai = Matrix([ai[i] for i in range(len(u_bfs))])              # form a vector

bi = sp.symbols(f'b:{len(v_bfs)}')                      # unknown constants
display(Math(f" \\text{{Unknown constants for $v$}}={latex(bi)}"))
bi = Matrix([bi[i] for i in range(len(v_bfs))])              # form a vector

ci = sp.symbols(f'c:{len(w_bfs)}')                      # unknown constants
display(Math(f" \\text{{Unknown constants for $w$}}={latex(ci)}"))
ci = Matrix([ci[i] for i in range(len(w_bfs))])              # form a vector

cs = ai.col_join(bi).col_join(ci)   # form a single vector of constants

ug = (ai.T@Matrix(u_bfs))[0]               # Form the general displacements
vg = (bi.T@Matrix(v_bfs))[0]
wg = (ci.T@Matrix(w_bfs))[0]
print(f"Total number of unknown constants:  {len(cs)}")
display(Math(f"\\text{{General expressions for displacements:}}"))
display(Math(f" $$u_g$$ ={latex(ug)}"))
display(Math(f" $$v_g$$ ={latex(vg)}"))
display(Math(f" $$w_g$$ ={latex(wg)}"))
```

Unknown constants for $u = (a_0,\ a_1,\ a_2,\ a_3,\ a_4,\ a_5,\ a_6,\ a_7)$

Unknown constants for $v = (b_0,\ b_1)$

Unknown constants for $w = (c_0,\ c_1,\ c_2,\ c_3,\ c_4)$

```
Total number of unknown constants: 15
```

General expressions for displacements:

$$u_g = a_0 + a_1 z + a_2 z^2 + a_3 z^3 + a_4 y^2 + a_5 y^2 z + a_6 x^2 + a_7 x^2 z$$

$v_g = b_0xy + b_1xyz$

$w_g = c_0x + c_1xz + c_2xz^2 + c_3xy^2 + c_4x^3$

The assumed displacements have a total of 15 unknown constants that need to be determined.

5.3.2 *Enforcing the displacement boundary conditions*

We assume that the displacements at $(0, 0, 0)$ are fixed as follows:

$$\begin{aligned} &u = v = w = 0 \\ &\frac{\partial u}{\partial z} = \frac{\partial v}{\partial x} = \frac{\partial w}{\partial y} = 0 \end{aligned} \tag{5.23}$$

These conditions impose six DBCs on the displacements. The conditions are enforced using the following snippet:

```
# Enforce displacement boundary conditions (DBCs):
dic0 = {x:0, y:0, z:0}                                          # DBCs
sln_cs_u0 = sp.solve([ug.subs(dic0), vg.subs(dic0), wg.subs(dic0),
 ug.diff(z).subs(dic0),vg.diff(x).subs(dic0),wg.diff(y).subs(dic0)],cs)

display(Math(f" \\text{{Conditions for fixed displacements}}=\
                                           {latex(sln_cs_u0)}"))
sln_cs = {**sln_cs_u0}      # update the solutions for unknown constants
ug = ug.subs(sln_cs).simplify()
vg = vg.subs(sln_cs).simplify()
wg = wg.subs(sln_cs).simplify()
display(Math(f" \\text{{General displacement $u$}}= {latex(ug)}")) #e.g.
```

Conditions for fixed displacements $= \{a_0 : 0,\ a_1 : 0\}$

General displacement $u = a_2z^2 + a_3z^3 + a_4y^2 + a_5y^2z + a_6x^2 + a_7x^2z$

We have found two of the constants using the DBCs.

5.3.3 *Enforcing stress boundary conditions*

The general expressions for the six stress components can be computed using Eq. (4.21) and the utility Python function `displacement2stress()` provided earlier. The snippet is as follows:

```
# Compute the general expressions for stresses from displacements:

U = sp.Matrix([ug, vg, wg])
σxxg, σyyg, σzzg, σyzg, σxzg, σxyg = displacement2stress(X, G, ν, U)

display(Math(f" \\text{{General expressions for stresses}}:" ))
display(Math(f" \\text{{$σ_{{xx}}$}}=\\small {latex(σxxg)}"))
display(Math(f" \\text{{$σ_{{yy}}$}}=\\small {latex(σyyg)}"))
display(Math(f" \\text{{$σ_{{xy}}$}}=\\small {latex(σxyg)}"))
display(Math(f" \\text{{$σ_{{zz}}$}}=\\small {latex(σzzg)}"))
display(Math(f" \\text{{$σ_{{yz}}$}}=\\small {latex(σyzg)}"))
display(Math(f" \\text{{$σ_{{xz}}$}}=\\small {latex(σxzg)}"))
```

General expressions for stresses:

$$\sigma_{xx} = Gx\left(\frac{4a_6\nu}{1-2\nu} + 4a_6 + \frac{2b_0\nu}{1-2\nu} + \frac{2c_1\nu}{1-2\nu} + z\left(\frac{4a_7\nu}{1-2\nu} + 4a_7 + \frac{2b_1\nu}{1-2\nu} + \frac{4c_2\nu}{1-2\nu}\right)\right)$$

$$\sigma_{yy} = Gx\left(\frac{4a_6\nu}{1-2\nu} + \frac{2b_0\nu}{1-2\nu} + 2b_0 + \frac{2c_1\nu}{1-2\nu} + z\left(\frac{4a_7\nu}{1-2\nu} + \frac{2b_1\nu}{1-2\nu} + 2b_1 + \frac{4c_2\nu}{1-2\nu}\right)\right)$$

$$\sigma_{xy} = Gy\left(2a_4 + b_0 + z\left(2a_5 + b_1\right)\right)$$

$$\sigma_{zz} = \frac{2Gx\left(-\nu\left(2a_6 + 2a_7z + b_0 + b_1z + c_1 + 2c_2z\right) + \left(c_1 + 2c_2z\right)\left(2\nu - 1\right)\right)}{2\nu - 1}$$

$$\sigma_{yz} = G\left(b_1xy + 2c_3xy\right)$$

$$\sigma_{xz} = G\left(2a_2z + 3a_3z^2 + a_5y^2 + a_7x^2 + c_0 + c_1z + c_2z^2 + c_3y^2 + 3c_4x^2\right)$$

These stresses contain unknown constants. We can now apply the stress boundary conditions (SBCs) for stresses in the x–y plane on the circumferential surface of the beam, using Eq. (4.25). To do this, we need to compute the direction cosines of the unit outward normals on the circumferential surface. For 3D beams with elliptic cross-sections, the cross-section is defined by the smooth one-piece function:

$$f(x, y) = \frac{x^2}{r_a^2} + \frac{y^2}{r_b^2} - 1 \tag{5.24}$$

where r_a and r_b are the lengths of the major and minor semi-axes of the ellipse, respectively. The outward normal of the ellipse is given by the gradient of $f(x, y)$:

$$\nabla f(x,y) = \begin{bmatrix} \frac{\partial f}{\partial x} & \frac{\partial f}{\partial y} \end{bmatrix}^\top = \begin{bmatrix} \frac{2x}{r_a^2} & \frac{2y}{r_b^2} \end{bmatrix}^\top \tag{5.25}$$

The **unit normal vector** becomes

$$\frac{\nabla f(x,y)}{|\nabla f(x,y)|} = \begin{bmatrix} \frac{r_b^2 x}{\sqrt{r_a^4 y^2 + r_b^4 x^2}} \\ \frac{r_a^2 y}{\sqrt{r_a^4 y^2 + r_b^4 x^2}} \end{bmatrix} \tag{5.26}$$

which represents the unit normal vector in Cartesian coordinates at any point (x, y) on the ellipse.

For the unit outward normal vector on the circumferential surface of the beam, we need to add the component n_3, which is simply 0 because the entire surface is parallel to the z-axis (see Fig. 4.3). We thus have

$$\mathbf{N}(x,y) = \begin{bmatrix} \frac{r_b^2 x}{\sqrt{r_a^4 y^2 + r_b^4 x^2}} \\ \frac{r_a^2 y}{\sqrt{r_a^4 y^2 + r_b^4 x^2}} \\ 0 \end{bmatrix} \tag{5.27}$$

The snippet for this computation is as follows:

```
# Compute the direction cosines at a point on the ellipse edge:

f_ellipse = (x**2/ra**2) + (y**2/rb**2) - 1   # equation of the ellipse

# Compute the normal vector N (gradient of f_ellipse):
N = sp.Matrix([sp.diff(f_ellipse, x), sp.diff(f_ellipse, y)])
magnitude_N = sp.sqrt(N[0]**2 + N[1]**2)                   # magnitude of N

N = sp.simplify(N/magnitude_N)                           # unit normal vector
N = N.col_join(sp.Matrix([sp.S.Zero]))                     # add n_3=0 to N
display(Math(f" \\text{{Unit normal vector on ellipse surface \
                                $\mathbf{{N}}^T$}}= {latex(N.T)}"))
```

Unit normal vector on ellipse surface $\mathbf{N}^T = \begin{bmatrix} \frac{r_b^2 x}{\sqrt{r_a^4 y^2 + r_b^4 x^2}} & \frac{r_a^2 y}{\sqrt{r_a^4 y^2 + r_b^4 x^2}} & 0 \end{bmatrix}$

With the unit outward normal obtained, we can enforce Eq. (4.25) on the circumferential surface, where $x = r_a\sqrt{1 - \frac{y^2}{r_b^2}}$. The snippet is as follows:

```
# Enforce SBCs over the circumferential surface of the beam:
x_ = sqrt(1 - y**2/rb**2)*ra # function for lower edge of cross-section

# SBCs on the edge:
SBC1 = sp.numer((N[0]*σxxg + N[1]*σxyg + N[2]*σxzg).subs(x,x_)\
                                        .simplify()).expand()
SBC2 = sp.numer((N[0]*σxyg + N[1]*σyyg + N[2]*σyzg).subs(x,x_)\
                                        .simplify()).expand()
SBC3 = sp.numer((N[0]*σxzg + N[1]*σyzg + N[2]*σzzg).subs(x,x_)\
                                        .simplify()).expand()
exprs = []
exprs.append(SBC1.coeff(y,2).coeff(z,1))
exprs.append(SBC1.coeff(y,2).coeff(z,0))
exprs.append(SBC1.coeff(z,1).coeff(y,0))
exprs.append(SBC1.coeff(y,0).coeff(z,0))

exprs.append(SBC2.coeff(y,1).coeff(sqrt(rb**2-y**2),1).coeff(z,1))
exprs.append(SBC2.coeff(y,1).coeff(sqrt(rb**2-y**2),1).coeff(z,0))

exprs.append(SBC3.coeff(sqrt(rb**2-y**2),1).coeff(z,2).coeff(y,0))
exprs.append(SBC3.coeff(sqrt(rb**2-y**2),1).coeff(z,1).coeff(y,0))
exprs.append(SBC3.coeff(y,2).coeff(sqrt(rb**2-y**2),1).coeff(z,0))
exprs.append(SBC3.coeff(sqrt(rb**2-y**2),1).coeff(y,0).coeff(z,0))
#exprs                                                      # lengthy
```

The SBCs must be satisfied over the entire circumferential surface at any point along the y- and z-coordinates. Therefore, all coefficients associated with the basis functions, such as

- y^2, y^1, y^0
- z^2, z^1, z^0
- $\sqrt{r_b^2 - y^2}$

and their combinations, must vanish. This leads to a new set of 10 conditions, which can now be used to solve for some of the unknown constants. The following snippet does this and follows up with updates:

```
# Use SBCs to solve for some of the unknown constants:
sln_SBC = sp.solve(exprs, cs)
#display(sln_SBC)
sln_cs = {**sln_cs_u0, **sln_SBC}          # update the unknown constants
sln_cs = {key: value.subs(sln_cs) for key, value in sln_cs.items()}
print(f"Current len of sln_cs = {len(sln_cs)}; Total cs = {len(cs)}")

# Using the current constants to update the stresses:
σ = sp.Matrix([σxxg,σyyg,σzzg,σyzg,σxzg,σxyg]).subs(sln_cs)\
                                              applyfunc(sp.simplify)
```

```
σxx, σyy, σzz, σyz, σxz, σxy = σ

display(Math(f" σ_{{xx}} ={latex(σxx)}; σ_{{yy}} ={latex(σyy)}; \
                                         σ_{{xy}} ={latex(σxy)}"))
display(Math(f" \\text{{General stress $σ_{{zz}}$:}} "))
display(Math(f" \scriptsize {latex(σzz.collect(ra).collect(rb))}"))
#display(Math(f" \\text{{General stress $σ_{{yz}}$}}={latex(σyz)}"))
#display(Math(f" \\text{{General stress $σ_{{xz}}$}}={latex(σxz)}"))
                                                             # Lengthy
# Using the current constants to update the displacements:
U = sp.Matrix([ug, vg, wg]).subs(sln_cs).applyfunc(sp.simplify)
#display(Math(f" \\text{{General displacement $U$}}={latex(U)}"))
                                                             # Lengthy
# Check how many constants are found and how many unknowns remain:
list_keys = list(sln_cs.keys())
cs_Eql = list(set(list(cs))-set(list_keys))
print(f"Constants yet to be found: {cs_Eql}")
```

```
Current len of sln_cs = 12; Total cs = 15
```

$\sigma_{xx} = 0; \sigma_{yy} = 0; \sigma_{xy} = 0$

$$\text{General stress } \sigma_{zz}: \frac{2Gx\left(r_a^2\left(c_1\nu^2 + c_1\nu + 4c_3z\nu + 4c_3z - 6c_4z\nu - 6c_4z\right) + r_b^2\left(-c_1\nu^2 - c_1\nu + 2c_3z\nu + 2c_3z\right)\right)}{\nu\left(r_a^2 - r_b^2\right)}$$

```
Constants yet to be found: [c1, c4, c3]
```

We made good progress. The undetermined constants have now been reduced to three. We will use the constants determined so far to update the stresses and displacements.

We found that the stress components $\sigma_{xx} = \sigma_{yy} = \sigma_{xy} = 0$ throughout the entire beam, despite only imposing zero conditions on the circumferential surfaces. Therefore, our focus is now solely on the three stress components in the z-plane: σ_{xz}, σ_{yz}, and σ_{zz}. We will display only the current σ_{zz}, as the other two components are quite lengthy. Readers can view them by uncommenting the relevant lines in the code.

5.3.4 *Enforcing force equilibrium conditions*

Next, we apply the equilibrium conditions for stresses on an arbitrary z-plane of the 3D beam:

$$
\begin{aligned}
\sum F_z &= 0: \iint_A \sigma_{zz} dx dy = 0 \\
\sum M_x &= 0: \iint_A \sigma_{zz} y dx dy = 0 \\
\sum M_y &= 0: \iint_A \sigma_{zz} x dx dy + P(L-z) = 0 \\
\sum F_x &= 0: \iint_A \sigma_{zx} dx dy - P = 0 \\
\sum F_y &= 0: \iint_A \sigma_{zy} dx dy = 0 \\
\sum M_z &= 0: \iint_A (\sigma_{zy} x - \sigma_{zx} y) dx dy = 0
\end{aligned}
\tag{5.28}
$$

Let us enforce these six equations:

```
# Enforce integral equilibrium equations (Ieq):
sln_cs = {**sln_cs_u0, **sln_SBC}                    # update constants

# update stresses
σ = sp.Matrix([σxxg, σyyg, σzzg, σyzg, σxzg, σxyg]).subs(sln_cs)\
                                            .applyfunc(sp.simplify)
σxx, σyy, σzz, σyz, σxz, σxy = σ

# Define the integral limits for x-direction:
x_limits = (x, -ra*sqrt(1-y**2/rb**2), ra*sqrt(1-y**2/rb**2))
exprs = []
Int = sp.integrate
exprs.append(Int(Int(σzz,   x_limits),(y,-rb,rb)))
exprs.append(Int(Int(σzz*y, x_limits),(y,-rb,rb)))
exprs.append(Int(Int(σzz*x, x_limits),(y,-rb,rb))+P*(L-z))
exprs.append(Int(Int(σxz,   x_limits),(y,-rb,rb))-P)
exprs.append(Int(Int(σyz,   x_limits),(y,-rb,rb)))
exprs.append(Int(Int((σyz*x-σxz*y),x_limits),(y,-rb,rb)))

sln_FBC = sp.solve(exprs, cs)
sln_FBC
sln_cs = {**sln_cs_u0, **sln_SBC, **sln_FBC}
sln_cs = {key: value.subs(sln_cs) for key, value in sln_cs.items()}
print(f"Current len of sln_cs = {len(sln_cs)}; Total cs = {len(cs)}")
#display(sln_cs)                                             # Lengthy
```

```
# update stresses
σ = sp.Matrix([σxxg,σyyg,σzzg,σyzg,σxzg,σxyg]).subs(sln_cs)\
                                        .applyfunc(sp.simplify)
σxx, σyy, σzz, σyz, σxz, σxy = σ
display(Math(f" \\text{{Final $σ_{{zz}}$}}=\scriptsize {latex(σzz)}"))
display(Math(f" \\text{{Current $σ_{{xz}}$}}=\scriptsize {latex(σxz)}"))
display(Math(f" \\text{{Current $σ_{{yz}}$}}=\scriptsize {latex(σyz)}"))

# update the displacements
U = sp.Matrix([ug, vg, wg]).subs(sln_cs).applyfunc(sp.simplify)
#display(U)                                              # lengthy
```

```
Current len of sln_cs = 14; Total cs = 15
```

$$\text{Final } \sigma_{zz} = \frac{4Px\left(-L+z\right)}{\pi r_a^3 r_b}$$

$$\text{Current } \sigma_{xz} = \frac{2\left(-3\pi G c_4 r_a^5 r_b^3 + 9\pi G c_4 r_a^5 r_b y^2 + 3\pi G c_4 r_a^3 r_b^3 x^2 + 2P r_a^4 + P r_a^2 r_b^2 - 2P r_a^2 x^2 + 3P r_a^2 y^2 - P r_b^2 x^2\right)}{3\pi r_a^3 r_b\left(r_a^2 + r_b^2\right)}$$

$$\text{Current } \sigma_{yz} = \frac{4xy\left(-3\pi G c_4 r_a^3 r_b^3 - P r_a^2 - 2P r_b^2\right)}{3\pi r_a^3 r_b\left(r_a^2 + r_b^2\right)}$$

We found updated stresses and displacements (now displayed), but there is still one unknown constant. However, we have used up all the available conditions. This occurs because the altered stresses may no longer satisfy the compatibility equations. Let us check it out:

To check whether the compatibility equations are satisfied, we need to verify that the strain components derived from these stresses satisfy the necessary conditions. If they do not, we will need to modify the stress or displacement assumptions, possibly by introducing additional higher-order terms or revising the boundary conditions.

```
# Check the compatibility conditions:
ε = sp.Matrix(stress2strain(E, ν, σ))   # compute strain using stresses
exprs = comptblty_check(E, ν, U, ε, σ, given = 'E')
display(Math(f" \\text{{$σ_{{xx}}$}}=\\scriptsize {latex(exprs)}"))
```

$$\sigma_{xx} = \left[0,\ 0,\ 0,\ 0,\ \frac{4\left(9\pi G c_4 r_a^5 r_b \nu + 9\pi G c_4 r_a^5 r_b + 3\pi G c_4 r_a^3 r_b^3 \nu + 3\pi G c_4 r_a^3 r_b^3 + P r_a^2 \nu + 4P r_a^2 - P r_b^2 \nu + 2P r_b^2\right)}{3\pi E r_a^3 r_b\left(r_a^2 + r_b^2\right)},\ 0\right]$$

As suspected, the fifth compatibility equation is violated. Let's enforce it:

```
sln_c4=sp.solve([exprs[4]], cs)
display(Math(f" \\text{{The last constant $c_{{4}}$}}=\
                                          \\small {latex(sln_c4)}"))
```

The last constant

$$c_4 = \left\{ c_4 : \frac{-Pr_a^2\nu - 4Pr_a^2 + Pr_b^2\nu - 2Pr_b^2}{9\pi G r_a^5 r_b \nu + 9\pi G r_a^5 r_b + 3\pi G r_a^3 r_b^3 \nu + 3\pi G r_a^3 r_b^3} \right\}$$

We can now update the constants, stresses and displacements.

```
sln_cs = {**sln_cs_u0,**sln_SBC,**sln_FBC,**sln_c4}  # update constants
sln_cs = {key: value.subs(sln_cs) for key, value in sln_cs.items()}
print(f" Current len of sln_cs = {len(sln_cs)}; Total cs = {len(cs)}")
#display(sln_cs)                                              # lengthy

# update stresses
σ = sp.Matrix([σxxg,σyyg,σzzg,σyzg,σxzg,σxyg]).subs(sln_cs)\
                                            .applyfunc(sp.simplify)
σxx, σyy, σzz, σyz, σxz, σxy = σ
display(Math(f" \\text{{Final $σ_{{zz}}$}}=\small {latex(σzz)}"))
display(Math(f" \\text{{Final $σ_{{xz}}$}}=\small {latex(σxz)}"))
display(Math(f" \\text{{Final $σ_{{yz}}$}}=\small {latex(σyz)}"))

# Final displacements:
U = sp.Matrix([ug, vg, wg]).subs(sln_cs).applyfunc(sp.simplify)
display(Math(f" \\text{{Final }} \mathbf{{U}}=\\small{latex(U)}"))
```

```
Current len of sln_cs = 15; Total cs = 15
```

Final $\sigma_{zz} = \dfrac{4Px\left(-L+z\right)}{\pi r_a^3 r_b}$

Final

$$\sigma_{xz} = \frac{2P\left(2r_a^4\nu + 2r_a^4 + r_a^2 r_b^2 - 2r_a^2x^2\nu - 2r_a^2x^2 + 2r_a^2y^2\nu - r_a^2y^2 - r_b^2x^2\right)}{\pi r_a^3 r_b\left(3r_a^2\nu + 3r_a^2 + r_b^2\nu + r_b^2\right)}$$

Final $\sigma_{yz} = \dfrac{4Pxy\left(-r_a^2\nu - r_a^2 - r_b^2\nu\right)}{\pi r_a^3 r_b\left(3r_a^2\nu + 3r_a^2 + r_b^2\nu + r_b^2\right)}$

$$\text{Final } \mathbf{U} = \begin{bmatrix} \frac{P\left(3Lx^2\nu - 3Ly^2\nu + 3Lz^2 - 3x^2z\nu + 3y^2z\nu - z^3\right)}{3\pi G r_a^3 r_b(\nu+1)} \\ \frac{2Pxy\nu(L-z)}{\pi G r_a^3 r_b(\nu+1)} \\ \frac{Px\left(-18Lr_a^2z - 6Lr_b^2z + 12r_a^4\nu + 12r_a^4 + 6r_a^2r_b^2 - r_a^2x^2\nu - 4r_a^2x^2 + 3r_a^2y^2\nu - 6r_a^2y^2 + 9r_a^2z^2 + r_b^2x^2\nu - 2r_b^2x^2 - 3r_b^2y^2\nu + 3r_b^2z^2\right)}{3\pi G r_a^3 r_b\left(3r_a^2\nu + 3r_a^2 + r_b^2\nu + r_b^2\right)} \end{bmatrix}$$

Finally, we obtained the entire set of solutions. These stresses are exactly the same as those obtained in Section 4.3, using the stress method. The distribution of σ_{xz} follows a parabolic pattern in both the x and y directions. The distribution of σ_{yz} is bilinear in both the x and y directions.

The displacements are newly obtained using the displacement method. The deflection u is cubic in the z direction and quadratic in both x and y. The displacement v (in the depth direction) is linear in z and quadratic in x and y. The displacement w (in the axial direction) is cubic in x and quadratic in both z and y.

The following code validates the solution obtained:

```
# Check on stress boundary conditions (SBCs) over the free surface:

SBCs=(SBC1.subs(sln_cs).simplify(), SBC2.subs(sln_cs).simplify(),\
                                    SBC3.subs(sln_cs).simplify())
print(f" Confirm free SBCs on the lower surface: {SBCs}")

# Confirmation free SBCs on the ellipse upper surface:
x_ = - sqrt(1 - y**2/rb**2)*ra                                  # SBCs:
SBC1 = sp.numer((N[0]*σxx + N[1]*σxy + N[2]*σxz).subs(x,x_)\
                                               .simplify()).expand()
SBC2 = sp.numer((N[0]*σxy + N[1]*σyy + N[2]*σyz).subs(x,x_)\
                                               .simplify()).expand()
SBC3 = sp.numer((N[0]*σxz + N[1]*σyz + N[2]*σzz).subs(x,x_)\
                                               .simplify()).expand()
print(f" Confirm free SBCs on the upper surface: {(SBC1, SBC2, SBC3)}")
```

```
Confirm free SBCs on the lower surface: (0, 0, 0)
Confirm free SBCs on the upper surface: (0, 0, 0)
```

```
# Check the displacement boundary conditions at (0, 0, 0):
u, v, w = U
u.subs(dic0), v.subs(dic0), w.subs(dic0), u.diff(z).subs(dic0), \
                        v.diff(x).subs(dic0), w.diff(y).subs(dic0)
```

(0, 0, 0, 0, 0, 0)

```
# Compute the stresses using the final displacements:
σxx, σyy, σzz, σyz, σxz, σxy = displacement2stress(X, G, ν, U)
display(Math(f" \\text{{Final $σ_{{zz}}$}}={latex(σzz.simplify())}"))
display(Math(f" \\text{{Current $σ_{{xz}}$}}={latex(σxz.simplify())}"))
display(Math(f" \\text{{Current $σ_{{yz}}$}}={latex(σyz.simplify())}"))
```

Final $\sigma_{zz} = \dfrac{4Px\left(-L+z\right)}{\pi r_a^3 r_b}$

Current

$$\sigma_{xz} = \frac{2P\left(2r_a^4\nu + 2r_a^4 + r_a^2 r_b^2 - 2r_a^2 x^2 \nu - 2r_a^2 x^2 + 2r_a^2 y^2 \nu - r_a^2 y^2 - r_b^2 x^2\right)}{\pi r_a^3 r_b \left(3r_a^2\nu + 3r_a^2 + r_b^2\nu + r_b^2\right)}$$

Current $\sigma_{yz} = \dfrac{4Pxy\left(-r_a^2\nu - r_a^2 - r_b^2\nu\right)}{\pi r_a^3 r_b\left(3r_a^2\nu + 3r_a^2 + r_b^2\nu + r_b^2\right)}$

We can also check the differential equilibrium equations, which were not used in the process of deriving these solutions. We would like to confirm that all these equations are satisfied:

```
# Check the differential equilibrium equations:
eqn1, eqn2, eqn3 = equilibrium_check(X, [σxx, σyy, σzz, σyz, σxz, σxy])
display(Math(f" \\text{{Is equilibrium Eq. 1 = 0? }}{latex(eqn1)}"))
display(Math(f" \\text{{Is equilibrium Eq. 2 = 0? }}{latex(eqn2)}"))
display(Math(f" \\text{{Is equilibrium Eq. 2 = 0? }}{latex(eqn3)}"))
```

Is equilibrium Eq. 1 = 0? 0

Is equilibrium Eq. 2 = 0? 0

Is equilibrium Eq. 2 = 0? 0

```
# Check on integral force boundary conditions:
FBC1 = (Int(Int(σzz,   x_limits),(y,-rb,rb))).simplify()
FBC2 = (Int(Int(σzz*y, x_limits),(y,-rb,rb))).simplify()
FBC3 = (Int(Int(σzz*x, x_limits),(y,-rb,rb))+P*(L-z)).simplify()
FBC4 = (Int(Int(σxz,   x_limits),(y,-rb,rb))-P).simplify()
FBC5 = (Int(Int(σyz,   x_limits),(y,-rb,rb))).simplify()
FBC6 = (Int(Int((σyz*x-σxz*y),x_limits),(y,-rb,rb))).simplify()
FBC1, FBC2, FBC3, FBC4, FBC5, FBC6
```

(0, 0, 0, 0, 0, 0)

Let us take a look at solutions at special locations:

```
display(Math(f" \\text{{At $x=y=0$: $σ_{{xz}}$}}=\
                                        {latex(σxz.subs({x:0,y:0}))}"))
display(Math(f" \\text{{At $x=y=0, ν=0$: $σ_{{xz}}$}}=\
                                   {latex(σxz.subs({x:0,y:0,ν:0}))}"))
display(Math(f" \\text{{For circular cross-section, at $x=y=0$: \
           $σ_{{xz}}$}}={latex(σxz.subs({x:0,y:0,rb:ra}).simplify())}"))
display(Math(f" \\text{{For circular cross-section, at $x=y=0$, ν=0: \
                   $σ_{{xz}}$}}={latex(σxz.subs({x:0,y:0,rb:ra,ν:0}))}"))
display(Math(f" \\text{{At $x=y=0$: $σ_{{yz}}$}}=\
                                        {latex(σyz.subs({x:0,y:0}))}"))
```

At $x = y = 0$: $\sigma_{xz} = \dfrac{2P\left(2r_a^4\nu + 2r_a^4 + r_a^2r_b^2\right)}{\pi r_a^3 r_b\left(3r_a^2\nu + 3r_a^2 + r_b^2\nu + r_b^2\right)}$

At $x = y = 0, \nu = 0$: $\sigma_{xz} = \dfrac{2P\left(2r_a^4 + r_a^2r_b^2\right)}{\pi r_a^3 r_b\left(3r_a^2 + r_b^2\right)}$

For circular cross-section, at $x = y = 0$: $\sigma_{xz} = \dfrac{P\,(2\nu + 3)}{2\pi r_a^2\,(\nu + 1)}$

For circular cross-section, at $x = y = 0$, $\nu = 0$: $\sigma_{xz} = \dfrac{3P}{2\pi r_a^2}$

At $x = y = 0$: $\sigma_{yz} = 0$

Lastly, let us compare the stress solution with that from the stress method, as given in Chapter 4:

```
# Compare the solution with that from the stress method (Chapter 4):
Iyy_=pi/4*ra**3*rb
σyz_= -P*x*y*(ra**2*(ν + 1) + rb**2*ν)/(Iyy_*(3*ra**2 + rb**2)*(ν + 1))
display(Math(f" \\text{{Stress $σ_{{yz}}$ same as that from stress \
                  method?}}=\\small{latex((σyz_-σyz).simplify()==0)}"))
```

Stress σ_{yz} same as that from stress method? = True

It is confirmed that all the solutions obtained are correct. Let us plot the results for shear stresses.

5.3.5 *Plot of the shear stresses on the cross-section*

```
from matplotlib import cm
import matplotlib.patches as patches

plt.rcParams['figure.dpi'] = 500
ra_, rb_ = 2., 1.
h, b, P_, ν_ = 2., 1., 1., 0.3                  # arbitrarily set values

xl, xr, dx = -h, h, 0.1
yl, yr, dy = -b, b, 0.05

X = np.arange(xl, xr+dx, dx)
Y = np.arange(yl, yr+dy, dy)
X, Y = np.meshgrid(X, Y)
xn, yn = X.shape
Szx = X*0; Szy = Y*0

dicE = {ra:ra_,rb:rb_,P:P_, ν:ν_}
np_σxz = lambdify((x,y), σxz.subs(dicE),'numpy')
np_σyz = lambdify((x,y), σyz.subs(dicE),'numpy')
for xk in range(xn):
    for yk in range(yn):
        Szx[xk, yk] = np_σxz(X[xk,yk],Y[xk,yk])
        Szy[xk, yk] = np_σyz(X[xk,yk],Y[xk,yk])

#Elliptic mask to-be-used in plot, print nothing outside the ellipse:
mask = (X/ra_)**2 + (Y/rb_)**2 >= 1
Szx[mask] = np.nan; Szy[mask] = np.nan

plt.ioff()
#plt.ion()                         # uncomment this to display the plots
mag = np.sqrt(Szx**2 + Szy**2)
fig = plt.figure(figsize=(8, 4))
ax = fig.add_subplot(131)
ax.quiver(Y,-X, Szy,-Szy*0, mag,scale=1,label='Component $5×σ_{zy}$',\
                                                            cmap=cm.jet)
ax.add_patch(patches.Ellipse((0,0), width=2*rb_, height=2*ra_, \
                               edgecolor='b', facecolor='none'))
plt.axis('equal')
plt.ylabel('Major radius $r_a$')
plt.xlabel('Minor radius $r_b$'); plt.legend()

ax = fig.add_subplot(132)
ax.quiver(Y,-X, Szx*0,-Szx, mag, scale=5,label='Componet $σ_{zx}$',\
                                                         cmap=cm.jet)
```

```
ax.add_patch(patches.Ellipse((0,0), width=2*rb_, height=2*ra_, \
                              edgecolor='b', facecolor='none'))
plt.axis('equal')
plt.xlabel('Minor radius $r_b$'); plt.legend()

ax = fig.add_subplot(133)
ax.quiver(Y,-X, Szy,-Szx, mag, scale=5,label='Shear Flow',cmap=cm.jet)
ax.add_patch(patches.Ellipse((0,0), width=2*rb_, height=2*ra_, \
                              edgecolor='b', facecolor='none'))
plt.axis('equal')
plt.xlabel('Minor radius $r_b$'); plt.legend()
plt.savefig('imagesSM/shearOnEllipseP.png', dpi=500) #save plot to a file
#plt.show()                          # uncomment this to display the plots
```

Figure 5.4 plots the distribution of the shear stresses on a cross-section of a 3D cantilever beam with an elliptic cross-section, where the major radius is in the vertical direction. The ratio of $r_a/r_b = 2$ (tall beam). The shear stress σ_{zy} (five times magnified) is shown on the left, σ_{zx} is in the middle, and the shear stress vector $[\sigma_{zx}; \sigma_{zy}]^\top$, representing the shear flow direction, is shown on the right. It is seen that σ_{zy} is much smaller than σ_{zx}. The shear flow is dominated by the component σ_{zx}, which is parabolically distributed along both the height and width. This distribution is largely the same as the shear force predicted by the thin beam theory for this tall beam case.

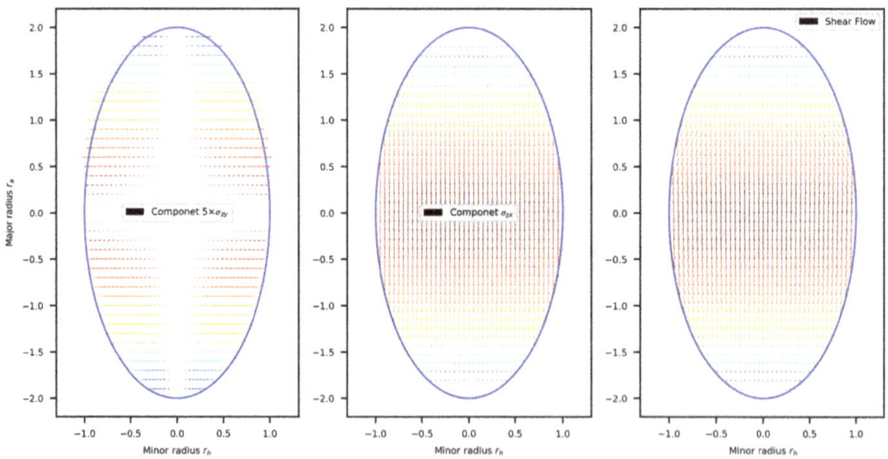

Figure 5.4. Distribution of the shear stresses on a cross-section of a 3D cantilever beam of elliptic cross-section. The major radius is the height (tall beam), and the ratio of $r_a/r_b = 2$.

Let us plot the same for a wide elliptic beam using the following snippet:

```
from matplotlib import cm
import matplotlib.patches as patches

plt.rcParams['figure.dpi'] = 500
ra_, rb_ = 1., 2.
h, b, P_, ν_ = 1., 2., 1., 0.3                            # arbitrarily set values

xl, xr, dx = -h, h, 0.05
yl, yr, dy = -b, b, 0.1

X = np.arange(xl, xr+dx, dx)
Y = np.arange(yl, yr+dy, dy)
X, Y = np.meshgrid(X, Y)
xn, yn = X.shape
Szx = X*0; Szy = Y*0

dicE = {ra:ra_,rb:rb_,P:P_, ν:ν_}
np_σxz = lambdify((x,y), σxz.subs(dicE),'numpy')
np_σyz = lambdify((x,y), σyz.subs(dicE),'numpy')
for xk in range(xn):
    for yk in range(yn):
        Szx[xk, yk] = np_σxz(X[xk,yk],Y[xk,yk])
        Szy[xk, yk] = np_σyz(X[xk,yk],Y[xk,yk])

#Elliptic mask to-be-used in plot, print nothing outside the ellipse:
mask = (X/ra_)**2 + (Y/rb_)**2 >= 1
Szx[mask] = np.nan; Szy[mask] = np.nan

plt.ioff()
#plt.ion()                                 # uncomment this to display the plots
mag = np.sqrt(Szx**2 + Szy**2)
fig = plt.figure(figsize=(6, 9))
ax = fig.add_subplot(311)
ax.quiver(Y,-X, Szy,-Szy*0, mag,scale=10,label='Componet $2×σ_{zy}$',\
                                                              cmap=cm.jet)
ax.add_patch(patches.Ellipse((0,0), width=2*rb_, height=2*ra_, \
                                edgecolor='b', facecolor='none'))
plt.axis('equal'); plt.legend()
plt.ylabel('Minor radius $r_b$')

ax = fig.add_subplot(312)
ax.quiver(Y,-X, Szx*0,-Szx, mag, scale=20,label='Componet $σ_{zx}$',\
                                                           cmap=cm.jet)
ax.add_patch(patches.Ellipse((0,0), width=2*rb_, height=2*ra_, \
```

```
                                edgecolor='b', facecolor='none'))
plt.axis('equal'); plt.legend()
plt.ylabel('Minor radius $r_b$')

ax = fig.add_subplot(313)
ax.quiver(Y,-X, Szy,-Szx, mag, scale=20,label='Shear Flow',cmap=cm.jet)
ax.add_patch(patches.Ellipse((0,0), width=2*rb_, height=2*ra_, \
                                edgecolor='b', facecolor='none'))
plt.axis('equal')
plt.xlabel('Major radius $r_a$')
plt.ylabel('Minor radius $r_b$'); plt.legend()
plt.savefig('imagesSM/shearOnEllipsePW.png', dpi=500) #save plot to a file
#plt.show()                          # uncomment this to display the plots
```

Figure 5.5 plots the distribution of the shear stresses on a cross-section of a 3D cantilever beam with an elliptic cross-section, where the minor radius is the height (wide beam). The ratio of $r_a/r_b = 2$. The shear stress σ_{zy} is shown at the top, σ_{zx} is in the middle, and the shear stress vector $[\sigma_{zx}\ \sigma_{zy}]^\top$ is shown at the bottom. In this case, it is observed that σ_{zy} is comparable to σ_{zx}. The shear flow is controlled by both components. The distribution of the stresses is entirely different from that predicted by the thin beam theory for this wide beam case.

It is found that: 1. The shear flow has a circulative component. 2. The largest shear stress may not be in the middle section but at the left and right edges for wide elliptic beams. This cannot be predicted by the thin beam theory. 3. There is significant shear stress (σ_{zy}) near the surface at approximately the 2-, 4-, 8-, and 10-o'clock directions. On the upper section, σ_{yz} points away from the x-axis, and on the lower section, σ_{yz} points toward the x-axis. Stress σ_{xz} always points downward. None of these characteristics can be predicted by the thin beam theory.

Based on our discussions, the results from the thin beam theory are only applicable to tall beams as an approximation.

5.4 Beams with rectangular cross-section

Consider the same problem shown in Fig. 5.6, which was studied in Section 4.5. The 3D beam is subjected to a downward force at the right end of the beam and is free from the body force. We would like to obtain the solutions for stresses in the beam in the Cartesian coordinate system by developing a displacement method for this 3D problem.

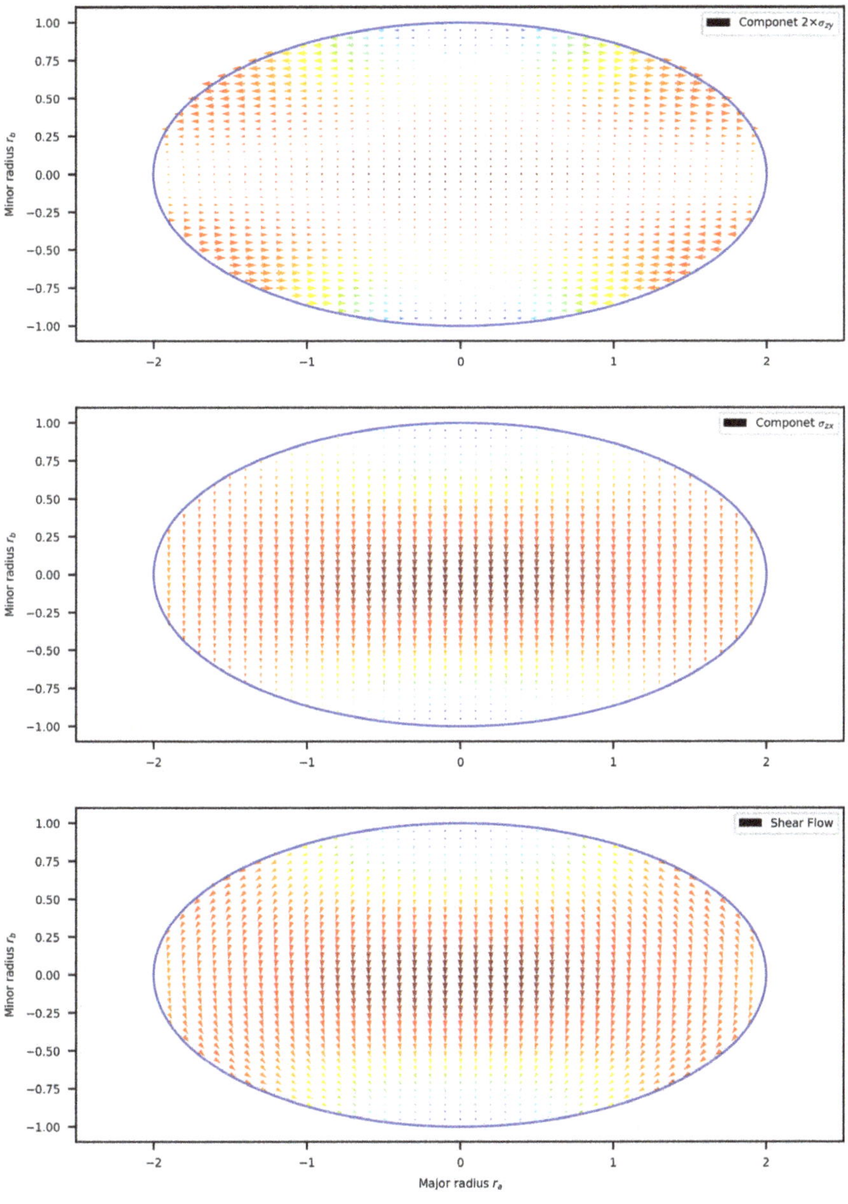

Figure 5.5. Distribution of the shear stresses on a cross-section of a 3D cantilever beam. The minor radius is the height (wide beam), and the ratio of $r_a/r_b = 2$.

5.4.1 *Constructing general displacements*

A general expression for displacements should consist of polynomial basis functions and other special basis functions depending on the type of problem.

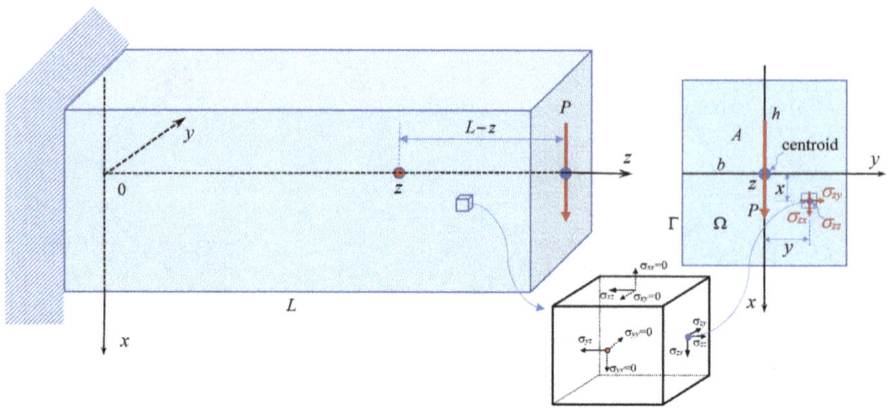

Figure 5.6. Schematic drawing of a 3D beam with a uniform rectangular cross-section subjected to a force at the right end. Three nonzero unknown stress components on the z-plane are shown and need to be determined.

5.4.1.1 *Use of polynomial basis functions*

The following code defines monomial bases, generates symbolic constants, and constructs these three displacement functions. In this process, we make use of the known symmetry properties for this problem:

1. The displacement function $u(x, y, z)$ is symmetric in x and antisymmetric in y.
2. The displacement function $v(x, y, z)$ is antisymmetric in both x and y.
3. The displacement function $w(x, y, z)$ is antisymmetric in x and symmetric in y.

These conditions are the same as those used for the elliptic beam problem studied in the previous section. Let's use them to filter out unnecessary basis functions. The snippet is as follows:

```
# Define necessary symbolic variables:
x, y, z = symbols("x, y, z ", real=True)
h, b, L, P, E, G, ν, Iyy = symbols("h,b,L,P,E,G,ν,I_{yy}",positive=True)
dic_I = {h**3:3*Iyy/(4*b)}                          # Iyy = (2b)(2h)^3/12
σxx,σyy,σzz,σyz,σxz,σxy = sp.symbols("σ_xx,σ_yy,σ_zz,σ_yz,σ_xz,σ_xy")
εxx,εyy,εzz,εyz,εxz,εxy   = sp.symbols("ε_xx,ε_yy,ε_zz,ε_yz,ε_xz,ε_xy")
n = symbols('n', integer=True, positive=True)
```

```
order = 3                # Set to use upto complete 3rd order monomials:
# Generate the complete set of the basis functions (monomials):
bfs_list = [x**i*y**j*z**k for i in range(order + 1)
                           for j in range(order + 1)
                           for k in range(order + 1) if i+j+k <= order]
bfs = sp.Matrix(bfs_list)          # form sympy vector with these bases
display(Math(f"\\text{{Complete 3rd order monomials in 3D, bfs_all:}}"))
display(Math(f"\small {latex(bfs_list)}"))

# Select basis functions (bfs) for u(x,y,z) using symmteric conditions:
def is_even_in_x_and_y(poly):
    return poly.subs(x, -x) == poly and poly.subs(y, -y) == poly
u_bfs = [poly for poly in bfs if is_even_in_x_and_y(poly)]
display(Math(f" \\text{{u_bfs}}={latex(u_bfs)}"))         # Bases for u

# Select bfs for v(x,y,z) using symmteric conditions:
def is_odd_in_x_and_y(poly):
    return -poly.subs(x, -x) == poly and -poly.subs(y, -y) == poly
v_bfs = [poly for poly in bfs if is_odd_in_x_and_y(poly)]
display(Math(f" \\text{{v_bfs}}={latex(v_bfs)}"))         # Bases for v

# Select bfs for w(x,y,z) using symmteric conditions:
def is_odd_in_x_and_even_y(poly):
    return -poly.subs(x, -x) == poly and poly.subs(y, -y) == poly
w_bfs = [poly for poly in bfs if is_odd_in_x_and_even_y(poly)]
display(Math(f" \\text{{w_bfs}}={latex(w_bfs)}"))         # Bases for w
```

Complete third-order monomials in 3D, bfs_all:

$$\left[1,\ z,\ z^2,\ z^3,\ y,\ yz,\ yz^2,\ y^2,\ y^2z,\ y^3,\ x,\ xz,\ xz^2,\ xy,\ xyz,\ xy^2,\right.$$
$$\left. x^2,\ x^2z,\ x^2y,\ x^3\right]$$

$$\text{u\_bfs} = \left[1,\ z,\ z^2,\ z^3,\ y^2,\ y^2z,\ x^2,\ x^2z\right]$$

$$\text{v\_bfs} = [xy,\ xyz]$$

$$\text{w\_bfs} = \left[x,\ xz,\ xz^2,\ xy^2,\ x^3\right]$$

Based on the number of basis functions to be used, we generate the necessary number of symbolic unknown constants. This must be done separately for each of the displacement components. The general expressions for these displacement components are the dot products of the basis functions and their corresponding constants.

```
# unknown constants for displacements:
ai = sp.symbols(f'a:{len(u_bfs)}')                          # unknown constants
display(Math(f" \\text{{Unknown coefficients for $u$}}={latex(ai)}"))
ai = Matrix([ai[i] for i in range(len(u_bfs))]) # form coeffs. in vector

bi = sp.symbols(f'b:{len(v_bfs)}')                          # unknown constants
display(Math(f" \\text{{Unknown coefficients for $v$}}={latex(bi)}"))
bi = Matrix([bi[i] for i in range(len(v_bfs))]) # form coeffs. in vector

ci = sp.symbols(f'c:{len(w_bfs)}')                          # unknown constants
display(Math(f" \\text{{Unknown coefficients for $w$}}={latex(ci)}"))
ci = Matrix([ci[i] for i in range(len(w_bfs))]) # form coeffs. in vector

cs = ai.col_join(bi).col_join(ci)   # form a single vector of constants

ug = (ai.T@Matrix(u_bfs))[0]            # Form the general displacements
vg = (bi.T@Matrix(v_bfs))[0]
```

Unknown coefficients for $u = (a_0,\ a_1,\ a_2,\ a_3,\ a_4,\ a_5,\ a_6,\ a_7)$

Unknown coefficients for $v = (b_0,\ b_1)$

Unknown coefficients for $w = (c_0,\ c_1,\ c_2,\ c_3,\ c_4)$

5.4.1.2 *Addition of special basis functions*

For our problem of a 3D cantilever beam with a rectangular cross-section and an end load, the displacement function $w(x, y, z)$ should include an additional harmonic term with an extra unknown constant. This is based on our experience from earlier studies on the same problem using the stress method. The additional term is in the following infinite summation form:

$$\Sigma w_n = \sum_{n=1}^{\infty} \frac{(-1)^n \cos\left(\frac{\pi n y}{b}\right) \sinh\left(\frac{\pi n x}{b}\right)}{n^3 \cosh\left(\frac{\pi h n}{b}\right)} \tag{5.29}$$

The snippet to define this term is as follows:

```
d0 = sp.symbols('d0')   # unknown constant for the biharmonic base Σψn
wn = sp.Function('w_n')(x, y)         # symbol of the base function w_n
Σwn = sp.Function('Σw_n')(x, y)     # symbol of Sum of the base function

# define the nth harmonic basis function (denoted with _):
wn_ = (-1)**n*sp.cos(n*sp.pi/b*y)*sp.sinh(n*sp.pi/b*x)\
                                        /(n**3*sp.cosh(n*sp.pi/b*h))
Sum = sp.summation
Σwn_ = Sum(wn_, (n, 1, oo))       # sum of these harmonic basis functions
display(Math(f"\\text{{Summation of terms $Σw_{{n}}$}}={latex(Σwn_)}"))
wg = (ci.T@Matrix(w_bfs))[0] + d0*Σwn_        # add to the monomial bases
display(Math(f"\\text{{Assumed displacements:}}"))
display(Math(f"\;\;\ \\text{{ $u(x,y,z)$}}={latex(ug)}"))
display(Math(f"\;\;\ \\text{{ $v(x,y,z)$}}={latex(vg)}"))
display(Math(f"\;\;\ \\text{{ $w(x,y,z)$}}=\small {latex(wg)}"))
```

$$\text{Summation of terms } \Sigma w_n = \sum_{n=1}^{\infty} \frac{(-1)^n \cos\left(\frac{\pi n y}{b}\right) \sinh\left(\frac{\pi n x}{b}\right)}{n^3 \cosh\left(\frac{\pi h n}{b}\right)}$$

Assumed displacements:

$$u(x, y, z) = a_0 + a_1 z + a_2 z^2 + a_3 z^3 + a_4 y^2 + a_5 y^2 z + a_6 x^2 + a_7 x^2 z$$

$$v(x, y, z) = b_0 xy + b_1 xyz$$

$$w(x, y, z) = c_0 x + c_1 xz + c_2 xz^2 + c_3 xy^2 + c_4 x^3 + d_0 \sum_{n=1}^{\infty} \frac{(-1)^n \cos\left(\frac{\pi n y}{b}\right) \sinh\left(\frac{\pi n x}{b}\right)}{n^3 \cosh\left(\frac{\pi h n}{b}\right)}$$

5.4.2 *Enforcing the displacement boundary conditions*

We assume that the displacements at $(0, 0, 0)$ are fixed. The equations are the same as those given in Eq. (5.23). This provides six DBCs for these displacements. The values for the component w at $(0, 0, 0)$ are all zero:

```
# Enforce DBCs, for w:
dic0 = {x:0, y:0, z:0}                                      # for DBCs
wg.subs(dic0), wg.diff(y).subs(dic0)     # 2 BCs automatically satisfied
```

$$\left(d_0 \sum_{n=1}^{\infty} 0,\ d_0 \sum_{n=1}^{\infty} 0\right)$$

Therefore, the zero condition for w is already satisfied (due to the symmetry of the basis). The conditions for u and v can be found using the following snippet:

```
# Enforce DBCs, for u and v:
sln_DBC = sp.solve([ug.subs(dic0), vg.subs(dic0),
            ug.diff(z).subs(dic0),   vg.diff(x).subs(dic0)], cs)
display(Math(f" \\text{{Conditions for fixed displacements}}=\
                                                {latex(sln_DBC)}"))
sln_cs = {**sln_DBC}   # update the solutions for the unknown constants
```

Conditions for fixed displacements $= \{a_0 : 0,\ a_1 : 0\}$

We have found two of the constants using these DBCs.

5.4.3 *Enforce stress boundary conditions*

The general expressions for the six stress components can be computed using the displacements and Eq. (4.21):

```
# Compute stresses using displacement-function:
e = ug.diff(x) + vg.diff(y) + wg.diff(z)   # volume strain (disp. div.)
σxxg = (2*G*(ν/(1-2*ν)*e+ug.diff(x))).expand().collect(G*x).collect(z)
σyyg = (2*G*(ν/(1-2*ν)*e+vg.diff(y))).expand().collect(G*x).collect(z)
σzzg = (2*G*(ν/(1-2*ν)*e+wg.diff(z))).simplify()
σyzg = (G*(wg.diff(y)+vg.diff(z)))
σxzg = (G*(wg.diff(x)+ug.diff(z))).simplify()
σxyg = (G*(vg.diff(x)+ug.diff(y))).expand().collect(G*y).collect(z)

display(Math(f" \\text{{General expressions for stresses}}:"))
display(Math(f" \\text{{$σ_{{xx}}$}}=\\scriptsize {latex(σxxg)}"))
display(Math(f" \\text{{$σ_{{yy}}$}}=\\scriptsize {latex(σyyg)}"))
display(Math(f" \\text{{$σ_{{xy}}$}}=\\scriptsize {latex(σxyg)}"))
display(Math(f" \\text{{$σ_{{zz}}$}}=\\scriptsize {latex(σzzg)}"))
display(Math(f" \\text{{$σ_{{yz}}$}}=\\scriptsize {latex(σyzg)}"))
display(Math(f" \\text{{$σ_{{xz}}$}}=\\scriptsize {latex(σxzg)}"))
```

General expressions for stresses:

$$\sigma_{xx} = Gx\left(\frac{4a_6\nu}{1-2\nu} + 4a_6 + \frac{2b_0\nu}{1-2\nu} + \frac{2c_1\nu}{1-2\nu}\right.$$

$$\left.+\, z\left(\frac{4a_7\nu}{1-2\nu} + 4a_7 + \frac{2b_1\nu}{1-2\nu} + \frac{4c_2\nu}{1-2\nu}\right)\right)$$

$$\sigma_{yy} = Gx\left(\frac{4a_6\nu}{1-2\nu} + \frac{2b_0\nu}{1-2\nu} + 2b_0 + \frac{2c_1\nu}{1-2\nu}\right.$$

$$\left. + z\left(\frac{4a_7\nu}{1-2\nu} + \frac{2b_1\nu}{1-2\nu} + 2b_1 + \frac{4c_2\nu}{1-2\nu}\right)\right)$$

$$\sigma_{xy} = Gy\left(2a_4 + b_0 + z\left(2a_5 + b_1\right)\right)$$

$$\sigma_{zz} = \frac{2Gx\left(-\nu\left(2a_6 + 2a_7z + b_0 + b_1z + c_1 + 2c_2z\right) + \left(c_1 + 2c_2z\right)\left(2\nu - 1\right)\right)}{2\nu - 1}$$

$$\sigma_{yz} = G\left(b_1xy + 2c_3xy + d_0\sum_{n=1}^{\infty} -\frac{(-1)^n \pi \sin\left(\frac{\pi ny}{b}\right)\sinh\left(\frac{\pi nx}{b}\right)}{bn^2\cosh\left(\frac{\pi hn}{b}\right)}\right)$$

$$\sigma_{xz} = G\left(a_1 + 2a_2z + 3a_3z^2 + a_5y^2 + a_7x^2 + c_0 + c_1z + c_2z^2\right.$$

$$\left. + c_3y^2 + 3c_4x^2 + \frac{\pi d_0 \sum_{n=1}^{\infty} \frac{(-1)^n \cos\left(\frac{\pi ny}{b}\right)\cosh\left(\frac{\pi nx}{b}\right)}{n^2\cosh\left(\frac{\pi hn}{b}\right)}}{b}\right)$$

We can now apply SBCs for stresses in the x–y plane on the circumferential surfaces of the beam:

$$\begin{aligned} &\text{On the top and bottom,} \quad x = \pm h : \sigma_{xx} = 0 \\ &\text{On the left and right,} \quad\quad y = \pm b : \sigma_{yy} = \sigma_{xy} = 0 \end{aligned} \tag{5.30}$$

These conditions must be satisfied for the x–y plane at any z. Thus, all the coefficients for z^1 and z^0 must vanish. This gives a new set of six conditions. We then use these conditions to solve for the constants:

```
SBCs = []
SBCs.append(σxxg.subs(x,h).coeff(G*h).coeff(z,1))               # Six SBCs
SBCs.append(σxxg.subs(x,h).coeff(G*h).coeff(z,0))
SBCs.append(σyyg.subs(y,b).coeff(G*x).coeff(z,1))
SBCs.append(σyyg.subs(y,b).coeff(G*x).coeff(z,0))
SBCs.append(σxyg.subs(y,b).coeff(G*b).coeff(z,1))
SBCs.append(σxyg.subs(y,b).coeff(G*b).coeff(z,0))
#display(expr1,expr2,expr3,expr4,expr5,expr6) #lengthy, one may display
```

```
sln_SBC = sp.solve(SBCs, cs)
display(sln_SBC)
sln_cs = {**sln_DBC, **sln_SBC}              # update the unknown constants

# Using the current constants to update the stresses:
σxx = σxxg.subs(sln_cs).simplify(); σyy = σyyg.subs(sln_cs).simplify ()
σzz = σzzg.subs(sln_cs).simplify(); σxy = σxyg.subs(sln_cs).simplify ()
σyz = σyzg.subs(sln_cs).simplify(); σxz = σxzg.subs(sln_cs).simplify ()

display(Math(f" σ_{{xx}} ={latex(σxx)}; σ_{{yy}} ={latex(σyy)}; \
                                          σ_{{xy}} ={latex(σxy)}"))
display(Math(f" \\text{{General stress $σ_{{zz}}$}}={latex(σzz)}"))
display(Math(f" \\text{{General stress $σ_{{yz}}$}}={latex(σyz)}"))
#display(Math(f" \\text{{General stress $σ_{{xz}}$}}={latex(σxz)}"))

# Using the current constants to update the displacements:
u = ug.subs(sln_cs).simplify(); v = vg.subs(sln_cs).simplify()
w = wg.subs(sln_cs).simplify()
display(Math(f" \\text{{General displacement $u$}}={latex(u)}"))
```

$$\left\{a_4 : \frac{c_1\nu}{2},\ a_5 : c_2\nu,\ a_6 : -\frac{c_1\nu}{2},\ a_7 : -c_2\nu,\ b_0 : -c_1\nu,\ b_1 : -2c_2\nu\right\}$$

$$\sigma_{xx} = 0; \sigma_{yy} = 0; \sigma_{xy} = 0$$

General stress $\sigma_{zz} = 2Gx\left(c_1\nu + c_1 + 2c_2z\nu + 2c_2z\right)$

General stress

$$\sigma_{yz} = -G\left(2c_2xy\nu - 2c_3xy + \frac{\pi d_0 \sum_{n=1}^{\infty} \frac{(-1)^n \sin\left(\frac{\pi ny}{b}\right)\sinh\left(\frac{\pi nx}{b}\right)}{n^2\cosh\left(\frac{\pi hn}{b}\right)}}{b}\right)$$

General displacement $u = a_2z^2 + a_3z^3 - \frac{c_1x^2\nu}{2} + \frac{c_1y^2\nu}{2} - c_2x^2z\nu + c_2y^2z\nu$

We have all the stresses and displacements updated using the constants found so far. We did not display all of them because some of these are too lengthy.

We found that the stress components $\sigma_{xx} = \sigma_{yy} = \sigma_{xy} = 0$ in the entire beam, although we only applied zero conditions on the circumferential surfaces. Thus, our focus is now only on the three stress components on the z-plane: σ_{xz}, σ_{yz}, and σ_{zz}.

5.4.4 *Enforce integral equilibrium conditions*

Next, we apply the integral equilibrium conditions (FBCs) given in Eq. (5.28) for stresses on a z-plane. Note that because σ_{zz} is odd in x, the first two equations are automatically satisfied. Also, since σ_{yz} is odd in x and odd in y, the fifth and sixth equations are also automatically satisfied. Therefore, we need to enforce the third and fourth equations, which needs to be done step by step.

Let us enforce the third equation now:

```
sln_cs = {**sln_DBC, **sln_SBC}
Int = sp.integrate
eqn3rd = Int(Int(σzzg.subs(sln_cs)*x, (x,-h,h)),(y,-b,b)) + P*(L-z)
eqn3_c1 = eqn3rd.expand().coeff(z,0)
eqn3_c2 = eqn3rd.expand().coeff(z,1)

sln_FBC = sp.solve([eqn3_c1, eqn3_c2], cs)
display(sln_FBC)
sln_cs = {**sln_DBC, **sln_SBC, **sln_FBC}

σzz = σzzg.subs(sln_cs).simplify()                    # update stresses
σyz = σyzg.subs(sln_cs).simplify()
σxz = σxzg.subs(sln_cs).expand().simplify()

display(Math(f" \\text{{Final $σ_{{zz}}$}}=\small {latex(σzz)}"))
#display(Math(f" \\text{{Current $σ_{{xz}}$}}={latex(σxz)}")) # lengthy
display(Math(f" \\text{{Current $σ_{{yz}}$}}=\small {latex(σyz)}"))

u = ug.subs(sln_cs).simplify()                # update the displacements
v = vg.subs(sln_cs).simplify()
w = wg.subs(sln_cs).simplify()
#display(u, v, w)                                             # lengthy
```

$$\left\{c_1 : -\frac{3LP}{8Gbh^3\nu + 8Gbh^3},\ c_2 : \frac{3P}{16Gbh^3\nu + 16Gbh^3}\right\}$$

$$\text{Final } \sigma_{zz} = \frac{3Px\left(-L+z\right)}{4bh^3}$$

$$\text{Current } \sigma_{yz} = \frac{2Gc_3xy\nu}{\nu+1} + \frac{2Gc_3xy}{\nu+1} - \frac{\pi Gd_0\sum_{n=1}^{\infty}\frac{(-1)^n\sin\left(\frac{\pi ny}{b}\right)\sinh\left(\frac{\pi nx}{b}\right)}{n^2\cosh\left(\frac{\pi hn}{b}\right)}}{b} - \frac{3Pxy\nu}{8bh^3\left(\nu+1\right)}$$

We found the final solution for the normal stress σ_{zz}. However, we still have unknown constants in σ_{yz} and σ_{xz}.

5.4.5 *Enforce boundary conditions for shear stresses*

Next, we need to enforce the following SBCs for σ_{xz} and σ_{yz} on the circumferential surfaces:

$$\begin{aligned} &\text{On the top and bottom,} \quad x = \pm h : \sigma_{xz} = 0 \\ &\text{On the left and right,} \quad y = \pm b : \sigma_{yz} = 0 \end{aligned} \tag{5.31}$$

The current expression for σ_{xz} on $x = h$ is very lengthy. The following snippet displays two of the terms in it:

```
display(σxz.subs(x,h).args[4].args[6]+σxz.subs(x,h).args[4].args[12])
```

$$16\pi G d_0 h^3 \nu \sum_{n=1}^{\infty} \frac{(-1)^n \cos\left(\frac{\pi n y}{b}\right)}{n^2} + 16\pi G d_0 h^3 \sum_{n=1}^{\infty} \frac{(-1)^n \cos\left(\frac{\pi n y}{b}\right)}{n^2}$$

These two infinite summation terms prevent SymPy from performing further derivation for us. We thus need to simplify them. From Eq. (5.32) obtained in the previous chapter, we have the following identity:

$$\sum_{n=1}^{\infty} \frac{(-1)^n \cos \frac{n\pi y}{b}}{n^2} = \frac{\pi^2}{12}\left(3\frac{y^2}{b^2} - 1\right) \tag{5.32}$$

Therefore, we can substitute the infinite summation using this identity.

These conditions must be satisfied for any z and any y. Thus, all the coefficients for monomials z^2, z^1, z^0, and y^2 must vanish. This gives another set of four expressions, which can be used to determine some of the remaining unknown constants. The snippet is as follows:

```
σxz_xh_= pi**2/12*(3*(y/b)**2-1)                    # the polynomial term
Sum_cos = Sum((-1)**n*cos(pi*n*y/b)/n**2, (n, 1, oo))   # the summation

# Substitution and collect the coefficients:
sln_cs = {**sln_DBC, **sln_SBC, **sln_FBC}
σxz = σxzg.subs(sln_cs).expand().simplify()
σxz_xh = σxz.subs({x:h}).subs(Sum_cos, σxz_xh_).expand().\
                                collect(z**2).collect(z).collect(y**2)
# 4 expressions:
SBC_xh = []
SBC_xh.append(σxz_xh.coeff(z**2))
SBC_xh.append(σxz_xh.coeff(z))
SBC_xh.append(σxz_xh.coeff(y**2))
SBC_xh.append(σxz_xh.coeff(y,0).coeff(z,0))
```

```
sln_SBC_xh = sp.solve(SBC_xh, cs)
sln_cs = {**sln_DBC, **sln_SBC, **sln_FBC, **sln_SBC_xh}
#display(sln_SBC_xh)

σyz = σyzg.subs(sln_cs).simplify()
σxz = σxzg.subs(sln_cs).expand().simplify()

display(Math(f"\\text{{General stress $σ_{{yz}}$}}=\small {latex(σyz)}"))
display(Math(f"\\text{{Shear stress $σ_{{xz}}|_{{x=h}}$}}=\
                    \small {latex(σxz.subs(x,h).simplify())}"))
u = ug.subs(sln_cs).simplify()
v = vg.subs(sln_cs).simplify()
w = wg.subs(sln_cs).simplify()
#u v, w                                              # Lengthy
```

General stress

$$\sigma_{yz} = -\frac{\pi G d_0 \sum_{n=1}^{\infty} \frac{(-1)^n \sin\left(\frac{\pi n y}{b}\right) \sinh\left(\frac{\pi n x}{b}\right)}{n^2 \cosh\left(\frac{\pi h n}{b}\right)} + \frac{\pi^3 G d_0 x y \nu}{2b^2(\nu+1)} + \frac{\pi^3 G d_0 x y}{2b^2(\nu+1)} + \frac{3Pxy\nu}{4h^3(\nu+1)}}{b}$$

Shear stress $\sigma_{xz}|_{x=h} = \dfrac{\pi G d_0 \left(\sum_{n=1}^{\infty} \frac{(-1)^n \cos\left(\frac{\pi n y}{b}\right)}{n^2} + \frac{\pi^2}{12} - \frac{\pi^2 y^2}{4b^2}\right)}{b}$

The expression for σ_{xz} is very lengthy. We display it on the top surface of the 3D beam. It is clear from Eq. (5.32) that σ_{xz} is zero, implying that the first equation in Eq. (5.31) is satisfied.

We are now ready to enforce the second equation in Eq. (5.31):

```
sln_cs = {**sln_DBC, **sln_SBC, **sln_FBC, **sln_SBC_xh}
σyz_yb = σyzg.subs(sln_cs).subs({y:b}).expand().collect(x).coeff(x)

sln_SBC_yb = sp.solve([σyz_yb], d0)
sln_cs = {**sln_DBC, **sln_SBC, **sln_FBC, **sln_SBC_xh, **sln_SBC_yb}
display(sln_SBC_yb)

σyz = σyzg.subs(sln_cs).simplify()
σxz = σxzg.subs(sln_cs).expand().simplify().expand()

display(Math(f"\\text{{General stress $σ_{{yz}}$}}=\small {latex(σyz)}"))
#display(Math(f" \\text{{General stress $σ_{{xz}}$}}={latex(σxz)}"))
# lengthy, display only one term:
display(Math(f" \\text{{Two terms in $σ_{{xz}}$}}=\
              \small {latex(σxz.args[0])}+...{latex(σxz.args[6])}+..."))
u = ug.subs(sln_cs).simplify()
v = vg.subs(sln_cs).simplify()
w = wg.subs(sln_cs).simplify()
#u, v, w                                              # Lengthy
```

$$\left\{d_0 : -\frac{3Pb^2\nu}{2\pi^3 G h^3 \nu + 2\pi^3 G h^3}\right\}$$

$$\text{General stress } \sigma_{yz} = \frac{3Pb\nu \sum_{n=1}^{\infty} \frac{(-1)^n \sin\left(\frac{\pi n y}{b}\right) \sinh\left(\frac{\pi n x}{b}\right)}{n^2 \cosh\left(\frac{\pi h n}{b}\right)}}{2\pi^2 h^3 (\nu + 1)}$$

$$\text{Two terms in } \sigma_{xz} = -\frac{24Pb^2\nu \sum_{n=1}^{\infty} \frac{(-1)^n \cos\left(\frac{\pi n y}{b}\right) \cosh\left(\frac{\pi n x}{b}\right)}{n^2 \cosh\left(\frac{\pi h n}{b}\right)}}{16\pi^2 b h^3 \nu + 16\pi^2 b h^3} + \cdots$$

$$-\frac{48\pi^2 G b c_4 h^5 \nu}{16\pi^2 b h^3 \nu + 16\pi^2 b h^3} + \cdots$$

5.4.6 *Force equilibrium on a cross-section*

We have found the final solution for σ_{yz}. The expression for σ_{xz} still contains one unknown constant c_4. We are ready to apply the fourth integral equilibrium condition (FBC) given in Eq. (5.28), which involves the applied end force P on a cross-section of the beam:

$$\sum F_x = 0 : \iint_A \sigma_{zx}\, dx\, dy = P \tag{5.33}$$

Since the expression for σ_{xz} contains a harmonic term with an infinite summation, the integration must be performed manually. The integration snippet is as follows:

```
#expr_harmonic_= σxz.expand().args[0]          # to get the harmonic term
harmonic = (-1)**n*cos(pi*n*y/b)*cosh(pi*n*x/b)/(n**2*cosh(pi*h*n/b))

harmonic_basis_I = pi*G*d0*Sum(Int(Int(harmonic,(x,-h,h)),(y,-b,b)),\
                                                            (n, 1, oo))
display(Math(f" \\text{{Integral of the harmonic term }}=\
                                          {latex(harmonic_basis_I)}"))
```

Integral of the harmonic term $= 0$

The integral of this infinite summation evaluates to zero. Therefore, the integral in Eq. (5.33) can be evaluated without this term.

```
expr_poly_ = (σxz - σxz.expand().args[0]).simplify() # exclude the sum
expr_poly = sp.integrate(sp.integrate(expr_poly_,(x,-h,h)),(y,-b,b)) -P

sln_FBC_P = sp.solve([expr_poly], cs)
sln_cs = {**sln_DBC, **sln_SBC, **sln_FBC, **sln_SBC_xh, **sln_SBC_yb,\
                                                          **sln_FBC_P}
#display(sln_FBC_P); display(sln_cs)

σyz_final = σyzg.subs(sln_cs).simplify()                       # update
σxz_final = σxzg.subs(sln_cs).expand().simplify().expand().\
            collect(8*pi**2*b*h**3).collect(3*h**2).collect(-3*x**2).\
                                     collect(pi**2*P).collect(ν+1)

display(Math(f" \\text{{Final stress $σ_{{zz}}$}}={latex(σzz)}"))
display(Math(f" \\text{{Final stress $σ_{{yz}}$}}={latex(σyz_final)}"))
display(Math(f" \\text{{Final stress $σ_{{xz}}$}}=\
             \scriptsize {latex(σxz_final.collect(ν+1).collect(3))}"))
u = ug.subs(sln_cs).simplify()
v = vg.subs(sln_cs).simplify()
w = wg.subs(sln_cs).simplify()
display(Math(f"\\text{{Final displacements, $u$}}=\small {latex(u)}"))
display(Math(f"\\text{{ $v$}}=\small {latex(v)}"))
display(Math(f"\\text{{ $w$}}=\scriptsize {latex(w.collect(sp.pi))}"))
```

$$\text{Final stress } \sigma_{zz} = \frac{3Px\left(-L+z\right)}{4bh^3}$$

$$\text{Final stress } \sigma_{yz} = \frac{3Pb\nu\sum_{n=1}^{\infty}\frac{(-1)^n\sin\left(\frac{\pi ny}{b}\right)\sinh\left(\frac{\pi nx}{b}\right)}{n^2\cosh\left(\frac{\pi hn}{b}\right)}}{2\pi^2h^3\left(\nu+1\right)}$$

$$\text{Final stress } \sigma_{xz} = \frac{-12Pb^2\nu\sum_{n=1}^{\infty}\frac{(-1)^n\cos\left(\frac{\pi ny}{b}\right)\cosh\left(\frac{\pi nx}{b}\right)}{n^2\cosh\left(\frac{\pi hn}{b}\right)} + \pi^2P\left(-b^2\nu+3y^2\nu+3\left(h^2-x^2\right)\left(\nu+1\right)\right)}{8\pi^2bh^3\left(\nu+1\right)}$$

$$\text{Final displacements, } u = \frac{P\left(3Lx^2\nu-3Ly^2\nu+3Lz^2-3x^2z\nu+3y^2z\nu-z^3\right)}{16Gbh^3\left(\nu+1\right)}$$

$$v = \frac{3Pxy\nu\left(L-z\right)}{8Gbh^3\left(\nu+1\right)}$$

$$w = -\frac{P\left(24b^3\nu\sum_{n=1}^{\infty}\frac{(-1)^n\cos\left(\frac{\pi ny}{b}\right)\sinh\left(\frac{\pi nx}{b}\right)}{n^3\cosh\left(\frac{\pi hn}{b}\right)} + \pi^3\left(6Lxz+2b^2x\nu-6h^2x\nu-6h^2x+x^3\nu+2x^3-3xy^2\nu-3xz^2\right)\right)}{16\pi^3Gbh^3\left(\nu+1\right)}$$

The solutions obtained are exactly the same as those in the previous chapter — Eq. (4.103) for the stresses and Eq. (4.117) for the displacements — though their forms are slightly different.

5.4.7 *Plot for cross-section deformation*

The following code computes the deformation or warping of the rectangular cross-section of a beam, using Eq. (4.117):

```
def section_warpe(b, h, E, ν, P, x, y, z, L, n_end = 21):
    '''Compute the warpe (displacement in the axial-direction) of a
       rectangular cross-section of a 3D cantilever with end load P.
    b:        half-height of the domain
    h:        half-width  of the domain
    ν:        Poisson's ratio
    E:        Young's modulus
    P:        constant - force P at the end of the cantilever
    n_end:    terms used in the series of the solution
    x, y, z: point (x,y) at cross-section at z.'''

    I = (2*b)*(2*h)**3/12
    C = P/(E*I)
    sum1 = 0.0
    for n in range(1, n_end):                          # the series
        sum1 += (-1)**n/(n**3*np.cosh(n*np.pi*h/b))\
                *np.cos(n*np.pi*x/b)*np.sinh(n*np.pi*y/b)

    poly = -L*x*z-b**2*x*ν/3+h**2*x*(ν+1)-x**3/6*(ν+2)+x/2*(y**2*ν+z**2)
    warpe = C*(-4*P*b**3*ν/sp.pi**3*sum1 + poly)
    return warpe
```

5.4.8 *Plot of sectional warping*

```
plt.rcParams['figure.dpi'] = 500
h, b, L, z = 2., 1., 1., 1/2                      # define geometry values
P, E, ν, n_end = 100, 1., 1/3, 21                      # define parameters

xl, xr, dx = -h, h, 0.09
yl, yr, dy = -b, b, 0.09
```

```
X = np.arange(xl, xr+dx, dx)
Y = np.arange(yl, yr+dy, dy)
X, Y = np.meshgrid(X, Y)
xn, yn = X.shape
ψ = X*0
W = X*0

for xk in range(xn):
    for yk in range(yn):
        W[xk, yk] = section_warpe(b, h, E, ν, P, X[xk,yk], Y[xk,yk],\
                                            z, L, n_end = n_end)
fig = plt.figure()       # Create a 3D plot
plt.ioff()
#plt.ion()                         # uncomment this to display the plots

ax = fig.add_subplot(111, projection='3d')
surf = ax.plot_surface(X,Y,W,rstride=1,cstride=1,cmap='viridis')
ax.view_init(elev=30, azim=85)            # Change elevation and azimuth
ax.zaxis.line.set_visible(False)
ax.yaxis.pane.fill = False            # Remove background pane for z-axis
ax.grid(False);
#plt.show()                        # uncomment this to display the plots
```

Figure 5.7 shows the warping (displacement component w) of a cross-section of a 3D cantilever beam with a rectangular cross-section and a height/base ratio of $h/b = 2$. The deformed cross-section is not a plane, indicating that the assumption of plane sections remaining plane in the thin beam theory does not hold for 3D beams.

5.4.9 *Plot shear stresses on rectangular cross-section*

The following code computes the shear stresses on a rectangular cross-section beam subjected to a downward force at its right end, using the formulas obtained earlier:

```
def shearS_rectP(b, h, nu, P, x, y, n_end = 21):
    '''Compute the shear stresses on a 3D rectangular cross-section
    beam under end force P.
    b:        half-height of the beam
    h:        half-width  of the beam
    P:        constant - force P at the end of the cantilever
    n_end:    terms used in the series of the solution
    x and y: point at which the solution to be found.

    Return: sigam_zx, sigam_zy
    '''
```

```
I = (2*b)*(2*h)**3/12  # second moment of area of the cross-section
C = nu*P/((1+nu)*I)                                      # coefficient

sum_x = 0.0;    sum_y = 0.0
for n in range(1, n_end):
    sum_x += (-1)**n/(n**2*np.cosh(n*np.pi*h/b))*\
              np.cosh(n*np.pi*x/b)*np.cos(n*np.pi*y/b)
    sum_y += (-1)**n/(n**2*np.cosh(n*np.pi*h/b))*\
              np.sinh(n*np.pi*x/b)*np.sin(n*np.pi*y/b)

s_zx = P*(h**2-x**2)/(2*I)+C/6*(3*y**2-b**2-12*b**2/np.pi**2*sum_x)
s_zy = 2*C*b**2/np.pi**2*sum_y

return s_zx, s_zy
```

Figure 5.7. Warping (displacement in the axial direction) of the cross-section of a 3D cantilever beam at mid-span, with a rectangular cross-section and height/base ratio of 2.

```
# Generate a plot of Shear Stresses on Rectangular Cross-section:
from matplotlib import cm
plt.rcParams['figure.dpi'] = 500
h, b, P, ν, n_end = 2., 1., 100, 0.3,  21      # arbitrarily set values
xl, xr, dx = -h, h, 0.2
yl, yr, dy = -b, b, 0.1

X = np.arange(xl, xr+dx, dx)
Y = np.arange(yl, yr+dy, dy)
X, Y = np.meshgrid(X, Y)
xn, yn = X.shape
Szx = X*0
Szy = Y*0

for xk in range(xn):
    for yk in range(yn):
        Szx[xk, yk],Szy[xk,yk] = shearS_rectP(b,h,ν,P,X[xk,yk],Y[xk,yk])

mag = np.sqrt(Szx**2 + Szy**2)
fig = plt.figure(figsize=(8, 4))                        # Create a 3D plot
plt.ioff()
#plt.ion()                          # uncomment this to display the plots
ax = fig.add_subplot(131)         #-90 degree rotate!  U_rot=V; V_rot=-U
ax.quiver(Y,-X, Szy,-Szy*0,mag,scale=4,label='$50×σ_{zy}$',cmap=cm.jet)
plt.axis('equal')
plt.ylabel('Height $h$')
plt.xlabel('Base $b$')
plt.legend()

ax = fig.add_subplot(132)
ax.quiver(Y,-X, Szx*0,-Szx, mag, scale=200,label='$σ_{zx}$',cmap=cm.jet)
plt.axis('equal')
plt.xlabel('Base $b$')
plt.legend()

ax = fig.add_subplot(133)
ax.quiver(Y,-X, Szy,-Szx, mag, scale=200,label='Shear Flow',cmap=cm.jet)
plt.axis('equal')  #plt.colorbar()
plt.xlabel('Base $b$')
plt.legend()
plt.savefig('imagesSM/shearOnRectTP.png', dpi=500)  # save plot to a file

#plt.show()                         # uncomment this to display the plots
```

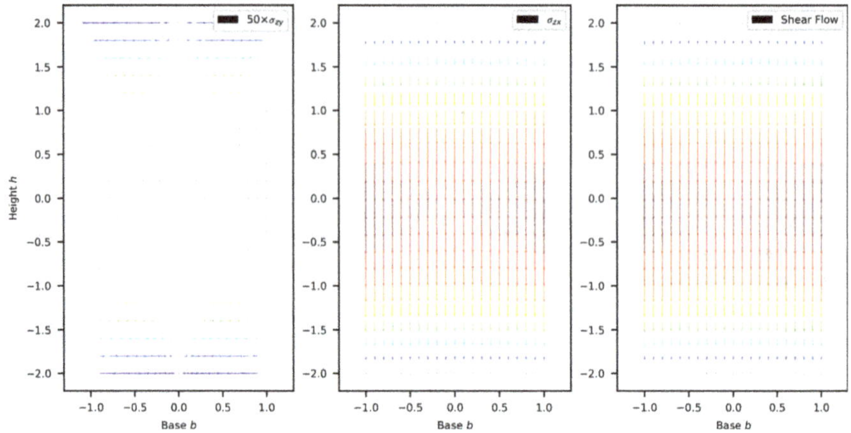

Figure 5.8. Distribution of the shear stresses on the cross-section of a 3D cantilever beam with a tall rectangular cross-section (height/base ratio of 2).

Figure 5.8 shows the distribution of shear stresses on the cross-section of a 3D cantilever beam with a rectangular cross-section and a height/base ratio of $h/b = 2$ (tall beam). The shear stress σ_{zy} (50 times magnified) is shown on the left, σ_{zx} in the middle, and the shear stress vector $[\sigma_{zx} \quad \sigma_{zy}]^\top$, representing the shear flow direction, is shown on the right. σ_{zy} is much smaller than σ_{zx}. The shear flow is dominated by the σ_{zx} component, which is parabolically distributed along the height. This distribution closely matches the shear force predicted by the thin beam theory for this tall beam.

Next, we plot the same for a wide beam using the following code snippet:

```
from matplotlib import cm
plt.rcParams['figure.dpi'] = 500
h, b, P, ν, n_end = 1., 2., 100, 0.3,  21      # arbitrarily set values
xl, xr, dx = -h, h, 0.1
yl, yr, dy = -b, b, 0.2

X = np.arange(xl, xr+dx, dx)
Y = np.arange(yl, yr+dy, dy)
X, Y = np.meshgrid(X, Y)#Y, -X )
xn, yn = X.shape
Szx = X*0
Szy = Y*0

for xk in range(xn):
    for yk in range(yn):
        Szx[xk, yk],Szy[xk,yk] = shearS_rectP(b,h,ν,P,X[xk,yk],Y[xk,yk])
```

```
mag = np.sqrt(Szx**2 + Szy**2)
fig = plt.figure(figsize=(6, 9))                         # Create a 3D plot
plt.ioff()
#plt.ion()                                # uncomment this to display the plots
ax = fig.add_subplot(311)              #-90 degree rotate!  U_rot=V; V_rot=-U
ax.quiver(Y, -X, Szy,-Szy*0, mag,scale=200,label='$σ_{zy}$',cmap=cm.jet)
plt.axis('equal')
plt.ylabel('Base $b$')
plt.legend()

ax = fig.add_subplot(312)
ax.quiver(Y,-X, Szx*0,-Szx, mag, scale=200,label='$σ_{zx}$',cmap=cm.jet)
plt.axis('equal')
plt.ylabel('Base $b$')
plt.legend()

ax = fig.add_subplot(313)
ax.quiver(Y,-X, Szy,-Szx, mag, scale=200,label='Shear Flow',cmap=cm.jet)
plt.axis('equal')  #plt.colorbar()
plt.xlabel('Height $h$')
plt.ylabel('Base $b$')
plt.legend()
plt.savefig('imagesSM/shearOnRectWP.png', dpi=500)  # save plot to a file

#plt.show()                                # uncomment this to display the plots
```

Figure 5.9 shows the distribution of shear stresses on the cross-section of a 3D cantilever beam with a rectangular cross-section and a height/base ratio of $h/b = 1/2$ (wide beam). The shear stress σ_{zy} is displayed at the top, σ_{zx} in the middle, and the shear stress vector $[\sigma_{zx} \quad \sigma_{zy}]^\top$ at the bottom. In this case, σ_{zy} is comparable to σ_{zx}, and the shear flow is influenced by both components. The stress distribution differs significantly from that predicted by the thin beam theory for this wide beam.

Key observations include the following:

1. The shear flow has a circulation component.
2. The largest shear stress does not occur at the midsection but rather at the left and right edges, which cannot be predicted by the thin beam theory.
3. Significant shear stresses σ_{yz} appear on both the top and bottom surfaces, particularly near the 2-, 4-, 8-, and 10-o'clock positions. In the upper section, σ_{yz} points away from the x-axis, while in the lower section, it points toward the x-axis. The stress σ_{xz} always points downward. None of these behaviors are captured by the thin beam theory.

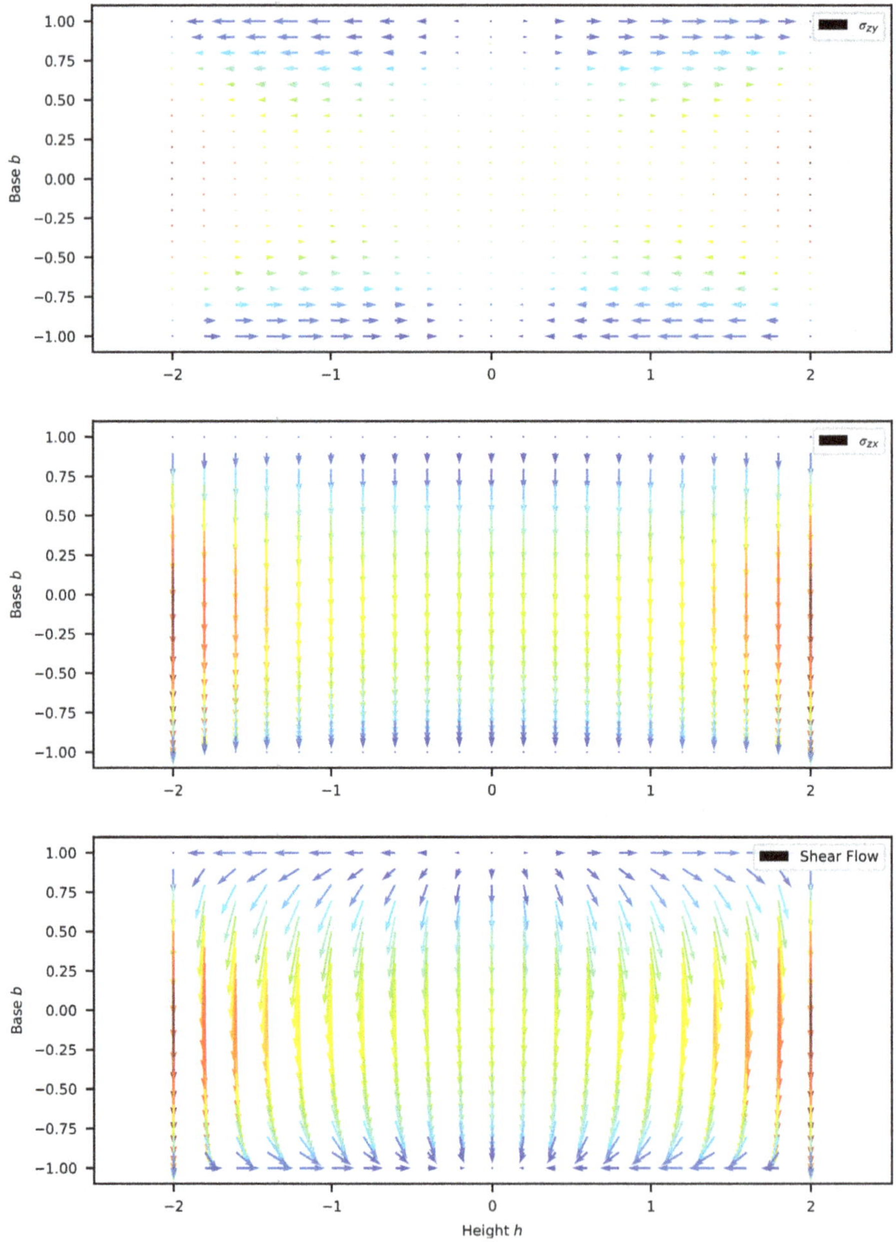

Figure 5.9. Distribution of the shear stresses on the cross-section of a 3D cantilever beam with a wide rectangular cross-section (height/base ratio of 1/2).

These findings confirm that the thin beam theory provides only an approximation and is applicable mainly to tall beams.

5.4.10 *Rigid body displacements*

The rigid body motion of a 3D solid can always be incorporated into the displacement solutions if needed. The formula is as follows:

$$\begin{aligned} u_0 &= a_0 - b_1 y - c_1 z \\ v_0 &= b_0 + b_1 x - c_2 z \\ w_0 &= c_0 + c_1 y + c_2 x \end{aligned} \tag{5.34}$$

where u_0, v_0, w_0 are the three components of rigid body displacement and $a_0, b_0, c_0, b_1, c_1, c_2$ are six arbitrary constants representing the possible rigid body motions.

Using the strain–displacement relations, one can confirm that these displacements do not contribute to strains or affect the stress solutions. They only alter the reference frame for measuring displacements.

The derivation of Eq. (5.34) simply identifies displacements that do not induce strain. These displacements must be linear since any higher-order terms would generate strain.

The following snippet computes these rigid body displacements:

```
x, y, z = sp.symbols("x, y, z", real = True)
bfs = sp.Matrix([1, x, y, z])     # form sympy vector with linear bases
u_bfs = v_bfs = w_bfs = bfs    # possible bases for linear displacements

ai = sp.symbols(f'a:{len(u_bfs)}')                    # unknown constants
bi = sp.symbols(f'b:{len(v_bfs)}')
ci = sp.symbols(f'c:{len(w_bfs)}')

ai = sp.Matrix(ai); bi = sp.Matrix(bi); ci = sp.Matrix(ci)
cs = ai.col_join(bi).col_join(ci)   # form a single vector of constants

u0 = (ai.T@Matrix(u_bfs))[0]         # all possible linear displacements
v0 = (bi.T@Matrix(v_bfs))[0]
w0 = (ci.T@Matrix(w_bfs))[0]

# Use strain-displacement relations to find zero strain contributors:
eqns =[u0.diff(x), v0.diff(y), w0.diff(z), v0.diff(z)+w0.diff(y),
       u0.diff(z)+w0.diff(x),  u0.diff(y)+v0.diff(x)]
s_d0 = sp.solve(eqns, cs)
display(Math(f"\\text{{Rigid body displacements: $u_0$}}\
                                        ={latex(u0.subs(s_d0))}"))
display(Math(f"\hspace{{4.8cm}}\\text{{$v_0$}}={latex(v0.subs(s_d0))}"))
display(Math(f"\hspace{{4.8cm}}\\text{{$w_0$}}={latex(w0.subs(s_d0))}"))
```

Rigid body displacements: $u_0 = a_0 - b_1 y - c_1 z$

$$v_0 = b_0 + b_1 x - c_2 z$$

$$w_0 = c_0 + c_1 x + c_2 y$$

5.5 Displacement potential functions

This section introduces techniques for solving 3D problems by representing displacements using displacement potential functions (DPFs) and displacement functions.

Assume that the displacement components can be derived from a potential function. The potential function, denoted as ψ, is defined in a 3D domain Ω, bounded by Γ, and is continuously differentiable. We can then express the displacement components as the derivatives of the displacement potential function:

$$u = \frac{1}{2G}\frac{\partial \psi}{\partial x}; \quad v = \frac{1}{2G}\frac{\partial \psi}{\partial y}; \quad w = \frac{1}{2G}\frac{\partial \psi}{\partial z} \tag{5.35}$$

If a single function ψ can be found, it allows us to express all three displacement components using a single scalar function. Once the potential function is determined, all displacement components — and consequently, all strains and stresses — can be obtained through differentiation.

5.5.1 *Conditions for displacement potential functions*

Using Eq. (5.35), it is clear that the Laplacian of the displacement components is the gradient of $\nabla^2\psi$. Additionally, e can be written as the Laplacian of ψ:

$$e = \frac{1}{2G}\nabla^2\psi \tag{5.36}$$

Thus, the gradient of e in Eq. (4.4) coincides with the gradient of $\nabla^2\psi$. Therefore, to satisfy the equilibrium in Eq. (4.19), the DPF must satisfy

$$\frac{\partial}{\partial x}\nabla^2\psi = 0; \quad \frac{\partial}{\partial y}\nabla^2\psi = 0; \quad \frac{\partial}{\partial z}\nabla^2\psi = 0 \tag{5.37}$$

This implies that the gradient of $\nabla^2\psi$ must vanish over the entire domain. Therefore, $\nabla^2\psi$ must be constant:

$$\nabla^2\psi = C \tag{5.38}$$

Equation (5.38) provides the condition for finding the DPF ψ.

Since ψ satisfies Eq. (5.38), and using Eq. (5.36), we find that e must be constant. This implies that the volume deformation of the solid is constrained. Moreover, if $\nabla^2\psi = 0$, then $e = 0$, indicating that the volume of the solid remains unchanged. This must be kept in mind.

5.5.2 *Some displacement potential functions*

Functions that satisfy Eq. (5.38) (with $C = 0$) are known as harmonic functions. Finding a proper ψ that satisfies Eq. (5.38) (with $C = 0$) for general solid mechanics problems is not easy when done manually. As discussed in Ref. [2], a function can be constructed using basis functions through a simple linear combination. Following are some harmonic basis functions that have been found so far and can be used to form the DPF $\psi(x, y, z)$:

1. **Polynomial harmonic bases:**

$$\begin{aligned}
&1,\ x,\ y,\ z,\ xz,\ xy,\ yz,\ x^2-y^2,\ y^2-z^2,\ x^2-z^2 && \leq \text{second-order}\\
&y^2z-\tfrac{z^3}{3},\ y^3-3yz^2,\ xyz,\ xy^2-xz^2,\\
&\quad x^2z-\tfrac{z^3}{3},\ x^2y-yz^2,\ x^3-3xz^2 && \text{third-order}\\
&y^3z-yz^3,\ y^4-6y^2z^2+z^4,\ xy^2z-\tfrac{xz^3}{3},\\
&\quad xy^3-3xyz^2,\ x^2yz-\tfrac{yz^3}{3}, x^2y^2-x^2z^2-y^2z^2+\tfrac{z^4}{3},\\
&\quad x^3z-xz^3,\ x^3y-3xyz^2,\ x^4-6x^2z^2+z^4 && \text{fourth-order}\\
&y^4z-2y^2z^3+\tfrac{z^5}{5},\ y^5-10y^3z^2+5yz^4,\ xy^3z-xyz^3,\\
&\quad xy^4-6xy^2z^2+xz^4,\ x^2y^2z-\tfrac{x^2z^3}{3}-\tfrac{y^2z^3}{3}+\tfrac{z^5}{15},\\
&\quad x^2y^3-3x^2yz^2-y^3z^2+yz^4,\ x^3yz-xyz^3,\\
&\quad x^3y^2-x^3z^2-3xy^2z^2+xz^4,\ x^4z-2x^2z^3+\tfrac{z^5}{5},\\
&\quad x^4y-6x^2yz^2+yz^4,\ x^5-10x^3z^2+5xz^4 && \text{fifth-order}\\
&\ldots\ldots
\end{aligned} \tag{5.39}$$

These polynomial bases are generated using **gr.create_nbf()** developed in Chapter 2. More can be generated using **gr.create_nbf()**, if needed.

Note that **gr.create_nbf()** can also be used to generate harmonic basis functions for 2D cases and was used in Ref. [7] for the construction of potential and stream functions.

2. **Rational harmonic basis:**

$$\begin{aligned}
&\frac{1}{\sqrt{x^2+y^2+z^2}} \\
&\frac{x}{z+\sqrt{x^2+y^2+z^2}}, \quad \frac{y}{z+\sqrt{x^2+y^2+z^2}} \\
&\frac{y}{x+\sqrt{x^2+y^2+z^2}}, \quad \frac{z}{x+\sqrt{x^2+y^2+z^2}} \\
&\frac{x}{y+\sqrt{x^2+y^2+z^2}}, \quad \frac{z}{y+\sqrt{x^2+y^2+z^2}}
\end{aligned} \tag{5.40}$$

3. **Logarithmic harmonic basis:**

$$\begin{aligned}
&\ln(x+\sqrt{x^2+y^2+z^2}) \\
&\ln(y+\sqrt{x^2+y^2+z^2}) \\
&\ln(z+\sqrt{x^2+y^2+z^2})
\end{aligned} \tag{5.41}$$

A linear combination of some of these basis functions may be used to form a displacement potential function. In our example problem given in the next section, we will use the second and third terms in Eq. (5.40).

5.5.3 *An example of generating polynomial DPFs*

The following is a snippet demonstrating the use of **gr.create_nbf()** to generate harmonic basis functions that can be used for constructing DPFs:

```
# Generate some polynomial harmonic basis functions for DPFs:
x, y, z = symbols('x, y, z')
X = sp.Matrix([x, y, z])
order = 6                                   # define the order of bases
DPFs = gr.create_nbf(order, X, dfn=2, prnt=True)
DPFs.T[:3]                                         # print out a few
```

```
Dimension=3, order=6, ∇^=2
No. of constants=28,  should be 28.0
Is  ^2 ψ=0? Matrix([[0]])! Number of bases=13
```

$$\left[y^5 z - \frac{10y^3z^3}{3} + yz^5,\ y^6 - 15y^4z^2 + 15y^2z^4 - z^6,\ xy^4z - 2xy^2z^3 + \frac{xz^5}{5}\right]$$

Note that the lower-order harmonic bases given in Eq. (5.39) cannot be fully generated using **gr.create_nbf()**. We can check whether these bases satisfy $\nabla^2\psi = 0$, using **gr.nabla_f()**:

```
bfs = Matrix([x*z, x*y, y*z, x**2-y**2, y**2-z**2, x**2-z**2])
bfs2 = gr.nabla_f(bfs, X, dfn=2, smplfy=2)                # Check ∇**2
bfs2.T
```

$$\begin{bmatrix}0 & 0 & 0 & 0 & 0 & 0\end{bmatrix}$$

The following snippet checks the rational and logarithmic basis functions for DPFs:

```
x, y, z = sp.symbols("x, y, z")
R = sp.sqrt(x**2 + y**2 + z**2)

basis = Matrix([sp.log(R+x), sp.log(R+y), sp.log(R+z)])              #Yes
#basis=Matrix([1/R,x/(R+z),y/(R+z),y/(R+x),z/(R+x),x/(R+y),z/(R+y)])#Yes
#basis = Matrix([R, R**2, x/R, y/R, sp.log(R)])                      # No

bfs2 = gr.nabla_f(basis, X, dfn=2, smplfy=1)                # Check ∇**2=0?
simplify(bfs2).T
```

$$\begin{bmatrix}0 & 0 & 0\end{bmatrix}$$

5.5.4 *Limitations of displacement potential functions*

The condition given in Eq. (5.38) for a displacement potential function ψ implies that the volume strain in a solid is constant throughout the problem domain. If the constant C is zero, the material is incompressible. Typical materials of this kind are rubber and soft biological tissues. In general, many materials do not necessarily exhibit constant volume strain. To avoid such artificial constraints on solids, we can use displacement potential functions in conjunction with the so-called displacement functions, such as the Love and Galerkin displacement functions.

5.6 Galerkin displacement functions

The Galerkin displacement functions (DFs) generally consist of three pieces of functions. The displacement components are written as follows:

$$
\begin{aligned}
u &= \frac{1}{2G}\left[2(1-\nu)\nabla^2\xi - \frac{\partial}{\partial x}\left(\frac{\partial \xi}{\partial x} + \frac{\partial \eta}{\partial y} + \frac{\partial \zeta}{\partial z}\right)\right] \\
v &= \frac{1}{2G}\left[2(1-\nu)\nabla^2\eta - \frac{\partial}{\partial y}\left(\frac{\partial \xi}{\partial x} + \frac{\partial \eta}{\partial y} + \frac{\partial \zeta}{\partial z}\right)\right] \\
w &= \frac{1}{2G}\left[2(1-\nu)\nabla^2\zeta - \frac{\partial}{\partial z}\left(\frac{\partial \xi}{\partial x} + \frac{\partial \eta}{\partial y} + \frac{\partial \zeta}{\partial z}\right)\right]
\end{aligned}
\tag{5.42}
$$

where $\xi(x, y, z)$, $\eta(x, y, z)$, and $\zeta(x, y, z)$ denote the three Galerkin displacement functions. Clearly, these are not potential functions of the displacement field.

Substituting Eq. (5.42) into the equilibrium equations written in displacements, Eq. (4.19), we obtain the conditions for these functions:

$$
\nabla^4\xi = 0; \quad \nabla^4\eta = 0; \quad \nabla^4\zeta = 0 \tag{5.43}
$$

These are homogeneous biharmonic equations, similar to those for the stress functions used in the stress methods, except it is now in 3D. There are many functions satisfying these conditions. Following are some examples.

5.6.1 *Some Galerkin displacement functions*

Functions satisfying Eq. (5.43) are harmonic in nature, similar to the stress function discussed earlier. Following are some additional known basis functions in 3D that can be used to form displacement functions $\xi(x, y, z)$, $\eta(x, y, z)$, or $\zeta(x, y, z)$ via linear combinations.

1. **Polynomial bases:**

$$\begin{aligned}
&1, x, y, z, xy, xz, yz, x^2, y^2, z^2, && \leq \text{ second-order}\\
&x^3, y^3, z^3, yzx^3, xyz^3, xy^3z, y^2(z^2 - x^2), x^2(z^2 - y^2), &&\\
&\quad y^3(x^2 - z^2), x^3(y^2 - z^2), z^3(x^2 - y^2) && \text{third-order}\\
&yz^3,\ y^2z^2 - \tfrac{z^4}{3},\ y^3z,\ y^4 - z^4,\ xz^3,\ xyz^2,\ xy^2z,\ xy^3, &&\\
&\quad x^2z^2 - \tfrac{z^4}{3},\ x^2yz,\ x^2y^2 - \tfrac{z^4}{3},\ x^3z,\ x^3y,\ x^4 - z^4 && \text{fourth-order}\\
&y^2z^3 - \tfrac{z^5}{5},\ y^3z^2 - yz^4,\ y^4z - \tfrac{z^5}{5},\ y^5 - 5yz^4, &&\\
&\quad xyz^3,\ xy^2z^2 - \tfrac{xz^4}{3},\ xy^3z,\ xy^4 - xz^4,\ x^2z^3 - \tfrac{z^5}{5}, &&\\
&\quad x^2yz^2 - \tfrac{yz^4}{3},\ x^2y^2z - \tfrac{z^5}{15},\ x^2y^3 - yz^4,\ x^3z^2 - xz^4, &&\\
&\quad x^3yz,\ x^3y^2 - xz^4,\ x^4z - \tfrac{z^5}{5},\ x^4y - yz^4,\ x^5 - 5xz^4 && \text{fifth-order}\\
&\ldots\ldots &&
\end{aligned} \tag{5.44}$$

Note that this includes all terms given in Eq. (5.39) because any function f that satisfies $\nabla^2 f = 0$ also satisfies $\nabla^4 f = 0$.

2. **Rational bases:**

$$\begin{aligned}
&\text{All terms in Eq. (5.40) and}\\
&R,\ R^2\\
&\frac{x}{R}, \frac{y}{R}, \frac{z}{R}
\end{aligned} \tag{5.45}$$

3. **Logarithmic basis:**

$$\begin{aligned}
&\text{All terms in Eq. (5.41) and}\\
&x\ln(R+z),\ y\ln(R+z),\ z\ln(R+z)\\
&x\ln(R+x),\ y\ln(R+x),\ z\ln(R+x)\\
&x\ln(R+y),\ y\ln(R+y),\ z\ln(R+y)
\end{aligned} \tag{5.46}$$

where $R = \sqrt{x^2 + y^2 + z^2}$. The write-up seems mostly clear.

Note that a displacement function can be formed using a linear combination of any of the basis terms. There is no need to use all of these terms, and $\xi(x, y, z)$, $\eta(x, y, z)$, or $\zeta(x, y, z)$ do not have to be the same. Often, necessary terms for a displacement function are handpicked based on experience.

5.6.2 *Example of generating DFs*

The following is a simple Python code that can be used for generating or examining displacement functions. We use $\xi(x, y, z)$ as an example:

```
x, y, z = symbols('x, y, z')
X = sp.Matrix([x, y, z])
order = 4                                    # define the order for BBFs
DFs = gr.create_nbf(order, X, dfn=4, prnt=True)
display(DFs.T[:11])                              # display sample terms

order = 5                                    # define the order for BBFs
DFs = gr.create_nbf(order, X, dfn=4, prnt=True)
display(DFs.T[:8])                               # display sample terms
```

```
Dimension=3, order=4, ∇^=4
No. of constants=15,  should be 15.0
Is ∇^4 ψ=0? Matrix([[0]])! Number of bases=14
```

$$\left[yz^3,\ y^2z^2 - \frac{z^4}{3},\ y^3z,\ y^4 - z^4,\ xz^3,\ xyz^2,\ xy^2z,\ xy^3,\ x^2z^2 - \frac{z^4}{3},\ x^2yz,\ x^2y^2 - \frac{z^4}{3}\right]$$

```
Dimension=3, order=5, ∇^=4
No. of constants=21,  should be 21.0
Is ∇^4 ψ=0? Matrix([[0]])! Number of bases=18
```

$$\left[y^2z^3 - \frac{z^5}{5},\ y^3z^2 - yz^4,\ y^4z - \frac{z^5}{5},\ y^5 - 5yz^4,\ xyz^3,\ xy^2z^2 - \frac{xz^4}{3},\ xy^3z,\ xy^4 - xz^4\right]$$

```
# Check whether a f satisfies  ∇^4 f = 0. Example:
DFs4 = gr.nabla_f(DFs, X, dfn=4, smplfy=2)
display(Math(f" \\text{{For polynomial DFs, $∇^4f = 0$?}} \
                                          {latex(DFs4.T[:18])}"))
# Check the rational and logarithmic DFs.
R = sp.sqrt(x**2 + y**2 + z**2)
#basis=Matrix([R, R**2, x/R, y/R, z/R])                      #∇^4=0, ∇^2≠0
basis=Matrix([x*sp.log(R+z),y*sp.log(R+z),z*sp.log(R+z)])  #∇^4=0,∇^2≠0

bfs4 = gr.nabla_f(basis, X, dfn=4, smplfy=1)                   # Check ∇^4
display(Math(f" \\text{{For rational and logarithmic DFs, $ ^4f=0$?}} \
                                 {latex(sp.simplify(bfs4).T)}"))
```

For polynomial DFs,

$\nabla^4 f = 0?$ [0, 0, 0, 0, 0, 0, 0, 0, 0, 0, 0, 0, 0, 0, 0, 0, 0, 0]

For rational and logarithmic DFs, $\nabla^4 f = 0?$ [0 0 0]

5.6.3 *Solution procedure using displacement functions*

The solution procedure using the Galerkin displacement functions is as follows:

1. Find the Galerkin displacement functions as linear combinations of basis functions that satisfy Eq. (5.43). Typically, the displacement functions are assumed based on the problem setting and known experience.
2. Use Eq. (5.42) to derive the displacements, involving only differentiations.
3. Apply the stress–displacement relations from Eq. (4.21) for the stresses, also involving only differentiations.
4. The constants in the displacement functions will appear in the displacement and stress expressions. Boundary conditions can then be used to determine these constants.

We demonstrate this process with the following example problem, where Python code is written to facilitate the solution. The same approach can be adapted (with modifications) for other similar problems.

5.7 Three-dimensional half-space with a shear force

5.7.1 *Problem setup*

Consider a three-dimensional (3D) half-space made of a uniform isotropic linear elastic material. This half-space is subjected to a point/concentrated shear force applied to its surface in a given direction, as shown in Fig. 5.10. The goal is to find analytic formulas for the displacements and stresses at any point in the 3D space. For simplicity and without loss of generality, we use Cartesian coordinates with the origin at the point where the force is applied and align the x-direction with the direction of the shear force.

This is a 3D problem that is not axially symmetric (unlike the point vertical force problem discussed in Chapter 5). Therefore, the solution

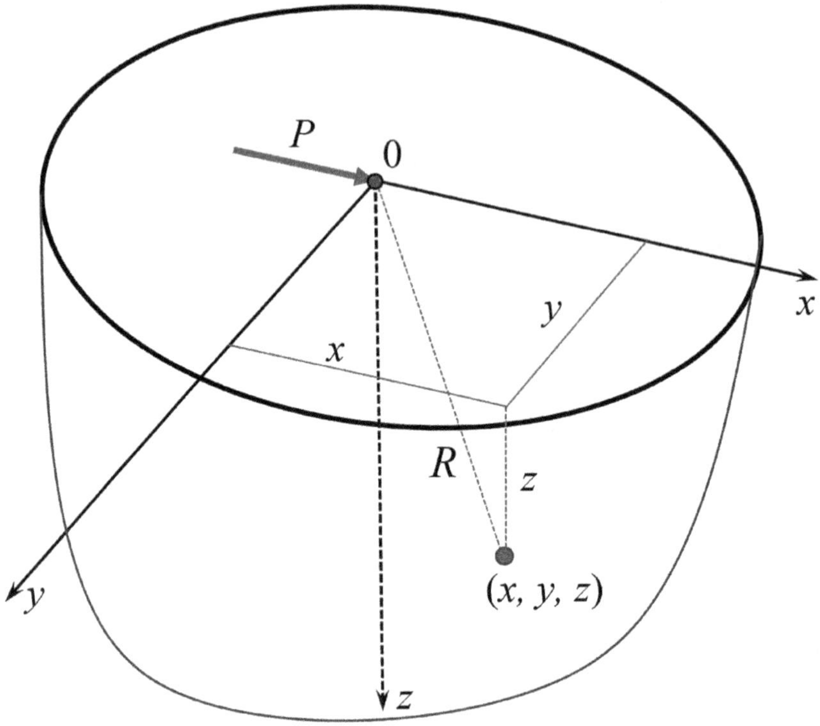

Figure 5.10. A three-dimensional half-space subjected to a shear force on its surface in the x-direction.

requires the use of displacement potential functions and Galerkin displacement functions.

The conditions for this problem are as follows:

1. Stress Boundary Conditions on the top surface:

$$\sigma_{zx} = \sigma_{zy} = \sigma_{zz} = 0 \quad \text{at } z = 0 \text{ and } R \neq 0 \tag{5.47}$$

2. Integral Equilibrium Conditions in the Domain:

Since the external force is a point force, the solution at the point of application is singular. The equilibrium conditions must be expressed in integral form, defined away from the singular point. For convenience, we choose an arbitrary plane parallel to the surface at a nonzero positive $z > 0$. The integrals of all stresses over this plane must satisfy the following equilibrium conditions:

$$\begin{aligned}
&\sum F_x = 0: \int_{-\infty}^{\infty}\int_{-\infty}^{\infty} \sigma_{zx} dxdy + P = 0 \\
&\sum F_y = 0: \int_{-\infty}^{\infty}\int_{-\infty}^{\infty} \sigma_{zy} dxdy = 0 \\
&\sum F_z = 0: \int_{-\infty}^{\infty}\int_{-\infty}^{\infty} \sigma_{zz} dxdy = 0 \\
&\sum M_x = 0: \int_{-\infty}^{\infty}\int_{-\infty}^{\infty} (y\sigma_{zz} - z\sigma_{zy}) dxdy = 0 \\
&\sum M_y = 0: \int_{-\infty}^{\infty}\int_{-\infty}^{\infty} (x\sigma_{zz} - z\sigma_{zx}) dxdy = 0 \\
&\sum M_z = 0: \int_{-\infty}^{\infty}\int_{-\infty}^{\infty} (y\sigma_{zx} - x\sigma_{zy}) dxdy = 0
\end{aligned} \tag{5.48}$$

5.7.2 *Choice of DFs and DPFs*

For this problem, we use the following displacement functions (DFs):

$$\begin{aligned}
\xi(x, y, z) &= c_0 R \\
\eta(x, y, z) &= 0 \\
\zeta(x, y, z) &= c_1 x \ln(R + z)
\end{aligned} \tag{5.49}$$

where $R = \sqrt{x^2 + y^2 + z^2}$. DFs have two coefficients c_0 and c_1.

Since the problem has six boundary conditions, as given in Eq. (5.47) and (5.48), we need additional coefficients. Thus, we introduce the second term from Eq. (5.40) as a basis function for the displacement potential function (DPF):

$$\psi(x, y, z) = c_2 \frac{x}{R + z} \tag{5.50}$$

This term adds the coefficient c_2. Therefore, we have a total of three coefficients in the assumed displacement functions.

The displacements will be computed by summing the displacements derived using Eqs. (5.42) and (5.35). We then follow the previously outlined procedure to find the solution and write the following Python code to implement this step by step:

Note that the expressions derived in this process can become quite lengthy and complicated, especially when using rational harmonic and logarithmic harmonic basis functions. The complexity tends to accumulate, which may result in overly intricate expressions. To mitigate this, we often manually simplify the expressions. While the details of the simplification process may not always be provided, an immediate follow-up check will be performed to ensure that the manually simplified expressions match the ones derived using SymPy.

5.7.3 *Techniques and Python code for solution*

Let's begin by defining the necessary variables.

```
G, ν, P = sp.symbols("G, ν, P", nonnegative =True)
x, y, z = sp.symbols("x, y, z")
c0, c1, c2 = sp.symbols("c0, c1, c2")

ux, uy, uz = sp.symbols("u_x, u_y, u_z")
σxx,σyy,σzz,σyz,σxz,σxy = sp.symbols("σ_xx,σ_yy,σ_zz,σ_yz,σ_xz,σ_xy")
```

5.7.3.1 *Step 1: Solutions from the DF*

The first step is to compute the displacements and the stresses σ_{zz}, σ_{zy}, and σ_{zx}, as we have boundary conditions for these stresses that will help determine the coefficients.

Begin by forming and checking the basis functions.

```
R = sp.sqrt(x**2 + y**2 + z**2)
ξ = c0*R                                           # displacement functions
ζ = c1*x*sp.log(R+z)

div_df=(ξ.diff(x,1)+ζ.diff(z,1)).simplify()  # divergence of disp. funs.
nab2ξ =(ξ.diff(x,2)+ξ.diff(y,2)+ξ.diff(z,2)).simplify()          # ∇^2ξ
#nab2η= η.diff(x,2)+η.diff(y,2)+η.diff(z,2)      # η = 0 for this problem
nab2ζ = (ζ.diff(x,2)+ζ.diff(y,2)+ζ.diff(z,2)).factor()          # ∇^2ζ

nab4ξ=(nab2ξ.diff(x,2)+nab2ξ.diff(y,2)+nab2ξ.diff(z,2)).simplify()  #∇^4
nab4ζ=(nab2ζ.diff(x,2)+nab2ζ.diff(y,2)+nab2ζ.diff(z,2)).simplify()
print(nab4ξ, nab4ζ)                              # check  ^4ξ=0,   ^4ζ=0 ?
```

```
0 0
```

It can be observed that the homogeneous biharmonic equation is satisfied for any values of c_0 and c_1.

Next, compute the displacement components by differentiating:

```
ux1 = (1/(2*G)*(2*(1-ν)*nab2ξ-div_df.diff(x,1))).simplify()
#ux1                    # 1: displacement from the displacement-function

uy1 = (1/(2*G)*(-div_df.diff(y,1))).simplify()                   # compute

uz1_=(1/(2*G)*(2*(1-ν)*nab2ζ-div_df.diff(z,1))).simplify()   # computed
uz1 = 1/(2*G)*(-4*x*c1*(ν-1)/((z+R)*R) + x*z*(c0+c1)/R**3) # simplified
# Check correctness of the simplified formula:
print(f" Is the simplified one correct ? {(uz1-uz1_).simplify()==0}")

display(Math(f" \\text{{Displacement from DF, $u_{{x1}}$}}=\
                                            {latex(ux1)}"))
display(Math(f" \\text{{Displacement from DF, $u_{{y1}}$}}=\
                                            {latex(uy1)}"))
display(Math(f" \\text{{Displacement from DF, $u_{{z1}}$}}=\
                                            {latex(uz1)}"))
```

```
Is the simplified one correct ? True
```

Displacement from DF,

$$u_{x1} = \frac{x^2 (c_0 + c_1) - (x^2 + y^2 + z^2)(4c_0(\nu - 1) + c_0 + c_1)}{2G(x^2 + y^2 + z^2)^{\frac{3}{2}}}$$

Displacement from DF, $u_{y1} = \dfrac{xy(c_0 + c_1)}{2G(x^2 + y^2 + z^2)^{\frac{3}{2}}}$

Displacement from DF, $u_{z1} = \dfrac{-\frac{4c_1 x(\nu-1)}{\left(z+\sqrt{x^2+y^2+z^2}\right)\sqrt{x^2+y^2+z^2}} + \frac{xz(c_0+c_1)}{(x^2+y^2+z^2)^{\frac{3}{2}}}}{2G}$

In the preceding code cell, we simplify the displacement u_{z1} to facilitate later derivation.

Now, compute the stresses σ_{zz}, σ_{zy}, and σ_{zx} using Eq. (4.21).

```
#Compute volume strain e (disp. divergence),using displacement-function:
e =(ux1.diff(x,1).simplify()+uy1.diff(y,1).simplify()+uz1.diff(z,1))

σzz1 = (2*G*(ν/(1-2*ν)*e+uz1.diff(z,1)))
σyz1 = (G*(uz1.diff(y,1)+uy1.diff(z,1)))
σxz1 = (G*(uz1.diff(x,1)+ux1.diff(z,1)))
#σzz1.simplify()              # 1: stresses from the DF, long expression
```

Note that these solutions from the potential function carry unknown constants c_0 and c_1.

5.7.3.2 *Step 2: Solutions from the DPF*

We compute the displacements and σ_{zz}, σ_{zy}, and σ_{zx} using the displacement potential function (DPF).

```
ψ = sp.symbols("ψ")                          # displacement potential function
ψ = c2*x/(R+z)
nab2ψ = ψ.diff(x,2)+ψ.diff(y,2)+ψ.diff(z,2)
#nab2ψ.simplify()                                            # check ∇^2ψ = 0?

ux2 = 1/(2*G)*ψ.diff(x,1)
#ux2                 # 2: solution from the displacement potential function

uy2 = 1/(2*G)*ψ.diff(y,1)
uz2 = 1/(2*G)*ψ.diff(z,1)

display(Math(f" \\text{{Displacement from DPF, $u_{{x2}}$}}=\
                                                  {latex(ux2)}"))
display(Math(f" \\text{{Displacement from DPF, $u_{{y2}}$}}=\
                                                  {latex(uy2)}"))
display(Math(f" \\text{{Displacement from DPF, $u_{{z2}}$}}=\
                                                  {latex(uz2)}"))
```

Displacement from DPF,

$$u_{x2} = \frac{-\frac{c_2 x^2}{\left(z+\sqrt{x^2+y^2+z^2}\right)^2 \sqrt{x^2+y^2+z^2}} + \frac{c_2}{z+\sqrt{x^2+y^2+z^2}}}{2G}$$

Displacement from DPF, $u_{y2} = -\dfrac{c_2 x y}{2G\left(z+\sqrt{x^2+y^2+z^2}\right)^2 \sqrt{x^2+y^2+z^2}}$

Displacement from DPF, $u_{z2} = \dfrac{c_2 x \left(-\frac{z}{\sqrt{x^2+y^2+z^2}} - 1\right)}{2G\left(z+\sqrt{x^2+y^2+z^2}\right)^2}$

```
# e =(ux2.diff(x,1)+uy2.diff(y,1)+uz2.diff(z,1))              # for checking
#e.simplify()                                                 # should be zero

σzz2 = ψ.diff(z,2)         # or: σzz2 = (2*G*(v/(1-2*v)*e+uz2.diff(z,1)))
σyz2 = ψ.diff(y,1).diff(z,1)
σxz2 = ψ.diff(x,1).diff(z,1)      # or (G*(uz2.diff(x,1)+ux2.diff(z,1)))

σzz2.simplify()  # 2: stresses from the displacement potential function
```

$$\frac{c_2 x \left(-\left(x^2+y^2\right)\left(x^2+y^2+z^2\right)+2\left(z+\sqrt{x^2+y^2+z^2}\right)\left(x^2+y^2+z^2\right)^{\frac{3}{2}}\right)}{\left(z+\sqrt{x^2+y^2+z^2}\right)^2\left(x^2+y^2+z^2\right)^{\frac{5}{2}}}$$

Note that these solutions from the potential function carry only the unknown constant c_2.

5.7.3.3 *Step 3: Use stress boundary conditions*

We can now solve for the constants using the SBCs on the surface of the half-space, which are given in Eq. (5.47).

```
sln = sp.solve([(σzz1+σzz2).subs({z:0}),(σxz1+σxz2).subs({z:0}),
                (σyz1+σyz2).subs({z:0})], [c0, c1, c2])
sln             # should be: c0 = c2/(2*(1-ν)*(1-2*ν)); c1=-c2/(2*ν-2)
```

$$\left\{c_0 : \frac{c_2}{4\nu^2-6\nu+2},\ c_1 : -\frac{c_2}{2\nu-2}\right\}$$

We have expressed both c_0 and c_1 in terms of c_2. We can now substitute c_0 and c_1 back into the solutions and express everything solely in terms of c_2. This is done in the following snippet. The solution expressions computed by SymPy are denoted with an underscore "_", and these can be quite complicated. Therefore, we performed a manual simplification to produce a simpler expression and then confirmed that the manually simplified expression is the same as the original ones found by the code.

```
#c0_c2 = sln[c0].simplify()                          # c0 expressed in c2
#c1_c2 =sln[c1]                    # c1 expressed in c2: c1_c2 = -c2/(2*ν - 2)
                                    # Note: 4ν^2-6ν + 2 = c2*(2*(1-ν)*(1-2*ν))
c0_c2 = c2/(2*(1-ν)*(1-2*ν))           # set the relationship of the coeffs.
c1_c2 = -c2/(2*ν - 2)

ux1c2_ = ux1.subs({c0:c0_c2, c1:c1_c2})   # compute displacements in c2
ux1c2 = -c2/(2*G*(2*ν-1))*(x**2/R**3 + 1/R)                    # simplified
# Check correctness of the simplified formula:
print(f" Is the simplified one correct ? "
      f"{(ux1c2_.simplify()-ux1c2.simplify()).simplify()==0}")
```

```
uy1c2_ = uy1.subs({c0:c0_c2, c1:c1_c2})
uy1c2 = -c2*x*y/(2*G*(2*ν - 1)*(x**2 + y**2 + z**2)**(3/2))
print(f" Is the simplified one correct ? "                    # check
      f"{(uy1c2_.simplify()-uy1c2.simplify()).simplify()==0}")

uz1c2_ = uz1.subs({c0:c0_c2, c1:c1_c2})
uz1c2=c2*x/((1-2*ν)*(1-ν)*4*G)*(z/R**3 + 4*(-ν+1)*(1-2*ν)/((R+z)*R)
                                + (1-2*ν)*z/R**3)
print(f" Is the simplified one correct ? "                    # check
      f"{(uz1c2_.simplify()-uz1c2.simplify()).simplify()==0}")

display(Math(f" \\text{{From DF with c2, $u_{{x1}}$}}=\
                                             {latex(ux1c2)}"))
display(Math(f" \\text{{From DF with c2, $u_{{y1}}$}}=\
                                             {latex(uy1c2)}"))
display(Math(f" \\text{{From DF with c2, $u_{{z1}}$}}=\
                                             {latex(uz1c2)}"))
```

```
Is the simplified one correct ? True
Is the simplified one correct ? True
Is the simplified one correct ? True
```

$$\text{From DF with c2, } u_{x1} = -\frac{c_2\left(\frac{x^2}{(x^2+y^2+z^2)^{\frac{3}{2}}} + \frac{1}{\sqrt{x^2+y^2+z^2}}\right)}{2G\,(2\nu - 1)}$$

$$\text{From DF with c2, } u_{y1} = -\frac{c_2 x y}{2G\,(2\nu - 1)\,(x^2 + y^2 + z^2)^{1.5}}$$

From DF with c2,

$$u_{z1} = \frac{c_2 x\left(\frac{z(1-2\nu)}{(x^2+y^2+z^2)^{\frac{3}{2}}} + \frac{z}{(x^2+y^2+z^2)^{\frac{3}{2}}} + \frac{(1-2\nu)(4-4\nu)}{\left(z+\sqrt{x^2+y^2+z^2}\right)\sqrt{x^2+y^2+z^2}}\right)}{4G\,(1 - 2\nu)\,(1 - \nu)}$$

5.7.3.4 *Combining the solutions*

Let us combine the solutions obtained from both the Galerkin displacement functions (DF) and the displacement potential function (DPF). Again, we performed manual simplifications and expressed the solution in simpler forms.

```
ux12 = ux1c2 + ux2                           # ux2 is already expressed in c0

uy12_ = uy1c2 + uy2                                              # computed
uy12=-c2/(2*G*(2*ν-1))*(x*y/R**3+x*y*(2*ν-1)/((z+R)**2*R)) # simplified
print(f" Is the simplified one correct ? "                          # check
      f"{(uy12_-uy12).factor().evalf().simplify().subs(1.0,1)==0}")

uz12_ = uz1c2 + uz2
uz12 = c2/(2*G*(2*ν - 1))*(-x*z/R**3 + (2*ν - 1)*x/((z+R)*R))
print(f" Is the simplified one correct ? "                          # check
      f"{(uz12_.simplify()-uz12.simplify()).simplify()==0}")

σxz_c2_=(G*(uz12.diff(x,1).simplify()+ux12.diff(z,1).simplify()))
σxz_c2 = 3*c2*x**2*z/((2*ν-1)*R**5)                            # simplified
print(f" Is the simplified one correct ? "                          # check
      f"{(σxz_c2_.simplify()-σxz_c2.simplify()).simplify()==0}")

display(Math(f" \\text{{Stress with c2, $σ_{{xz}}$}}={latex(σxz_c2)}"))
```

```
Is the simplified one correct ? True
Is the simplified one correct ? True
Is the simplified one correct ? True
```

Stress with c2, $\sigma_{xz} = \dfrac{3c_2x^2z}{(2\nu - 1)\left(x^2 + y^2 + z^2\right)^{\frac{5}{2}}}$

Note that simplifying a given expression manually can be quite challenging. Helpful SymPy methods include **simplify()**, **factor()**, **expand()**, **sp.apart()**, **collect()**, etc. One effective trick is to define a new variable name and use it to substitute long terms in an expression. The following is an example:

```
q = symbols('q', positive=True)
dict_ = {x**2 + y**2 + z**2: q**2}          # replace x**2+y**2+z**2 by q
temp_ = σxz_c2_.subs(dict_).simplify()             # σxz_c2_ is very long
temp_ = temp_.subs(q**2, x**2 + y**2 + z**2)           #  substitute back

display(Math(f" \\text{{simplified expression}}:\\small{latex(temp_)}"))
```

simplified expression: $\dfrac{3c_2x^2z}{(2\nu - 1)\left(x^2 + y^2 + z^2\right)^{\frac{5}{2}}}$

which is used in the previous snippet. Using a similar approach, simpler formulas can often be derived.

5.7.3.5 *Use of integral equilibrium condition*

We now determine c_2 using the first equation in Eq. (5.48):

$$\sum F_x = 0 : \int_{-\infty}^{\infty} \int_{-\infty}^{\infty} \sigma_{zx}\, dx\, dy + P = 0 \tag{5.51}$$

To avoid the singular point, we set $z = 1$ for σ_{zx}. Since Eq. (5.51) must be satisfied at any z, any choice of z except 0 is acceptable. Since this is a double integral over the entire x–y plane, $(-\infty, \infty) \times (-\infty, \infty)$, and the integrand is symmetric with respect to x and y, we integrate over the first quadrant $(0, \infty) \times (0, \infty)$ and then multiply the result by 4. The domain-reduced double integral can then be carried out in two stages. First, let's integrate along the y-axis:

```
σxz_z1 = σxz_c2.subs(z,1)
σxz_intx = sp.integrate(σxz_z1, (y, 0, oo))          # integrate along y
σxz_intx
```

$$\frac{2c_2x^2}{\left(x^2+1\right)^2(2\nu-1)}$$

Next, integrate along the x-axis:

```
σxz_int =(4*sp.integrate(σxz_intx, (x, 0, oo))).evalf() # int. along x
display(Math(f" \\text{{Stress $σ_{{xz}}$ integrated}}=\
                                         {latex(σxz_int)}"))
```

$$\text{Stress } \sigma_{xz} \text{ integrated} = \frac{6.28318530717959c_2}{2.0\nu - 1.0}$$

Note that $2\pi = 6.28318530717959$; we can substitute this value back, and then solve for c_2 using Eq. (5.51). With c_2 found, c_0 and c_1 can all be determined by substitutions.

```
σxz_int= (2*sp.pi)*c2/(2*ν-1)                           # write in pi value
sln_c2 = sp.solve(σxz_int+P, c2)[0]
sln_c2                                           # solution for coefficient c2
display(Math(f" \\text{{Coefficient $c_2$ found}}= {latex(sln_c2)}"))
c0_ = c0_c2.subs(c2,sln_c2)
display(Math(f" \\text{{Coefficient $c_0$ found}}= {latex(c0_)}"))
c1_ = c1_c2.subs(c2,sln_c2)
display(Math(f" \\text{{Coefficient $c_1$ found}}= {latex(c1_)}"))
```

$$\text{Coefficient } c_2 \text{ found} = \frac{P\,(1-2\nu)}{2\pi}$$

$$\text{Coefficient } c_0 \text{ found} = \frac{P}{2\pi\left(2-2\nu\right)}$$

$$\text{Coefficient } c_1 \text{ found} = -\frac{P\left(1-2\nu\right)}{2\pi\left(2\nu-2\right)}$$

5.7.3.6 *Final set of solutions*

Now that all the coefficients have been found, we can substitute these values into the formulas to obtain the complete set of solutions.

Displacement Solutions:

```
#R_= symbols('R', positive=True)  # R:symbol (not variable) in formulas
sln_c2 = (1-2*ν)*P/(2*sp.pi)
dict_c2 = {c2:sln_c2}
R = sp.symbols("R", positive=True)
dict_R = {x**2 + y**2 + z**2: R**2}

ux = ux12.subs(dict_c2)
ux_ = ux.subs(dict_R).simplify()
display(Math(f" \\text{{Final displacement $u_x$}}= {latex(ux_)}"))

uy = uy12.subs(dict_c2)
uy_ = uy.subs(dict_R).simplify()
display(Math(f" \\text{{Final displacement $u_y$}}= {latex(uy_)}"))

uz = uz12.subs(dict_c2)
uz_ = uz.subs(dict_R).simplify()
display(Math(f" \\text{{Final displacement $u_z$}}= {latex(uz_)}"))
```

Final displacement

$$u_x = \frac{P\left(R^2\left(2\nu-1\right)\left(-R\left(R+z\right)+x^2\right)+\left(R+z\right)^2\left(R^2+x^2\right)\right)}{4\pi G R^3\left(R+z\right)^2}$$

$$\text{Final displacement } u_y = \frac{Pxy\left(R^2\left(2\nu-1\right)+\left(R+z\right)^2\right)}{4\pi G R^3\left(R+z\right)^2}$$

$$\text{Final displacement } u_z = -\frac{Px\left(R^2\left(2\nu-1\right)-z\left(R+z\right)\right)}{4\pi G R^3\left(R+z\right)}$$

Stress Solutions: The stress components can be obtained by differentiating the displacement solutions using Eq. (4.21). The corresponding code snippets are provided as follows:

```
# use displacement expression ux,..., NOT ux_,...:
# Compute the volume strain (disp. divergence)
e =(ux.diff(x,1) + uy.diff(y,1) + uz.diff(z,1))
σxx = (2*G*(ν/(1-2*ν)*e + ux.diff(x,1)))
#σxx.factor() #numerator, denominator = σxx.as_numer_denom()

σyy = (2*G*(ν/(1-2*ν)*e + uy.diff(y,1)))
σzz = (2*G*(ν/(1-2*ν)*e + uz.diff(z,1)))
σyz = ( G*(uy.diff(z,1) + uz.diff(y,1)))
σxy = ( G*(ux.diff(y,1) + uy.diff(x,1)))
```

Note: Due to the complexity of the displacement functions, the resulting stress expressions can be quite intricate and lengthy. Therefore, careful and laborious simplification — both using SymPy and manual methods — is necessary to reduce the expressions as much as possible. We will not delve into this lengthy process further. Instead, we summarize the displacement and stress solutions in the following more readable formulas, as presented in Ref. [6]. Given that this textbook was published nearly 50 years ago, it's likely that these formulas were derived by hand, although the exact origin remains uncertain:

$$
\begin{aligned}
u_x &= \frac{P}{4\pi G}\left\{\frac{2(1-\nu)R+z}{R(R+z)}+\frac{[2R(\nu R+z)+z^2]x^2}{R^3(R+z)^2}\right\}\\
u_y &= \frac{P}{4\pi G}\left\{\frac{[2R(\nu R+z)+z^2]xy}{R^3(R+z)^2}\right\}\\
u_z &= \frac{P}{4\pi G}\left\{\frac{(1-2\nu)x}{R(R+z)}+\frac{xz}{R^3}\right\}\\
\sigma_{xx} &= \frac{Px}{2\pi R^3}\left\{\frac{1-2\nu}{(R+z)^2}\left(R^2-y^2-\frac{2Ry^2}{R+z}\right)-\frac{3x^2}{R^2}\right\}\\
\sigma_{yy} &= \frac{Px}{2\pi R^3}\left\{\frac{1-2\nu}{(R+z)^2}\left(3R^2-x^2-\frac{2rx^2}{R+z}\right)-\frac{3y^2}{R^2}\right\}\\
\sigma_{zz} &= -\frac{3Pxz^2}{2\pi R^5};\quad \sigma_{zy}=-\frac{3Pxyz}{2\pi R^5};\quad \sigma_{zx}=-\frac{3Px^2z}{2\pi R^5}\\
\sigma_{xy} &= \frac{Py}{2\pi R^3}\left\{\frac{1-2\nu}{(R+z)^2}\left(-R^2+x^2+\frac{2rx^2}{R+z}\right)-\frac{3x^2}{R^2}\right\}
\end{aligned}
\tag{5.52}
$$

Note: The variable R is defined as $R=\sqrt{x^2+y^2+z^2}$. It is important to mention that obtaining precise and concise formulas can be challenging, even with the assistance of SymPy. To verify the accuracy of the hand written formulas against those computed using the code provided above, one may use the following code snippets as examples:

```
# An example: check the correctness of displacement ux:
ux_ = P/(4*sp.pi*G)*((2*(1 - ν)*R + z)/(R*(R + z))+  # textbook formula
      (2*R*(ν*R + z) + z**2)*x**2/(R**3*(R + z)**2))

dictR = {R**2: x**2 + y**2 + z**2}
print(f" Is the simplified one correct ? "         # substitute and check
      f"{(ux-ux_.subs(dictR)).simplify()==0}")
```

```
Is the simplified one correct ? True
```

```
# An example: check the correctness of stresses:
σxx_ = P*x/(2*sp.pi*R**3)*(((1-2*ν)/(R + z)**2)*(R**2 - y**2 - \
                        2*R*y**2/(R + z))-3*x**2/R**2)  # textbook formulas
σzz_ = -3*P*x*z**2/(2*sp.pi*R**5)

print(f" Is the simplified one correct ? "                      # check
      f"{(σxx - σxx_.subs(dictR)).simplify()==0}")
print(f" Is the simplified one correct ? "                      # check
      f"{(σzz - σzz_.subs(dictR)).simplify()==0}")
#executed in 47.8s, finished 09:30:21 2025-02-21
```

```
Is the simplified one correct ? True
Is the simplified one correct ? True
```

It is evident from Eq. (5.52) that the three stresses on the top surface of the half-space, σ_{zz}, σ_{zy}, and σ_{zx}, are all zero at $z = 0$. One can easily verify that the equilibrium conditions in Eq. (5.48) are satisfied by simple inspection based on the symmetry of the stresses and by substituting σ_{zz}, σ_{zy}, and σ_{zx} into Eq. (5.48). Notably, the integration of any antisymmetric function over a symmetric domain results in zero.

5.7.4 *Forces in other directions*

If the shear force is applied in the y-direction, we can obtain the solution in exactly the same manner. All that is required is to use the following displacement functions instead:

$$\xi(x, y, z) = 0, \quad \eta(x, y, z) = c_0 R, \quad \zeta(x, y, z) = c_1 y \ln(R + z) \tag{5.53}$$

along with the displacement potential function:

$$\psi(x, y, z) = c_2 \frac{y}{R + z} \tag{5.54}$$

In fact, there is no need to repeat the entire solution process. We can simply take the solution for the shear force in the x-direction and swap the x and y coordinates.

Readers may consider what displacement functions and displacement potential functions to use if the point force is a vertical force in the z-direction. Such a problem is axially symmetric and is easier to solve in the cylindrical coordinate system, as shown in Ref. [6].

It is worth noting that obtaining analytical closed forms for stresses is preferred, provided the expressions are not excessively long. However, it is not as critical as having closed forms for displacements. This is because stresses can be easily obtained through differentiation and can always be computed efficiently with numerical methods for exact solutions (to machine accuracy), as long as the closed forms for displacements are available.

5.8 Numerical stress analysis for given displacements

As observed, when the displacement functions are overly complicated, deriving formulas for stresses can become a laborious task, even with SymPy's assistance. In such cases, it is often more efficient to obtain solutions using numerical methods. This section demonstrates this process. We will utilize the codes from Ref. [8] with modifications to accomplish this task.

We will consider the same half-space solid with point forces applied on its top surface. In this study, we will use the analytical displacement solutions derived in the previous section by superimposing the displacement solutions for P_x, P_y, and P_z. This solution is classical and can also be found in various books and papers, such as the textbook [6, 9]. The full set of formulas for the displacements is given by [9]

$$\begin{aligned}
u_x &= \frac{(1+\nu)}{2\pi E}\left\{\left[\frac{xz}{R^3} - \frac{(1-2\nu)x}{R(R+z)}\right]P_z + \frac{2(1-\nu)R+z}{R(R+z)}P_x \right. \\
&\quad \left. + \frac{[2R(\nu R+z)+z^2]x}{R^3(R+z)^2}(xP_x+yP_y)\right\} \\
u_y &= \frac{(1+\nu)}{2\pi E}\left\{\left[\frac{yz}{R^3} - \frac{(1-2\nu)y}{R(R+z)}\right]P_z + \frac{2(1-\nu)R+z}{R(R+z)}P_y \right. \\
&\quad \left. + \frac{[2R(\nu R+z)+z^2]y}{R^3(R+z)^2}(xP_x+yP_y)\right\} \\
u_z &= \frac{(1+\nu)}{2\pi E}\left\{\left[\frac{2(1-\nu)}{R} - \frac{z^2}{R^3}\right]P_z + \left[\frac{1-2\nu}{R(R+z)} + \frac{z}{R^3}\right](xP_x+yP_y)\right\}
\end{aligned} \tag{5.55}$$

where E is Young's modulus, $E = 2(1+\nu)G$, $R = \sqrt{x^2+y^2+z^2}$, and $P_i (i = x, y, z)$ are the three components of the point force vector acting at the origin $(0, 0, 0)$. The Python code defining these displacements is as follows:

```
#The following snippet allows to format the graphs.
repo="https://raw.githubusercontent.com/nicoguaro/matplotlib_styles/master"
style = repo + "/styles/minimalist.mplstyle"
plt.style.use(style)

# Define symbolic variables:
x, y, z, R, E, ν, Px, Py, Pz = symbols('x y z R E ν P_x P_y P_z')

# Displacements in x, y, z directions
ux = (1+ν)/(2*pi*E)*((x*z/R**3  - (1-2*ν)*x/(R*(R+z)))*Pz +
     (2*(1-ν)*R+z)/(R*(R+z))*Px +
     ((2*R*(ν*R+z) + z**2)*x)/(R**3*(R+z)**2)*(x*Px + y*Py))

uy = (1+ν)/(2*pi*E)*((y*z/R**3  - (1-2*ν)*y/(R*(R+z)))*Pz +
     (2*(1-ν)*R+z)/(R*(R+z))*Py +
     ((2*R*(ν*R+z) + z**2)*y)/(R**3*(R+z)**2)*(x*Px + y*Py))

uz = (1+ν)/(2*pi*E)*((2*(1-ν)/R + z**2/R**3)*Pz +
     ((1-2*ν)/(R*(R+z)) + z/R**3)*(x*Px + y*Py))
display(Math(f" \\text{{$u_x$}}= {latex(ux)}"))    # display an example
```

$$u_x = \frac{(\nu+1)\left(\frac{P_x(R(2-2\nu)+z)}{R(R+z)} + P_z\left(-\frac{x(1-2\nu)}{R(R+z)} + \frac{xz}{R^3}\right) + \frac{x(P_x x + P_y y)\left(2R(R\nu+z)+z^2\right)}{R^3(R+z)^2}\right)}{2\pi E}$$

Without loss of generality, we can set $P_y = 0$, which means that we purposely set x-axis along with the force direction.

```
ux = ux.subs(Py, 0); uy = uy.subs(Py, 0); uz = uz.subs(Py, 0)
display(Math(f" \\text{{$u_x (P_y=0)$ }}= {latex(ux)}"))
```

$$u_x(P_y = 0) = \frac{(\nu+1)\left(\frac{P_x(R(2-2\nu)+z)}{R(R+z)} + \frac{P_x x^2\left(2R(R\nu+z)+z^2\right)}{R^3(R+z)^2} + P_z\left(-\frac{x(1-2\nu)}{R(R+z)} + \frac{xz}{R^3}\right)\right)}{2\pi E}$$

We can then construct the displacement vector and analyze its properties, including its singularity at the origin and behavior at infinity.

```
# Form displacement vector
u = Matrix([ux, uy, uz])

display(Math(f" \\text{{Displacements at the infinity}}=\
                                          {latex(u.limit(R, oo).T)}"))
display(Math(f" \\text{{Displacements at the origin}}=\
               {latex(sp.simplify(u.subs({x:0, y:0}).limit(z, 0)).T)}"))
```

Displacements at the infinity $= \begin{bmatrix} 0 & 0 & 0 \end{bmatrix}$

$$\text{Displacements at the origin} = \begin{bmatrix} \frac{P_x(1-\nu^2)}{\pi ER} & 0 & \frac{P_z(1-\nu^2)}{\pi ER} \end{bmatrix}$$

It can be observed that the displacements vanish at infinity and exhibit a $1/r$ singularity at the origin. The displacement component in the y-direction on the $x = z$ plane ($y = 0$) is zero.

Now, let's compute the strains using the displacements and then the stresses.

5.8.1 *Strains and stresses*

We compute the strain and stress tensors using the symmetric gradient (`vector.sym_grad()`) of the displacement and the strain-to-stress function (`solids.strain()`), both imported from `continuum_mechanics.solids`. The following source code demonstrates how to compute stresses from strains using Eq. (4.3):

```
def strain_stress(strain, parameters):              # parameters in a tuple
    '''source code for strain_stress() from continuum_mechanics.'''
    λ, G = parameters                                          #Lamé's constants
    strain_trace = strain.trace()
    stress = Matrix(3, 3, lambda i, j:
                    λ*np.eye(3)[i,j]*strain_trace+2*G*strain[j,i])
    return stress
```

Let us use it to compute the stress tensor:

```
# compute the strain and stresses:
λ = E*ν/((1 + ν)*(1 - 2*ν))
G = E/(2*(1 + ν))
u = Matrix([ux, uy, uz]).subs(R, (x**2+y**2+z**2)**sp.S.Half)

strain = sym_grad(u)  # comput the strain using symmetric gradient of u
stress = strain_stress(strain, [λ, G])
#stress                       # One may take a look at this (very lengthy)
```

In the previous code cell, we used Eq. (4.8) to compute the strains.

The resulting expressions for the stresses are lengthy. Therefore, it is more practical to compute the stresses for a specific set of parameters at fixed coordinates.

```
simplify(stress.subs({x: 1, y: 1, z:1, E: 1, ν: sp.S(1)/4, Px: 0}))
```

$$\left[\begin{matrix}\frac{P_z\left(-6.75-3.75\sqrt{3}\right)}{\pi\left(90.0+54.0\sqrt{3}\right)} & \frac{P_z\left(-27-13\sqrt{3}\right)}{72\pi\left(5+3\sqrt{3}\right)} & \frac{P_z\left(-2\sqrt{3}-3\right)}{18\pi\left(\sqrt{3}+2\right)}\\ \frac{P_z\left(-27-13\sqrt{3}\right)}{72\pi\left(5+3\sqrt{3}\right)} & \frac{P_z\left(-6.75-3.75\sqrt{3}\right)}{\pi\left(90.0+54.0\sqrt{3}\right)} & \frac{P_z\left(-2\sqrt{3}-3\right)}{18\pi\left(\sqrt{3}+2\right)}\\ \frac{P_z\left(-2\sqrt{3}-3\right)}{18\pi\left(\sqrt{3}+2\right)} & \frac{P_z\left(-2\sqrt{3}-3\right)}{18\pi\left(\sqrt{3}+2\right)} & \frac{P_z\left(-9.0-5.0\sqrt{3}\right)}{\pi\left(90.0+54.0\sqrt{3}\right)}\end{matrix}\right]$$

If we examine the solutions in the x–z plane, we can set $y = 0$ and find the expressions for the stresses:

```
sij = simplify(stress.subs({y:0, E:1, ν:sp.S(1)/4, Px:0, Py:0, Pz:1}))
sij[:,0]                              # print the first column to take a look
```

$$\left[\begin{matrix}\frac{0.25x^6-2.25x^4z^2-1.25x^4z\sqrt{x^2+z^2}-2.25x^2z^4-2.5x^2z^3\sqrt{x^2+z^2}+0.25z^6+0.25z^5\sqrt{x^2+z^2}}{\pi\left(1.0x^8+5.0x^6z^2+2.0x^6z\sqrt{x^2+z^2}+9.0x^4z^4+6.0x^4z^3\sqrt{x^2+z^2}+7.0x^2z^6+6.0x^2z^5\sqrt{x^2+z^2}+2.0z^8+2.0z^7\sqrt{x^2+z^2}\right)}\\ 0\\ \frac{3xz^2\left(-z-\sqrt{x^2+z^2}\right)}{2\pi\left(x^6+3x^4z^2+x^4z\sqrt{x^2+z^2}+3x^2z^4+2x^2z^3\sqrt{x^2+z^2}+z^6+z^5\sqrt{x^2+z^2}\right)}\end{matrix}\right]$$

5.8.2 *Plots of displacement distribution*

We are now ready to visualize the distribution of the displacement and stress solutions. The first step is to create mesh grids for these plots. The mesh grid allows us to evaluate and visualize the displacement and stress fields over the spatial domain.

```
x_mesh, z_mesh = np.mgrid[-1.:1.:400j, 0.001:1.1:200j]
print(f"Mesh grid for plots: {x_mesh.shape}")

ui = sp.simplify(u.subs({y:0, E:1, ν:sp.S(1)/4, Px:0, Py:0, Pz:1}))
display(Math(f" \\text{{Typic displacement function}}={latex(ui.T)}"))
```

```
Mesh grid for plots: (400, 200)
```

$$\text{Typic displacement function} = \left[\begin{matrix}\frac{5x\left(-x^2-z^2+2z\left(z+\sqrt{x^2+z^2}\right)\right)}{16\pi\left(x^2+z^2\right)^{\frac{3}{2}}\left(z+\sqrt{x^2+z^2}\right)} & 0 & \frac{5\left(3x^2+5z^2\right)}{16\pi\left(x^2+z^2\right)^{\frac{3}{2}}}\end{matrix}\right]$$

The following utility function can be used to plot the solutions:

```
def field_plot(expr, x_mesh, z_mesh, c_norm=0.5, title=''):
    """Plot the field, by Guarín-Zapata, Nicolás, under MIT license.
    at https://github.com/nicoguaro/continuum_mechanics
    """
    # Lambdify the function
    var = [x, z]
    arr = [x_mesh, z_mesh]
    vals = OrderedDict(zip(var, arr)).values()
    expr_fun = lambdify(var, expr, "numpy")
    expr_mesh = expr_fun(*vals)

    # Determine extrema
    vmin = np.min(expr_mesh)
    vmax = np.max(expr_mesh)
    print("Minimum value in the domain: {:g}".format(vmin))
    print("Maximum value in the domain: {:g}".format(vmax))
    vmax = max(np.abs(vmax), np.abs(vmin))
    #print(f"Log10(vmax)={vmax},{float(vmax)}")
    vmax = float(vmax)

    fig = plt.figure(dpi=500)#; plt.ioff()
    fig = plt.gcf()
    levels = np.logspace(-1, np.log10(vmax), 20)
    levels = np.hstack((-levels[-1::-1], [0], levels))
    cbar_ticks = ["{:.2g}".format(level) for level in levels]
    cont = plt.contourf(x_mesh, z_mesh, expr_mesh, levels=levels,
           cmap="seismic", norm=colors.SymLogNorm(c_norm)) #,base=10))
    cbar = fig.colorbar(cont, ticks=levels[::4])
    cbar.ax.set_yticklabels(cbar_ticks[::4], fontsize=5)
    plt.axis("image")
    plt.gca().invert_yaxis()
    plt.xticks(fontsize=5)
    plt.yticks(fontsize=5)
    plt.xlabel(r"$x$")
    plt.ylabel(r"$z$")
    plt.title(title)
    #print("Figure DPI:", fig.get_dpi())
    return cont
```

```
from collections import OrderedDict
from matplotlib import colors
plt.ioff()
#plt.ion()                        # uncomment this to display the plots
field_plot(ui.norm(), x_mesh, z_mesh, c_norm=0.5, title=r"u$_{norm}$")
plt.savefig('imagesSM/DispNorm.png', dpi=500)             # save to file
#plt.show()                       # uncomment this to display the plots
```

```
Minimum value in the domain: 0.277018
Maximum value in the domain: 120.716
```

Displacement norm distribution

The displacement vector in a solid represents the movement of points, and its norm is a scalar function that measures the magnitude of this movement. The distribution of the displacement norm in the half-space subjected to a unit vertical point force at the top surface is shown in Fig. 5.11. The plot clearly demonstrates the singularity at the location of the point load, where the displacement magnitude is largest. As expected, the displacement magnitude decays rapidly as we move away from the point force.

```
# Plot the displacement component ux over the $x-z$ plane:
plt.figure()
plt.ioff()
#plt.ion()                          # uncomment this to display the plots
field_plot(ui[0], x_mesh, z_mesh, c_norm = 0.1, title=r"$u_x$")
plt.savefig('imagesSM/Disp_ux.png', dpi=500)               # save to file
#plt.show();                        # uncomment this to display the plots
```

```
Minimum value in the domain: -8.05779
Maximum value in the domain: 8.05779
```

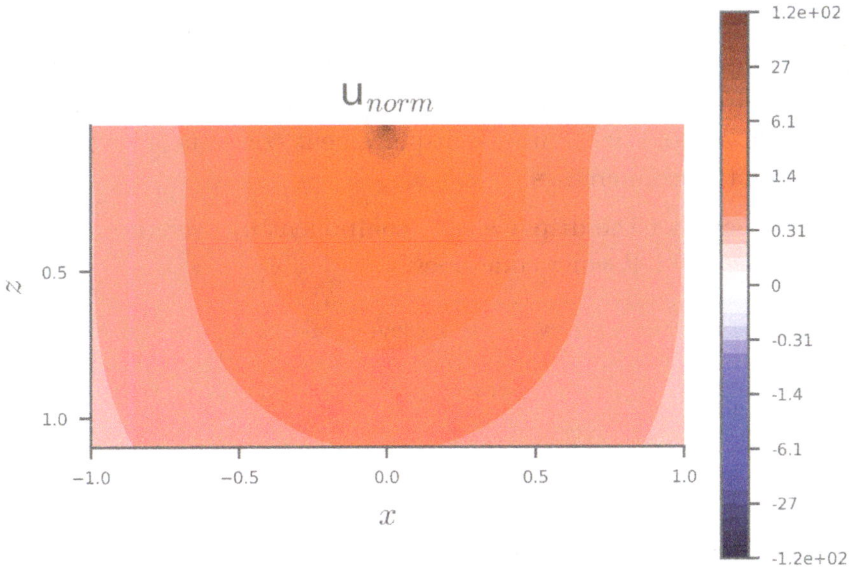

Figure 5.11. The distribution of the displacement norm in the half-space subjected to a unit vertical point force at the top surface.

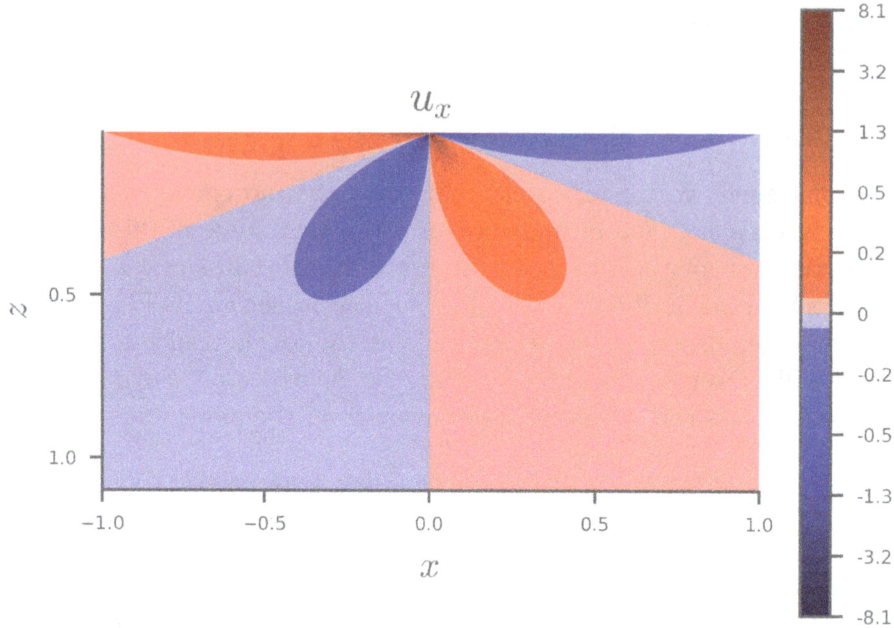

Figure 5.12. Distribution of the displacement u_x in the x–z plane in the half-space subjected to a unit vertical point force at the top surface.

Distribution of displacement component u_x

The distribution of the displacement component u_x in the half-space, subjected to a unit vertical point force at the top surface, is shown in Fig. 5.12. The plot clearly illustrates the singularity at the point where the load is applied, with a rapid decay in the displacement component as the distance from the point force increases.

Readers may plot the displacement component u_z over the x–z plane by uncommenting the following code lines.

```
# Plot the displacement component uz over the $x-z$ plane:
#plt.figure()
#field_plot(ui[2], x_mesh, z_mesh, c_norm = 0.5, title=r"u$_z$")
#plt.show()
```

5.8.3 *Plots of stress distributions*

Now, let's plot the distribution of stresses in the x–z plane. Since stresses form a tensor and are stored in a matrix, we need to access the individual stress components by referring to their respective row and column indices. The snippet is as follows:

```
# Plot norm stress σxx (or σ11) on $y=0$ plane:
plt.figure()
plt.ioff()
#plt.ion()                        # uncomment this to display the plots
row, col = 0, 0
field_plot(sij[row, col], x_mesh, z_mesh, c_norm=0.1,
                               title=r"$\sigma_{%i%i}$"%(row+1, col+1))
plt.savefig('imagesSM/Stress_xx.png', dpi=500)             # save to file
#plt.show();                        # uncomment this to display the plot
```

```
Minimum value in the domain: -12989.6
Maximum value in the domain: 227.629
```

The distribution of the stress component σ_{xx} in the half-space subjected to a unit vertical point force at the top surface is plotted in Fig. 5.13. The singularity at the point load is clearly shown. It decays rapidly as the distance from the point force increases.

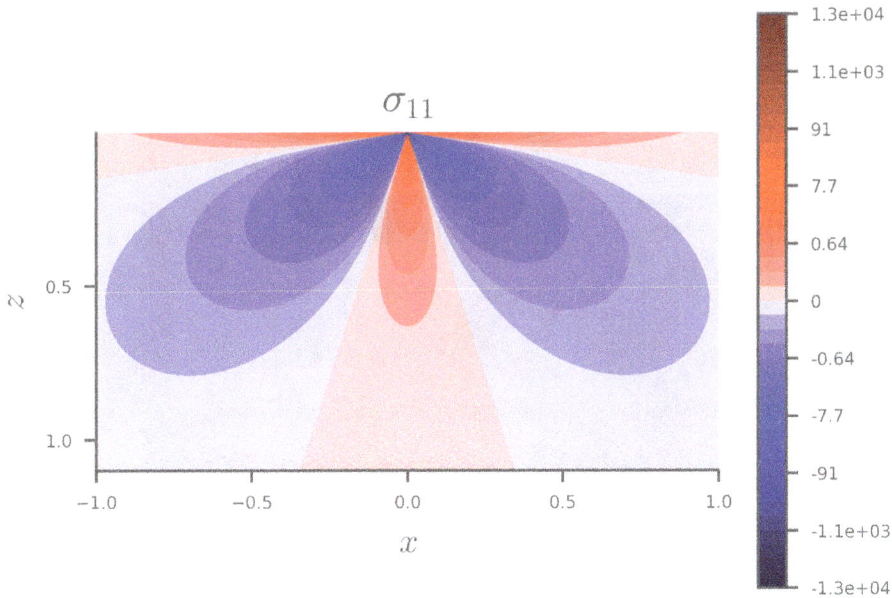

Figure 5.13. Distribution of the stress component σ_{xx} in the x–z plane in the half-space subjected to a unit vertical point force at the top surface.

```
# Plot norm stress σyy (or σ22) on $y=0$ plane:
plt.figure()
plt.ioff()
#plt.ion()                              # uncomment this to display the plots
row, col = 1, 1
field_plot(sij[row, col], x_mesh, z_mesh, c_norm=0.1,
                              title=r"$\sigma_{%i%i}$"%(row+1, col+1))
plt.savefig('imagesSM/Stress_yy.png', dpi=500)                # save to file
#plt.show();                              # uncomment this to display the plot
```

```
Minimum value in the domain: -3923.73
Maximum value in the domain: 678.422
```

The distribution of the stress component σ_{yy} in the half-space subjected to a unit vertical point force at the top surface is plotted in Fig. 5.14. The stress is singular at the point of the load and decays rapidly as the distance from the point increases.

```
# Shear stresses σxy (or σ12) and σyz (or σ33) is zero on $y=0$ plane:
row, col = 1, 0
print('\u03C3_%i%i='%(row+1, col+1),sij[1,0])
row, col = 2, 1
print('\u03C3_%i%i='%(row+1, col+1),sij[1,0])
```

```
σ_21= 0
σ_32= 0
```

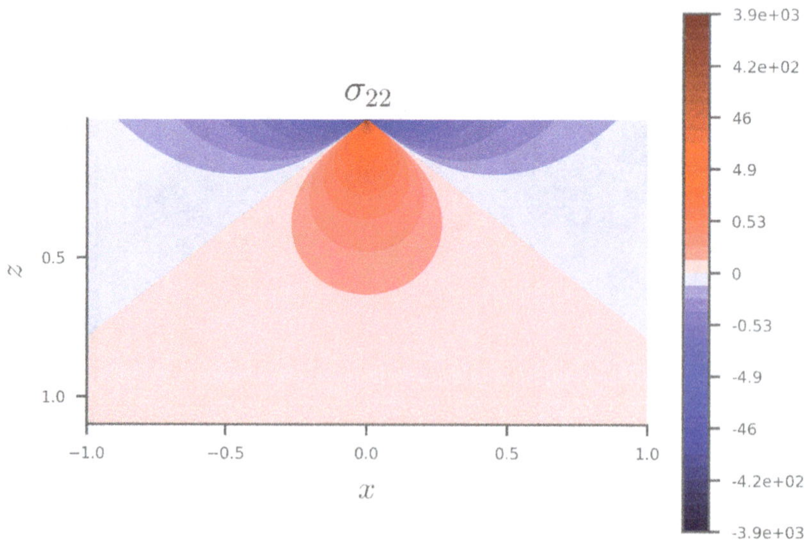

Figure 5.14. Distribution of the stress component σ_{yy} in the x–z plane in the half-space subjected to a unit vertical point force at the top surface.

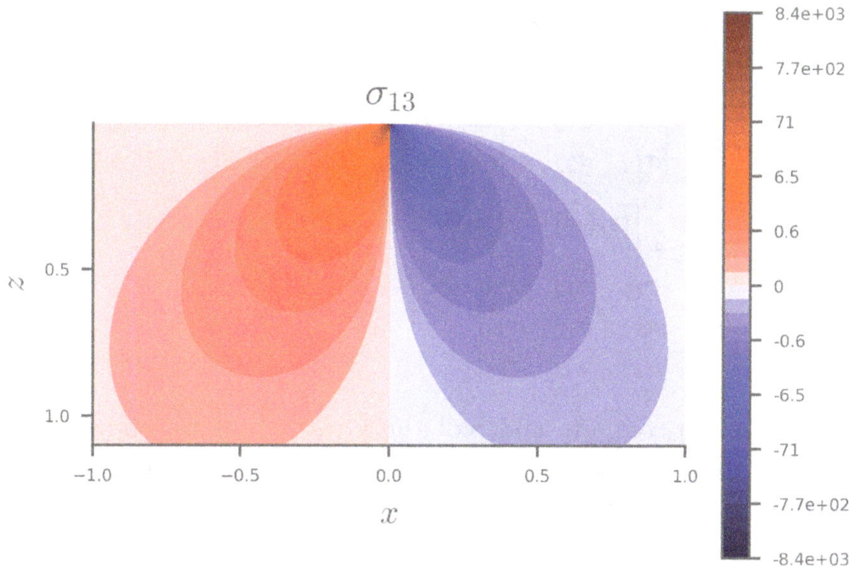

Figure 5.15. Distribution of the stress component σ_{xz} in the x–z plane in the half-space subjected to a unit vertical point force at the top surface.

```
# Plot shear stress σxz (or σ13) on $y=0$ plane:
plt.figure()
plt.ioff()
#plt.ion()                         # uncomment this to display the plots
row, col = 0, 2
field_plot(sij[row, col], x_mesh, z_mesh, c_norm=0.1,
                              title=r"$\sigma_{%i%i}$"%(row+1, col+1))
plt.savefig('imagesSM/Stress_xz.png', dpi=500)          # save to file
#plt.show();                         # uncomment this to display the plot
```

```
Minimum value in the domain: -8364.43
Maximum value in the domain: 8364.43
```

The distribution of the stress component σ_{xz} in the half-space subjected to a unit vertical point force at the top surface is plotted in Fig. 5.15. It is singular at the point where the load is applied and decays rapidly as the distance from the point increases. The largest values occur around the 45° lines.

We can also plot the three stress invariants for a given stress tensor σ. The detailed definition of these invariants is as follows [1]:

$$
\begin{aligned}
I_1 &= \sigma_{11} + \sigma_{22} + \sigma_{33} \\
I_2 &= \sigma_{11}\sigma_{22} + \sigma_{11}\sigma_{33} + \sigma_{22}\sigma_{33} - \sigma_{12}^2 - \sigma_{13}^2 - \sigma_{23}^2 \\
&= \begin{vmatrix} \sigma_{11} & \sigma_{12} \\ \sigma_{12} & \sigma_{22} \end{vmatrix} + \begin{vmatrix} \sigma_{11} & \sigma_{13} \\ \sigma_{13} & \sigma_{33} \end{vmatrix} + \begin{vmatrix} \sigma_{22} & \sigma_{23} \\ \sigma_{23} & \sigma_{33} \end{vmatrix} \\
&= \frac{1}{2}\left[(\mathrm{tr}(\boldsymbol{\sigma}))^2 - \mathrm{tr}\left(\boldsymbol{\sigma}^2\right)\right] \\
I_3 &= \begin{vmatrix} \sigma_{11} & \sigma_{12} & \sigma_{13} \\ \sigma_{12} & \sigma_{22} & \sigma_{23} \\ \sigma_{12} & \sigma_{23} & \sigma_{33} \end{vmatrix}
\end{aligned} \tag{5.56}
$$

The snippet to compute and plot the invariants is as follows. Here, we plot only I_3. Readers can plot any other invariant by uncommenting the corresponding code lines. First, let's compute these invariants using the stress components:

```
# Plot three invariants on $y=0$ plane:
I1 = sp.S(1)/3 * sij.trace()

I2 = sij[0,0]*sij[1,1] + sij[0,0]*sij[2,2] + sij[1,1]*sij[2,2] \
    -sij[0,1]**2 - sij[0,2]**2 - sij[1,2]**2
I2_= sp.S(1)/2 * (sij.trace()**2 - (sij**2).trace())
print(f" I2 = I2_? {abs((I2-I2_).subs({x:1, y:1, z:1}).evalf())<1e-12}")

I3 = sij.det()
```

```
I2 = I2_? True
```

```
# Plot the first invariant, I3, on $y=0$ plane:
plt.figure()
plt.ioff()
#plt.ion()                         # uncomment this to display the plots

#field_plot(I1, x_mesh, z_mesh, c_norm=0.1, title=r"$I_1$")
#plt.savefig('imagesSM/Stress_I1.png', dpi=500)          # save to file

#field_plot(I2, x_mesh, z_mesh, c_norm=0.1, title=r"$I_2$")
#plt.savefig('imagesSM/Stress_I2.png', dpi=500)          # save to file

field_plot(I3, x_mesh, z_mesh, c_norm=0.1, title=r"$I_3$")
plt.savefig('imagesSM/Stress_I3.png', dpi=500)           # save to file
#plt.show();                        # uncomment this to display the plot
```

```
Minimum value in the domain: -4.54869e+09
Maximum value in the domain: 1.04419e+11
```

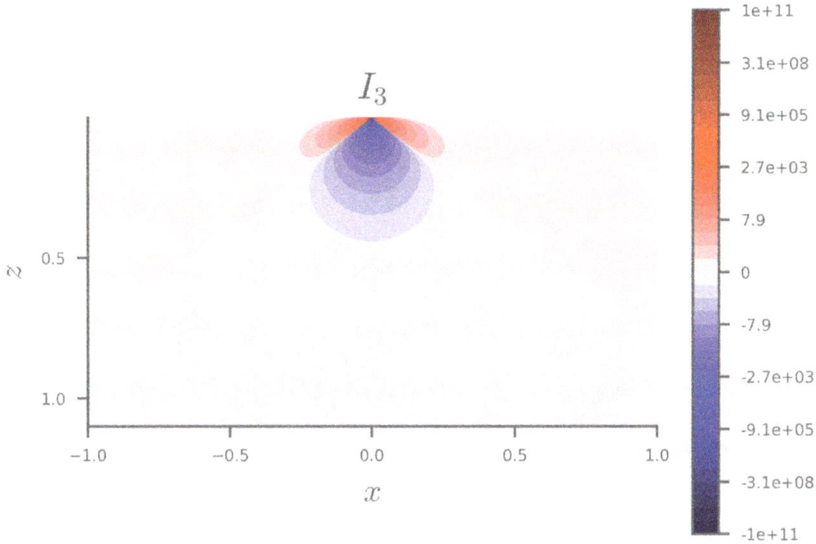

Figure 5.16. Distribution of the stress invariant I_3 in the x–z plane in the half-space subjected to a unit vertical point force at the top surface.

The distribution of the stress invariant I_3 in the half-space subjected to a unit vertical point force at the top surface is plotted in Fig. 5.16. The stress invariant is also singular at the point of the load.

Finally, we will compute and plot the von Mises stress, defined as

$$\sigma_{vm} = \sqrt{\frac{1}{2}\left[(\sigma_{11}-\sigma_{22})^2 + (\sigma_{22}-\sigma_{33})^2 + (\sigma_{33}-\sigma_{11})^2 + 6(\sigma_{12}^2 - \sigma_{13}^2 - \sigma_{23}^2)\right]} \tag{5.57}$$

The snippet to compute and plot it is as follows:

```
# Compute and plot the von Mises stress:

Mises = sqrt(((sij[0,0]-sij[1,1])**2 + (sij[1,1]-sij[2,2])**2 +
              (sij[2,2]-sij[0,0])**2 +
               6*(sij[0,1]**2 + sij[1,2]**2 + sij[0,2]**2))/2)
plt.figure()
plt.ioff()
#plt.ion()                          # uncomment this to display the plots
field_plot(Mises, x_mesh, z_mesh, c_norm=0.1, title=r"$von Mises$")
plt.savefig('imagesSM/Stress_Mises.png', dpi=500)
#plt.show();                           # uncomment this to display the plot
```

```
Minimum value in the domain: 0.0496518
Maximum value in the domain: 17255.1
```

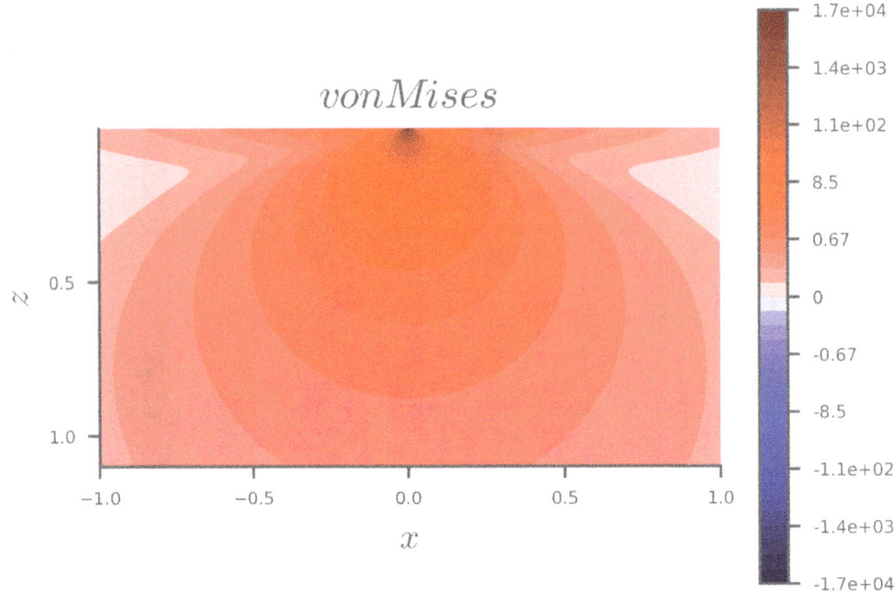

Figure 5.17. Distribution of the von Mises stress in the x–z plane in the half-space subjected to a unit vertical point force at the top surface.

The distribution of the von Mises stress in the half-space subjected to a unit vertical point force at the top surface is plotted in Fig. 5.17. The stress is also singular at the point of the load, where the material will yield first.

5.9 Line load on the half-space surface

5.9.1 *Setting of the problem*

When a solution (stress or displacement) is found for the point load, one can easily obtain solutions for any type of distributed load using superposition. The point load solution is also known as Green's function, which offers a fundamental solution; solutions to other types of loads can be found by integration. If the load function multiplied by Green's function is integrable, we can find the solution analytically. Otherwise, it can be found via numerical integration.

Consider an example of a vertical line load q uniformly distributed on the half-space surface along the x-axis over a length of $[-l/2, l/2]$ centered at $x = 0$, as shown in Fig. 5.18.

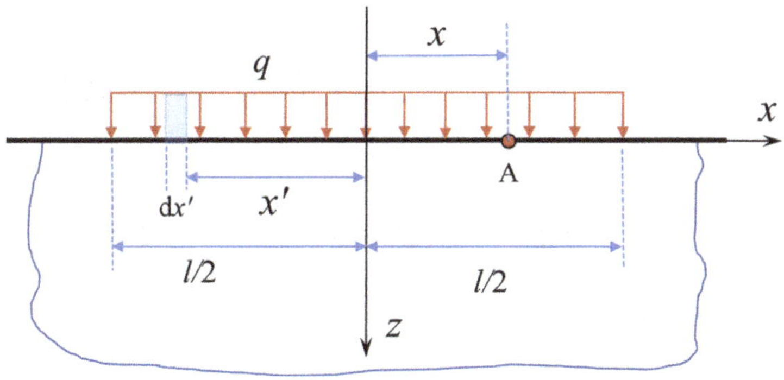

Figure 5.18. A distributed load q on the surface of the half-space.

5.9.2 *Integral formula over the line load*

The displacements at point A, located at x in the half-space, can be computed by making use of solutions for the point force. Let u be the infinitesimal displacement at x resulting from a unit point force $P_z = 1$. Thus, du will be the differential displacement caused by the differential small force $\mathrm{d}P = q\,\mathrm{d}x$. The total displacement u at x is then computed via integration.

Since q is uniformly distributed, it is symmetric with respect to the z-axis. The fields of the displacement components should be either symmetric u_z or antisymmetric u_x. Consider first the symmetric displacement u_z; the displacement resulting from the distributed line load should be

$$d_z(x) = q \int_{-l/2}^{l/2} u_z(x - x')\, dx' \tag{5.58}$$

where x' is the integral parameter. This is a typical convolution integral. Introducing a new parameter $\xi = x - x'$, we have $d\xi = -dx'$. The foregoing equation becomes

$$d_z(x) = -q \int_{x+l/2}^{x-l/2} u_z(\xi)\, d\xi \tag{5.59}$$

Assuming the antiderivative of u_z is denoted as U_z, the foregoing integral can be found as [10]

$$d_z(x) = q\big(U_z(x + l/2) - U_z(x - l/2)\big) \tag{5.60}$$

Next, consider the antisymmetric displacement u_x. The distributed load q on the left side of A results in positive u_x for a positive q, and that on the right side results in a negative u_x. Therefore, the integration should be split into two parts:

$$d_x(x) = -q\int_{-l/2}^{x} u_x(x-x')dx' + q\int_{x}^{l/2} u_x(x-x')dx' \tag{5.61}$$

or

$$d_x(x) = q\int_{x+l/2}^{0} u_x(\xi)d\xi + q\int_{0}^{x-l/2} u_x(\xi)d\xi \tag{5.62}$$

Assuming the antiderivative of u_x is denoted as U_x, the foregoing integral can be found as

$$d_x(x) = -q\big[U_x(x+l/2) + U_x(x-l/2)\big] \tag{5.63}$$

Therefore, all we need is to find the antiderivative of the displacement component in question, which can be conveniently done using Sympy.

We can now write the following Python code to perform this calculation and generate the plot:

```
# Use the displacement vector u obtained earlier, and set parameters as:

ui = sp.simplify(u.subs({y:0, E:1, ν:sp.S(1)/4, Px:0, Py:0, Pz:1}))
display(Math(f" \\text{{ Fundamental solution for u}}={latex(ui.T)}"))
```

Fundamental solution for $u = \left[\begin{matrix}\frac{5x\left(-x^2-z^2+2z\left(z+\sqrt{x^2+z^2}\right)\right)}{16\pi\left(x^2+z^2\right)^{\frac{3}{2}}\left(z+\sqrt{x^2+z^2}\right)} & 0 & \frac{5\left(3x^2+5z^2\right)}{16\pi\left(x^2+z^2\right)^{\frac{3}{2}}}\end{matrix}\right]$

```
# Integrate ui(x) to obtain the antiderivatives Ui(x):
Ui = integrate(ui,(x))

l = symbols('l')                          # coverage length of the line load
l_h = 1                                   # half-length of the line-load
dx_asym = (-Ui[0].subs(x,x+l/2) + Ui[0].subs(x,x-l/2)).subs(l,l_h)
dz_sym  = ( Ui[2].subs(x,x+l/2) - Ui[2].subs(x,x-l/2)).subs(l,l_h)
display(Math(f" \\text{{Displacement by a line load $d_{{x}}$}}:"))
display(Math(f" \\scriptsize {latex(dx_asym.simplify())}"))
```

Displacement by a line load d_x:

$$\frac{5z}{4\pi\sqrt{4z^2+(2x+1)^2}} - \frac{5z}{4\pi\sqrt{4z^2+(2x-1)^2}} - \frac{5\log\left(z+\frac{\sqrt{4z^2+(2x-1)^2}}{2}\right)}{16\pi} + \frac{5\log\left(z+\frac{\sqrt{4z^2+(2x+1)^2}}{2}\right)}{16\pi}$$

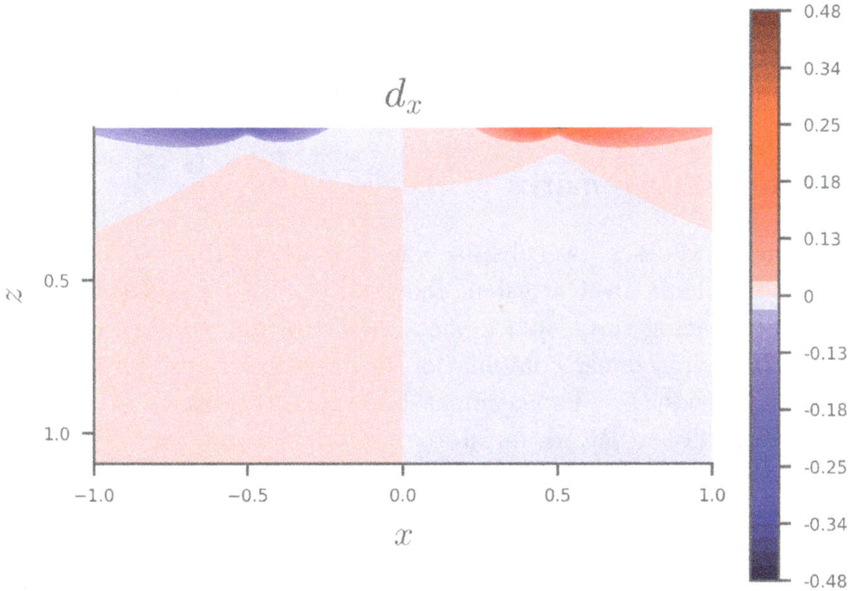

Figure 5.19. Distribution of the displacement u_x in the x–z plane in the half-space subjected to a unit vertical line force over $x \in [-l/2, l/2]$ on the top surface.

```
# Plot the results, for Pz, ux is antisymmetric: uz is symmetric:
plt.figure()
plt.ioff()
#plt.ion()                          # uncomment this to display the plots
#field_plot(dz_sym, x_mesh, z_mesh, c_norm=0.1, title=r"$d_z$")
field_plot(dx_asym, x_mesh, z_mesh, c_norm=0.5, title=r"$d_x$")
plt.savefig('imagesSM/LineLoad_ux.png', dpi=500)
#plt.show();                         # uncomment this to display the plot
```

The distribution of the displacement u_x in the half-space subjected to a unit vertical line force on the top surface is plotted in Fig. 5.19.

5.10 Review questions

1. Assume a linearly distributed body force defined in Eq. (5.19). Find the solutions for displacements and stresses for the same half-space problem studied in Section 5.1, using the displacement method.
2. Consider a 3D beam with an elliptical cross-section. Derive formulas for the general form of displacements using the DBCs given in Eq. (5.23).
3. Consider a 3D half-space subjected to a shear force in the y-direction. Write out the solutions for displacements and stresses using Eq. (5.52).

4. Consider a 3D half-space subjected to a shear force in the y-direction. Derive the solutions for displacements and stresses using Eqs. (5.53) and (5.54), following the similar procedure outlined in Section 5.7.

5.11 Concluding remarks

This chapter discusses two displacement methods for solving 3D solid mechanics problems in Cartesian coordinates. The basic solution strategy involves treating the displacements as the primary unknown functions, assuming them as a linear combination of basis functions. Various types of boundary conditions — displacement boundary conditions (DBCs), stress boundary conditions (SBCs), and force boundary conditions (FBCs) — are then used to determine the unknown constants. The operations in our displacement method involve differentiation, definite integration, and algebraic manipulation. In theory, this method is always workable, straightforward, and powerful for 3D problems, which inherently present challenges. The condition is that a proper set of basis functions can be found.

The first method is proposed for the first time in this chapter to the best of the author's knowledge. It uses a polynomial basis function enriched by a harmonic basis in an infinite series and is applied to 3D beams with finite dimensions. A set of Python code snippets is provided for solving these types of problems. The solution process can be automated by appropriately handling the harmonic infinite series term.

The second method is classical and can be found in multiple sources. It uses a displacement potential function along with displacement functions, applied to a 3D half-space problem with an infinite domain. Since these functions involve rational functions, the derivation process becomes more complicated and often requires manual operations for simplification. A set of Python code snippets is also provided for solving these types of problems.

Note that the sequence of imposing various types of boundary conditions may vary depending on convenience during the derivation process. Additionally, the differential equilibrium equations in either stresses or displacements can be used. The key is to ensure that all of these conditions are satisfied.

References

[1] G.R. Liu, *Mechanics of Materials: Formulations and Solutions with Python*, World Scientific, New Jersey, 2024.
[2] G.R. Liu, *Numbers and Functions: Theory, Formulation, and Python Codes*, World Scientific, New Jersey, 2024.

[3] G.R. Liu and S.S. Quek, *The Finite Element Method: A Practical Course*, Butterworth-Heinemann, New York, 2013.

[4] G.R. Liu and T.T. Nguyen, *Smoothed Finite Element Methods*, Taylor & Francis Group, New York, 2010.

[5] G.R. Liu, *Mesh Free Methods: Moving Beyond the Finite Element Method*, Taylor & Francis Group, New York, 2010.

[6] Z.L. Xu, *Elasticity*, Vols. 1 & 2, People's Publisher, China, 1979.

[7] G.R. Liu, *Vector Calculus: Formulation, Applications, and Python Codes*, World Scientific, New Jersey, 2025.

[8] N. Guarín-Zapata, *Continuum_Mechanics: Continuum Mechanics Calculations in Python*, 2020. Online https://github.com/nicoguaro/continuum_mechanics.

[9] L.D. Landau, E.N. Lifšic, E.M. Lifshitz *et al.*, *Theory of Elasticity: Volume 7*, Elsevier, London, 1986.

[10] G.R. Liu, *Calculus: A Practical Course with Python*, World Scientific, New Jersey, 2025.

Chapter 6

Layered Solids

Let us import the necessary external modules or dependencies for later use in this chapter.

```
# Place cursor in this cell, and press Ctrl+Enter to import dependencies.
import sys                          # for accessing the computer system
sys.path.append('../grbin/')   # Change to the directory in your system

from commonImports import *          # Import dependencies from '../grbin/'
import grcodes as gr                    # Import the module of the author
importlib.reload(gr)                 # When grcodes is modified, reload it

init_printing(use_unicode=True)         # For latex-like quality printing

np.set_printoptions(precision=4,suppress=True,    # Digits in print-outs
                    formatter={'float_kind': lambda x: f"{x:10.3e}"})
```

The layered half-space is an idealized solid domain often used in many practical problems in nature and engineering, such as structures near the Earth's surface or solids coated with a thin film. This chapter studies two typical problems that can be reduced to 1D problems, which can then be solved using displacement methods to obtain exact closed-form solutions.

6.1 A finite layer on half-space

Consider a layer of solid material with finite thickness d, perfectly bonded on top of a three-dimensional (3D) half-space. The layer is subjected to a distributed force q over its entire top surface, as well as a vertical body force b_1

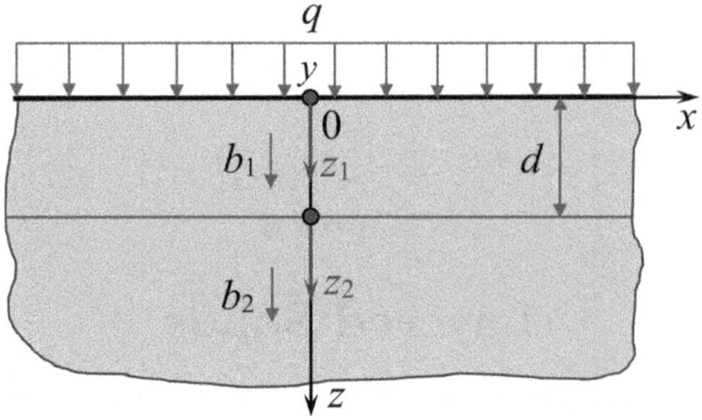

Figure 6.1. Schematic view of a cross-section of a 3D layered half-space, subjected to a surface pressure and a body force in the vertical direction.

within the layer and a body force b_2 within the half-space, as shown in Fig. 6.1. The material constants for the top layer are denoted as E_1 and ν_1, which can be used to compute G_1 and λ_1. The material constants for the half-space are E_2 and ν_2, which can be used to compute G_2 and λ_2. Our goal is to determine the stress and displacement distributions in the layered half-space.

6.1.1 *Analyzing the problem*

Since the pressure is uniformly distributed over the entire surface of the 3D layered half-space, and the gravitational forces are invariant in any plane parallel to the x–y plane, any vertical plane is symmetric. Therefore, the displacements u and v will be zero throughout the entire domain:

$$u(x, y, z) = 0; \quad v(x, y, z) = 0 \tag{6.1}$$

The only nonzero displacement is w, which does not vary in any x–y plane. Therefore, w is a function of z only. The first two equations of equilibrium (Eq. 4.16) are automatically satisfied, and the third equation becomes

$$\frac{\partial^2 w}{\partial z^2} = -\frac{b_z(z)}{2G + \lambda} \tag{6.2}$$

It is evident that this 3D half-space problem is effectively a 1D problem, similar to the 1D strain problem discussed in Section 5.1. The key difference is that we now have two sub-domains for z: one for the top layer and one for the half-space. Equations (6.1) and (6.2) apply to both sub-domains.

To simplify the formulation, we use two separate z-coordinates. For the layer, we introduce z_1, with its origin at the top surface of the layer. For the half-space, we introduce z_2, with its origin at the top surface of the half-space. The solution to this problem is then given by the solution from Section 5.1 and is expressed as follows.

6.1.2 *Solution for displacements and stresses*

We now rewrite the full set of final solutions in the layers. The displacement solution is

$$\begin{aligned} w &= \frac{1}{2G+\lambda}\left[q(d-z_1)+\frac{b_1}{2}(d-z_1)^2\right], \quad 0 \le z_1 \le d \text{ in the top layer} \\ w &= \frac{1}{2G+\lambda}\left[-q(z_2)+\frac{b_2}{2}(z_2)^2\right], \quad 0 \le z_2 \le \infty \text{ in the half-space} \end{aligned} \tag{6.3}$$

Here, the reference point for the displacement is set at $z_1 = d$ or $z_2 = 0$.

The stress solutions are as follows. In the top layer $0 \le z_1 \le d$:

$$\begin{aligned} \sigma_{xx} &= \sigma_{yy} = -\frac{\nu_1}{1-\nu_1}(b_1 z_1 + q) \\ \sigma_{zz} &= -(b_1 z_1 + q) \\ \sigma_{yz} &= 0; \quad \sigma_{xz} = 0; \quad \sigma_{xy} = 0 \end{aligned} \tag{6.4}$$

In the half-space $0 \le z_2 \le \infty$:

$$\begin{aligned} \sigma_{xx} &= \sigma_{yy} = -\frac{\nu_2}{1-\nu_2}(b_2 z_2 + q + q_L) \\ \sigma_{zz} &= -(b_2 z_2 + q + q_L) \\ \sigma_{yz} &= 0; \quad \sigma_{xz} = 0; \quad \sigma_{xy} = 0 \end{aligned} \tag{6.5}$$

where $q_L = b_1 d$, which is the additional pressure on the top surface of the half-space from the top layer.

It is seen that all these stresses in the layered half-space are linear functions of z, and hence the strains will also be linear functions of z. Therefore, the compatibility equations are automatically satisfied. For positive b_1, b_2, and q, the solid is compressed. Further, if $b_1 = b_2 = 0$, the stress and strains are all constants.

6.2 Thin beams on foundation

6.2.1 *Setting of the problem*

Consider a thin beam of infinite length on a two-dimensional (2D) elastic foundation, as studied in Ref. [2]. It is subjected to a vertical force P, as shown in Fig. 6.2. Let the origin 0 of the Cartesian coordinates be at the centroid of the beam's cross-section. The vertical load P is applied at $x = 0$. The bottom surface of the beam is perfectly connected to the surface of the foundation. Under these settings, the beam will deform, leading to a displacement field in the foundation, and a distributed force will develop between the beam and the foundation. We aim to find a solution to this problem, including the following:

1. Deflection, rotation of the beam.
2. Moment, shear force, and stresses on the beam's cross-section.
3. Interfacial distributed force between the beam and foundation.

We assume small deformation, and all materials involved are linear elastic. The beam is thin, so the Euler–Bernoulli beam theory holds. Additionally, all displacements are small. Formulas developed for beams in Chapter 8 of Ref. [1] are applicable.

We also assume that all variables are independent of z (the depth direction). Clearly, the solution depends on the stiffness of the beam and the foundation.

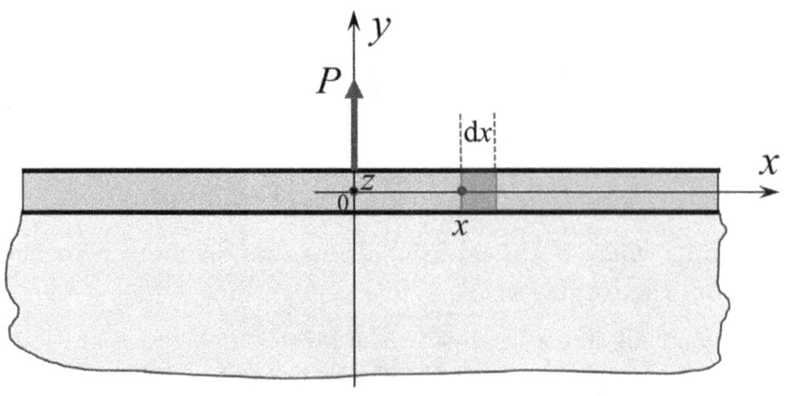

Figure 6.2. Schematic drawing of a thin beam on a two-dimensional (2D) elastic foundation, subjected to a vertical point force P. The depth (dimension in the z-direction) for both the beam and foundation is unit.

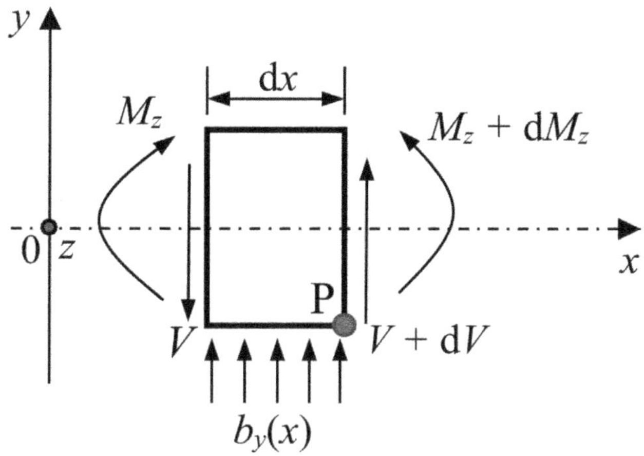

Figure 6.3. A free-body diagram taken from a beam on a foundation.

6.2.2 *Strategy for analyzing this problem*

We use the usual method of taking an infinitely small free body with length dx from the beam, as shown in Fig. 6.2. The diagram of the free body with all these forces involved is shown in Fig. 6.3, where all the force-related variables are defined.

Assume the beam is subjected to a distributed body force $b_y(x)$, which is a function of x. Based on the Euler–Bernoulli beam theory, the **slope** of the deflection of the beam is the same as the rotation of the cross-section of the beam with respect to the z-axis, which gives

$$\theta = \frac{\partial v}{\partial x} \tag{6.6}$$

where v denotes the deflection of the beam and θ is the rotation of the cross-section of the beam. Using the strain–displacement relation and Hooke's law, the moment M_z (about the z-axis) on the cross-section of the beam can be computed using the second derivative of the deflection:

$$M_z = EI_z \frac{d^2 v}{dx^2} \tag{6.7}$$

where E is the Young's modulus of the beam material and I_z is the **second moment of area** of the cross-section of the beam about the z-axis. Using the equilibrium conditions, the shear force can be computed using

$$V = -EI_z \frac{d^3 v}{dx^3} \tag{6.8}$$

The deflection of the beam is governed by the following differential equation:

$$b_y = EI_z \frac{d^4 v}{dx^4} \tag{6.9}$$

The detailed derivation of these foregoing equations can be found in Chapter 8 of Ref. [1]. In the following study, we simply use these equations. It is important to note that all these important variables (rotation, moment, shear force, and body force) depend on the deflection of the beam. In other words, if the deflection $v(x)$ is found, we find everything. This is a typical feature of a displacement method.

However, in our problem, the distributed body force b_y on the beam is unknown. It is the interfacial force between the beam and the foundation, which in turn depends on the deflection of the beam.

Assume the interfacial force and the deflection of the beam are linearly related, we then have

$$b_y = -kv \tag{6.10}$$

in which k is the elastic spring constant of the foundation per unit depth, and it is positive. A list of such constants for different types of foundations, such as sand and clay soils, is given in Table 10.1 in Ref. [2]. The data were sourced from another resource, a full book on foundation analysis and design [3].

Using this spring constant k to model the effects of the solid beneath the beam, the foundation can be any type, including half-space, bulky solids with finite depth, or another stiff beam, as long as k is available.

The minus sign in Eq. (6.10) appears because a positive deflection v (in the y-direction) results in the foundation pulling (in the negative y-direction) on the beam.

6.2.3 *Governing equation for beams on foundation*

Using Eqs. (6.10) and (6.9), the governing equation for the beam on an elastic foundation is obtained as

$$EI_z \frac{d^4 v}{dx^4} = -kv \tag{6.11}$$

Note that the unknown beam deflection appears on both sides of the equation. This equation can be simplified as

$$\frac{d^4v}{dx^4} = -4\beta^4 v \tag{6.12}$$

where β is a constant given by

$$\beta = \sqrt[4]{\frac{k}{4EI_z}} \tag{6.13}$$

It is a ratio of the stiffness of the foundation and the beam, and clearly it is positive. It carries a unit of [1/L]. When β and k are known, the bending stiffness (per length) EI_z is

$$EI_z = \frac{k}{4\beta^4} \tag{6.14}$$

Equation (6.11) is a fourth-order differential equation (DE). It is quite special: the fourth-order derivatives of the deflection function are proportional to the function itself. This implies that the deflection must be some kind of biharmonic function. Let us use SymPy to find the solution.

6.2.4 *General solution for beam deflection*

We write the following code to find the solution for the deflection of the beam on an elastic half-space foundation with spring constants k and subjected to a point force P.

We define first symbolic variables and the differential equation (6.12) in SymPy.

```
β, k = sp.symbols('β, k', positive=True)
c0, c1 = symbols('c0, c1')                          # integration constant
x, P = symbols('x, P')
v = sp.Function('v')(x)                               # deflection function
deq = sp.diff(v,(x,4)) + 4*β**4*v     # define the differential equation
display(Math(f" \\text{{ Governing equation for the beam: }}\
                                        {latex(deq)}=0"))
```

Governing equation for the beam: $4\beta^4 v(x) + \frac{d^4}{dx^4}v(x) = 0$

```
# Solve the Governing DE for deflection v(x)

sp.dsolve(deq,v)
```

$$v(x) = (C_1 \sin(x\beta) + C_2 \cos(x\beta))\, e^{-x\beta} + (C_3 \sin(x\beta) + C_4 \cos(x\beta))\, e^{x\beta}$$

We obtained the general solution for the deflection of the beam. It is indeed a biharmonic function and contains four integral constants. We rewrite the solution in the following slightly more readable form:

$$v(x) = e^{-x\beta}(c_0 \sin \beta x + c_1 \cos \beta x) + e^{x\beta}(c_2 \sin \beta x + c_3 \cos \beta x) \tag{6.15}$$

We also replaced the four free variables C_i generated by SymPy with the unknown constants c_i defined earlier, for easy management from now on.

6.2.5 *Infinite beam subject to a point force*

We consider an infinite beam subjected to a point force. We choose the origin of the coordinates at the neutral axis of the beam and right under the point force, as shown in Fig. 6.2. In this case, the problem is symmetric with respect to the y-axis, so we need to consider only half of it. Also, the deflection shall go to zero when $x \to \infty$. Therefore, we can only take the terms with $e^{-x\beta}$ for the right half of the beam with $x \geq 0$ because β is a positive constant. The general solution is written as

$$v(x) = e^{-x\beta}(c_0 \sin \beta x + c_1 \cos \beta x) \tag{6.16}$$

To determine these two constants in the general Eq. (6.16), we follow the following procedure.

6.2.5.1 *Determine the unknown constants*

The slope of the deflection v, or the rotation of the beam cross-section, can be found by differentiating the deflection once (see Chapter 8 of Ref. [1]).

```
# Re-define the deflection function:
v   = sp.exp(-β*x)*(c0*sp.sin (β*x) + c1*sp.cos(β*x))

dvdx = diff(v, x)                          # the slope of the deflection
display(Math(f" \\text{{ Slope: }}\\frac{{dv}}{{dx}}= {latex(dvdx)}"))
```

$$\text{Slope: } \frac{dv}{dx} = -\beta \left(c_0 \sin\left(x\beta\right) + c_1 \cos\left(x\beta\right)\right) e^{-x\beta}$$
$$+ \left(c_0\beta \cos\left(x\beta\right) - c_1\beta \sin\left(x\beta\right)\right) e^{-x\beta}$$

Due to the symmetry, the slope of the deflection v must be zero at $x = 0$. We thus have the following condition:

```
dvdx_0 = dvdx.subs({x:0})
display(Math(f" \\text{{ Zero slop condition at $x=0$: }}\
                                       {latex(dvdx_0)}=0"))
```

Zero slop condition at $x = 0$: $c_0\beta - c_1\beta = 0$

This zero-slope condition at $x = 0$ gives $c_0 = c_1$. Therefore, Eq. (6.16) becomes

$$v(x) = c_0 e^{-x\beta}(\sin\beta x + \cos\beta x) \tag{6.17}$$

We now have only one constant, c_0, to be determined.

Next, using the equilibrium condition, the integral of the distributed force on the beam (from the foundation) should be balanced by the external force P applied to the beam. This gives

$$2\int_0^\infty kv dx = P \tag{6.18}$$

which leads to

$$c_0 = \frac{P\beta}{2k} \tag{6.19}$$

This can be confirmed by carrying out the integration using the following snippet:

```
v  = c0*sp.exp(-β*x)*(sp.sin (β*x) + sp.cos(β*x))
k, β = sp.symbols('k, β', positive=True)
P_  = 2*integrate(k*v,(x,0,oo))
display(Math(f" \\text{{ Equilibrium condition for the beam: }}\
                                                    {latex(P_)}=P"))
```

Equilibrium condition for the beam: $\dfrac{2c_0k}{\beta} = P$

6.2.5.2 *Final solution for a point force*

Using Eqs. (6.17) and (6.19), we obtain the final solution for the deflection for our problem:

```
v = (P*β)/(2*k)*sp.exp(-β*x)*(sp.sin (β*x) + sp.cos(β*x))
```

The solution for the deflection for $x \geq 0$ becomes

$$v(x) = \underbrace{\frac{P\beta}{2k}}_{amplitude} \underbrace{e^{-\beta x}}_{decay} \underbrace{(\sin \beta x + \cos \beta x)}_{sinusoidal} \tag{6.20}$$

The **deflection** has an important feature: it is **sinusoidal**, modulated by a fast exponential **decay** function, and has an **amplitude**. This is a typical, well-behaved function found in many engineering problems. It has a number of predictive features. The deflection and all the variables related to its derivatives (rotation, moment, shear force, and interfacial force) will decay exponentially.

Using the symmetry property of the deflection with respect to y, the solution for the deflection over the entire 1D domain can be written as follows:

$$v(x) = \begin{cases} \dfrac{P\beta}{2k} e^{-x\beta}(\sin \beta x + \cos \beta x) & \text{for } x \geq 0 \\ \dfrac{P\beta}{2k} e^{x\beta}(-\sin \beta x + \cos \beta x) & \text{for } x < 0 \end{cases} \tag{6.21}$$

From the foregoing equation, we find the following additional features for the beam deflection:

1. It has the smallest roots at $\pm\frac{3\pi}{4}$, at which $v = 0$. Let us denote it as

$$L_0 = \frac{3\pi}{4\beta}, \quad \text{where } v(L_0) = 0 \tag{6.22}$$

2. The peak deflection is $\frac{P\beta}{2k}$ at $x = 0$, where P is applied. Let us denote it as

$$v_{\text{max0}} = \frac{P\beta}{2k} \tag{6.23}$$

3. The deflection decays exponentially at a rate of $e^{-x\beta}$. At $x = \pm\frac{3\pi}{2\beta}$, the amplitude of the deflection is less than 0.9% of v_{max0}:

```
ratio = (v.subs(x,3*sp.pi/2/β)/v.subs(x,0)).evalf()
display(Math(f" \\text{{ The ratio of deflection}}={latex(ratio)}"))
```

The ratio of deflection $= -0.00898329102112943$

Therefore, if a 0.9% deflection error is the criterion, a beam with a full length larger than $3\pi/\beta$ can be treated as an infinitely long beam. Let us denote this critical length as

$$L_c = \frac{3\pi}{2\beta} \tag{6.24}$$

A beam longer than $2L_c$ can be treated as an infinite beam. The error in the amplitude of the deflection is less than 0.9%.

4. At $x = \pm\frac{9\pi}{8\beta}$, the amplitude of the deflection is approximately 3.8% of $v_{\max 0}$.

```
ratio = (v.subs(x,9*sp.pi/8/β)/v.subs(x,0)).evalf()
display(Math(f" \\text{{ The ratio of deflection}}={latex(ratio)}"))
```

The ratio of deflection $= -0.0381247448354074$

Note that because of the discontinuous nature of the point force, the solution can only be given in piecewise functions. Extra care is needed when coding these types of piecewise functions.

After the deflection is obtained, other variables can be derived using different orders of derivatives of the deflection function (see Chapter 8 of Ref. [1]), which can be easily obtained using the following snippet:

```
θ = diff(v, x).simplify()
display(Math(f" \\text{{ Rotation of the cross-section of the beam:\
                          }}θ=\\frac{{dv}}{{dx}}= {latex(θ)}"))
```

Rotation of the cross-section of the beam: $\theta = \frac{dv}{dx} = -\frac{P\beta^2 e^{-x\beta}\sin(x\beta)}{k}$

We obtain the formula for **rotation** over the entire 1D domain:

$$\theta(x) = \frac{dv(x)}{dx} = \begin{cases} -\dfrac{P\beta^2}{k}e^{-\beta x}\sin\beta x & \text{for } x \geq 0 \\ \dfrac{P\beta^2}{k}e^{\beta x}\sin\beta x & \text{for } x < 0 \end{cases} \tag{6.25}$$

Here, we used the antisymmetry property of the rotation. We note the following:

1. The rotation will be zero at $x = 0$, where the deflection reaches its maximum, as expected. It will be zero periodically at $x = \pm\frac{n\pi}{\beta}$. Its next zero occurs at $x = \pm\frac{\pi}{\beta}$, where the deflection reaches the second peak.

2. The maximum (or minimum) rotation is $\theta_{\max} = \pm\frac{P\beta^2}{\sqrt{2}e^{\frac{\pi}{4}}k}$, occurring at $x = \pm\frac{\pi}{4\beta}$.
3. At $x = L_c$, the amplitude of the rotation is approximately 2.8% of $\theta_{\max}$.

```
θ_max = θ.subs(x,sp.pi/4/β)
display(Math(f" \\text{{ The maximum rotation of the beam cross-section\
                                                    }}={latex(θ_max)}")

ratio = (θ.subs(x,3*sp.pi/2/β)/θ.subs(x,sp.pi/4/β)).evalf()
display(Math(f" \\text{{ The ratio of rotation}}={latex(ratio)}"))
```

The maximum rotation of the beam cross-section $= -\frac{\sqrt{2}P\beta^2}{2ke^{\frac{\pi}{4}}}$

The ratio of rotation $= -0.0278640701953884$

The **moment** on the cross-section is given by the second derivative of the deflection, using Eq. (6.7):

```
EI = sp.symbols('EI', positive=True)
Mz  = (EI*diff(v, (x,2))).subs({EI:k/(4*β**4)})
display(Math(f" \\text{{ Moment on the cross-section of the beam:\
                        }} M_z =EI\\frac{{d^2v}}{{dx^2}}= {latex(Mz)}"))
```

Moment on the cross-section of the beam:
$M_z = EI\frac{d^2v}{dx^2} = \frac{P\left(\sin\left(x\beta\right) - \cos\left(x\beta\right)\right)e^{-x\beta}}{4\beta}$

We obtain

$$M_z = EI_z\frac{d^2v}{dx^2} = \begin{cases} \dfrac{P}{4\beta}e^{-x\beta}(\sin\beta x - \cos\beta x) & \text{for } x \geq 0 \\ \dfrac{P}{4\beta}e^{x\beta}(-\sin\beta x - \cos\beta x) & \text{for } x < 0 \end{cases} \tag{6.26}$$

Here, we used the symmetry property of the moment. From the foregoing equation, we find the following additional features for the moment:

1. The moment has the smallest roots at $\pm\frac{\pi}{4}$, at which $M_z = 0$, where the rotation is maximum (or minimum).
2. The peak moment is $\frac{P}{4\beta}$ at $x = 0$, where P is applied. Let us denote it as

$$M_{\max 0} = \frac{P}{4\beta} \tag{6.27}$$

3. The moment decays exponentially at a rate of $e^{-x\beta}$. At L_c, the amplitude of the moment is less than 0.9% of $M_{\max 0}$.

```
ratio = (Mz.subs(x,3*sp.pi/2/β)/Mz.subs(x,0)).evalf()
display(Math(f" \\text{{ The ratio of moment}}={latex(ratio)}"))
```

The ratio of moment $= 0.00898329102112943$

The **shear force** on the cross-section is given by the third derivative of the deflection, using Eq. (6.8):

```
Vy = (-EI*diff(v, (x,3))).subs({EI:k/(4*β**4)})
Vy = Vy.simplify()
Vy
```

$$-\frac{Pe^{-x\beta}\cos(x\beta)}{2}$$

We obtain:

$$V_y = -EI_z\frac{d^3v}{dx^3} = \begin{cases} -\dfrac{P}{2}e^{-\beta x}\cos\beta x & \text{for } x \geq 0 \\ \dfrac{P}{2}e^{\beta x}\cos\beta x & \text{for } x < 0 \end{cases} \tag{6.28}$$

Here, we used the antisymmetry property of the shear force. From the foregoing equation, we observe the following features of the shear force:

1. The shear force has its smallest roots at $x = \pm\frac{\pi}{2\beta}$, where the moment is at its maximum or minimum.
2. The peak shear force is given by $V_{\max} = \frac{P}{2}$ at $x = 0$, where the point force P is applied. A discontinuity (jump) of magnitude P occurs at $x = 0$ due to the concentrated force.
3. The shear force decays exponentially at a rate of $e^{-x\beta}$.
4. At $x = \pm\frac{9\pi}{8\beta}$, the amplitude of the shear force is approximately 2.7% of $V_{\max}$.

```
ratio = (Vy.subs(x,9*sp.pi/8/β)/Vy.subs(x,0)).evalf()
display(Math(f" \\text{{ The ratio of shear force}}={latex(ratio)}"))
```

The ratio of shear force $= -0.0269582656041234$

The **interfacial force** on the beam is given by the fourth derivative of the deflection, using Eq. (6.9):

```
qy = (EI*diff(v, (x,4))).subs({EI:k/(4*β**4)})
qy
```

$$\frac{P\beta\left(-\sin\left(x\beta\right)-\cos\left(x\beta\right)\right)e^{-x\beta}}{2}$$

which equals $-kv$, and we recovered Eq. (6.10). This shows that our solution is verified as correct. Note that once the deflection is known, obtaining the interfacial distributed force is straightforward: simply multiply Eq. (6.21) by the foundation stiffness k and reverse the sign.

Equations (6.21, 6.25–6.28) are the full set of solutions for a beam on an elastic foundation with spring constant k. We now summarize the following:

1. All solutions decay exponentially.
2. A beam longer than $2L_c$ can be treated as an infinite beam with a solution amplitude error of less than 1%.
3. The amplitudes of these solutions are $\frac{P\beta}{2k}$, $\frac{P\beta^2}{k}$, $\frac{P}{4\beta}$, and $\frac{P}{2}$, respectively, for deflection, rotation, moment, and shear force.
4. There are four significant reference points:

$$\begin{aligned}
&\frac{\pi}{4}: \text{ zero moment, largest rotation}\\
&\frac{\pi}{2}: \text{ zero shear force, largest moment}\\
&\frac{3\pi}{4}: \text{ zero deflection}\\
&\pi: \text{ zero shear stress, second-largest moment}\\
&\frac{3\pi}{2}: L_c\beta, \text{ half-length for an infinite beam with less than 1\% error}
\end{aligned} \tag{6.29}$$

5. Since this solution is for a point force, it is a **fundamental solution**. Solutions to other distributed forces can be obtained via integration using the fundamental solution.

6.2.5.3 *Plots of the fundamental solutions*

We now use the following code to plot the distribution of all these solutions:

```
slns = [v, θ, Mz, Vy]         # put all the solution expression in a list
s_Amp = [P*β/(2*k), P*β**2/k, P/(4*β), P/2]              # The amplitudes
```

In order to plot the distribution profile of these solutions, we set all involved variables to unity.

```
# Set all factors unit for plots:
slns1 = [slns[i].subs({P:1,β:1,k:1}) for i in range(len(slns))]
print(f"  Solutions as function of x:")
slns1
```

```
Solutions as function of x:
```

$$\left[\frac{(\sin(x)+\cos(x))\,e^{-x}}{2},\ -e^{-x}\sin(x),\ \frac{(\sin(x)-\cos(x))\,e^{-x}}{4},\ -\frac{e^{-x}\cos(x)}{2}\right]$$

With this setting, the amplitudes of the solutions can be computed using:

```
s_Amp1 = [s_Amp[i].subs({P:1,β:1,k:1}) for i in range(len(s_Amp))]
display(Math(f" \\text{{The amplitudes of the solution functions}}=\
                                                  {latex(s_Amp1)}"))
```

$$\text{The amplitudes of the solution functions} = \left[\frac{1}{2},\ 1,\ \frac{1}{4},\ \frac{1}{2}\right]$$

The solutions are now only a function of x. We may, for example, determine some of their maximum and minimum values.

```
gr.maxminS0(slns1[0], x, title="deflection")
gr.maxminS0(slns1[1], x, title="rotation")
```

```
Stationary points for deflection: [0, 3.14159265]
At x=0, local maximum deflection=0.50000000
At x=3.14159265, local minimum deflection=-0.02160696

At x=[0, 0], Max deflection=0.50000000

Stationary points for rotation: [0.78539816]
At x=0.78539816, local minimum rotation=-0.32239694

At x=[0], Max rotation=0
```

```
gr.maxminS0(slns1[2], x, title="moment")
gr.maxminS0(slns1[3], x, title="shear force")
```

```
Stationary points for moment: [1.57079633, 4.71238898]
At x=1.57079633, local maximum moment=0.05196989
At x=4.71238898, local minimum moment=-0.00224582

At x=[1.57079633], Max moment=0.05196989

Stationary points for shear force: [-0.78539816]
At x=-0.78539816, local minimum shear force=-0.77544160

At x=[1], Max shear force=-0.09938306
```

Let us plot the distribution curves for all solutions.

```
title = [["Deflection", "Rotation"], ["Moment", "Shear force"]]

for i in range(len(title)):
    gr.plot2curveS(slns1[2*i:2*(i+1)], xL=0., xR=5., title=title[i])

#fig.tight_layout()
#plt.savefig('imagesSM/beam_cq.png', dpi=500)    # save the plot to file
plt.show()
```

```
Maximum Deflection value=5.000e-01, at x=0.0
Maximum Rotation value=1.393e-02, at x=3.93
Extreme Deflection values=[ 5.000e-01 -2.161e-02],at x=[ 0.000e+00  3.140e+00]

Critical Rotation values=[-0.000e+00 -6.893e-05],at x=[ 0.000e+00  3.140e+00]

Deflection values at boundary =(0.5, -0.0022749400837603654)
Rotation values at boundary =(-0.0, 0.006461180938816702)

Maximum Moment value=5.197e-02, at x=1.57
Maximum Shear force value=3.351e-02, at x=2.36
Extreme Moment values=[-2.246e-03],at x=[ 4.710e+00]

Critical Shear force values=[ 1.076e-05],at x=[ 4.710e+00]

Moment values at boundary =(-0.25, -0.002093120427528168)
Shear force values at boundary =(-0.5, -0.0009556503856479853)
```

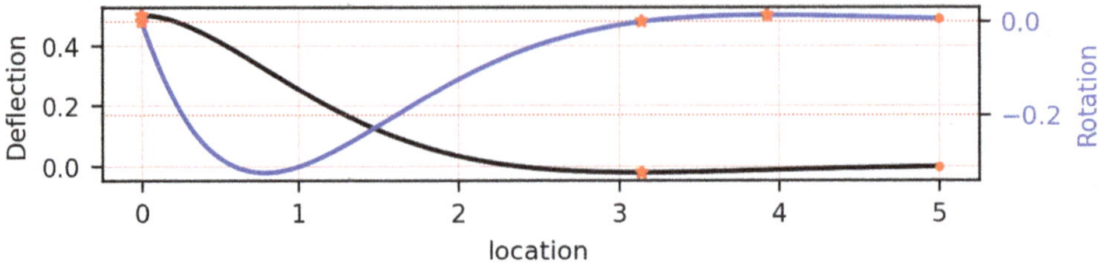

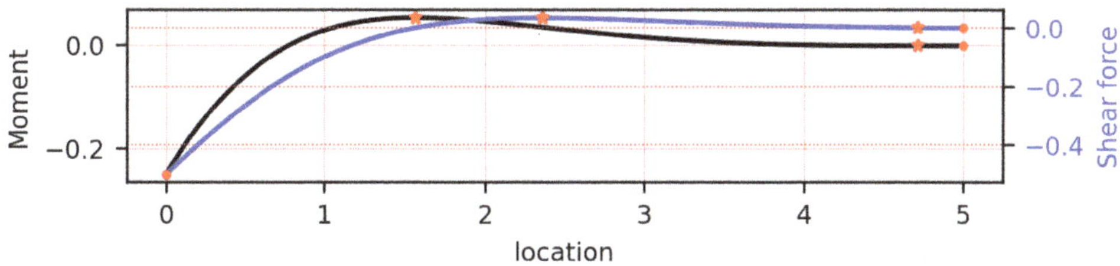

Note that our code may miss some extreme values. In general, finding all extreme values for a given function can be challenging. One may explore better Python tools for this purpose. The following is a NumPy-based approach, which can serve as an alternative for numerical analysis.

6.2.5.4 *NumPy code for numerical computation*

SymPy is useful for deriving formulas. Once the formulas are obtained, it is often more convenient, efficient, and robust to compute solutions numerically, especially for complex cases. For example, when dealing with multiple loads or distributed loads with complicated functions. We write the following function in NumPy for such purposes:

```
def beamOnFndn_s(x, P, β, k, side='R'):
    '''compute the numerical solutions in a beam on foundation of
    spring k, subject to point force P. β = sqrt[4](k/(4*EI))
     if side="R" solution in the right side, otherwise left side of P.
    Return: array v, t, M, V -deflection, rotation, moment, shear force.
    '''
    bx = β*x
    cbx = np.cos(bx); sbx = np.sin(bx)
    if side == "R":
        ebx = np.exp(-bx)
        v = (P*β)/(2*k)* ebx*( sbx+cbx)    # deflection
        t = -(P*β**2)/k* ebx*sbx           # rotation
        M = P/(4*β)*     ebx*( sbx-cbx)    # moment
        V = -P/2*        ebx*cbx           # shear force
    elif side == "L":
        ebx = np.exp( bx)
        v = (P*β)/(2*k)* ebx*(-sbx+cbx)    # deflection
        t = -(P*β**2)/k* ebx*sbx           # rotation
        M = P/(4*β)*     ebx*(-sbx-cbx)    # moment
        V =  P/2*        ebx*cbx           # shear force
    else: print('set side to "R" or "L"')
    return np.array([v, t, M, V]) #(v, t, M, V)
```

6.2.5.5 *Example: Solution over 1D domain*

```
# Set parameters for the problem:
P = -1.0; β = 1.0; k = 1.0                                # All unit values

# The amplitudes of the solutions:
s_Amp = [abs(P)*β/(2*k), abs(P)*β**2/k, abs(P)/(4*β), abs(P)/2]
print(f"The amplitudes of the solutions: {s_Amp}")

# Compute the solution and plot the results:
labels=['deflection','rotation','moment','shear force']
xL = -5; xR = 5; dx = 0.1

x_dataL = np.arange(xL, 0.,    dx)  # domain for computation and plot
x_dataR = np.arange(0., xR+dx, dx)
x_data  = np.concatenate((x_dataL, x_dataR), axis=0)

slnL  = beamOnFndn_s(x_dataL, P, β, k, side="L")
slnR  = beamOnFndn_s(x_dataR, P, β, k, side="R")

slnx = [np.concatenate((slnL[i], slnR[i])) for i in range(len(slnL))]

f_data = slnx #[vx, tx, Mx, Vx]
f_ij=[[1, 1, 1, "xl", "yl"]]                          # define figure grids
fij=f_ij[0]                                           # only one figure

plt.ioff()
#plt.ion()                          # uncomment this to display the plots
fig_s = plt.figure()
gr.plot_fij(fig_s, x_data, f_data, labels, '$x$', '$v$', fij)

# Key locations:
kLL = np.array([-np.pi/4,-np.pi/2,-np.pi,-3*np.pi/4,-3*np.pi/2   ])
kLR = np.array([ np.pi/4, np.pi/2, np.pi, 3*np.pi/4, 3*np.pi/2, 0])
plt.scatter(kLL, [0]*len(kLL), c='r', marker='*', s=20)
plt.scatter(kLR, [0]*len(kLR), c='r', marker='*', s=20)

kSL  = beamOnFndn_s(kLL, P, β, k, side="L")    # solutions at key locs.
kSR  = beamOnFndn_s(kLR, P, β, k, side="R")
plt.scatter([kLL[0],kLR[0]], [kSL[1,0],kSR[1,0]],c='r', marker='o',s=20)
plt.scatter([kLL[1],kLR[1],kLR[-1]], [kSL[2,1],kSR[2,1],kSR[2,-1]],
            c='g', marker='o',s=20)
plt.scatter([kLR[-1],kLR[-1]],[kSR[0,-1],kSR[3,-1]],
            c='b', marker='o',s=20)

plt.savefig('imagesSM/beamOnFndn_sP.png',dpi=500,bbox_inches='tight')
#plt.show()                         # uncomment this to display the plots
```

```
The amplitudes of the solutions: [0.5, 1.0, 0.25, 0.5]
```

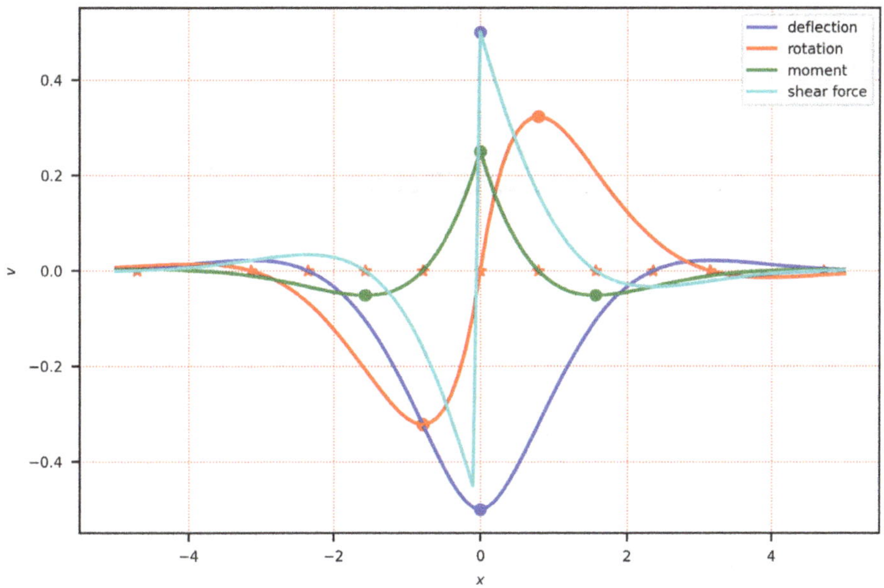

Figure 6.4. Fundamental solutions for a beam on elastic foundation subjected to a unit point force at $x = 0$. Solutions at key locations are marked. All parameters involved are set to unit.

Figure 6.4 plots the fundamental solutions for a beam on an elastic foundation. The numerical results confirm the following points:

1. The values of the solutions are significantly large near the point force.
2. The solution curves are sinusoidal and decay rapidly with x.
3. Key values are found at the key locations, as given in Eq. (6.29).
4. All values beyond $\pm\frac{3\pi}{2}$ or $\pm L_c$ are very small.

6.2.6 *Beam under distributed loading*

6.2.6.1 *Solution strategy, fundamental solutions*

Consider distributed loading p applied on the beam surface over the span of $[-L/2, L/2]$, as shown in Fig. 6.5. For convenience, we choose the origin of the x-axis at the center of the distributed loading. We aim to find solutions for deflection, rotation, moment, and shear force in the beam.

With the fundamental solution found in Eqs. (6.21–6.28), we can now use the superposition technique to obtain a solution to arbitrarily distributed loading. This is done by considering $p(x)dx$ as a small point force, and the solution should be an integral over the span of the distributed loading.

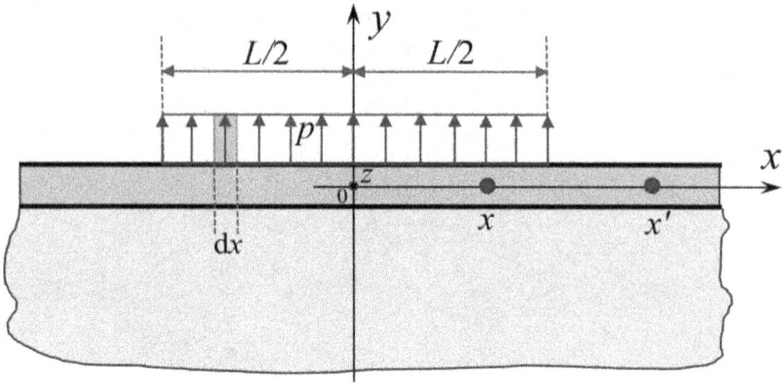

Figure 6.5. Schematic drawing of a thin beam on an elastic foundation, subjected to a vertical distributed force p that may not necessarily be uniform.

Note that the force is, in general, discontinuous at $x = \pm L/2$. Therefore, the integral needs to be done piecewise.

When the observation point x is within the span of the distributed loading, there are two contributions: one from the left-side loading, for which we shall use the fundamental solution for $x \geq 0$; another from the right-side loading, for which we shall use the fundamental solution for $x < 0$.

When the observation point x is beyond the span of the distributed loading, say at x', there is only one contribution from the left-side loading, and we need only to use the fundamental solution corresponding to $x \geq 0$.

Let us use the deflection solution as an example to detail the computation process.

Assume first that the observation point x is within the span of the distributed loading. The integrals become

$$v(x) = \int_0^{L/2+x} \left[\frac{p(x)\beta}{2k} e^{-x\beta}(\sin \beta x + \cos \beta x)\right] dx + \int_{x-L/2}^{0} \left[\frac{p(x)\beta}{2k} e^{x\beta}(-\sin \beta x + \cos \beta x)\right] dx \quad (6.30)$$

If the observation point is within the span of the distributed loading, say at x', we need only one integral, as follows:

$$v(x) = \int_{x-L/2}^{x+L/2} \left[\frac{p(x)\beta}{2k} e^{-x\beta}(\sin \beta x + \cos \beta x)\right] dx \quad (6.31)$$

For other solutions, rotation, moment, and shear force, the formulations are exactly the same. The only change needed is to replace the fundamental solution accordingly. If $p(x)$ is a simple function, such as a constant or linear, the foregoing integrals can be evaluated analytically, using, for example, SymPy. The analytical expressions obtained can be quite lengthy.

Since $p(x)$ can be, in general, a complicated function, the above integrations can be done numerically using the Gaussian integration technique that is readily available in NumPy, with integral error accessible. In addition, it is very fast. Our previously given code function, **beamOnFndn_s()**, can be used for the integration.

6.2.6.2 *Example: Beams on foundation, uniform load*

Consider an example given in the textbook [2] (Example 10.4). A thin beam made of wood with Young's modulus of $E = 10.0$ GPa has a rectangular cross-section with height $h = 200$ mm and base $b = 100$ mm. It is on a soil foundation with spring constant of $k_0 = 0.040$ N/mm^3. A uniformly distributed load $p = -35.0$ N/mm covers a span of $L = 3.61$ m (see Fig. 6.5).

1. Determine the maximum deflection in the beam and the distributed force between the beam and the foundation.
2. Determine the maximum moment and the maximum normal stress in the beam.
3. Plot the distribution of the deflection, rotation, moment, and shear force.

Solution:

We write the following Python code to find the solutions. First, preparation of data:

```
b = .100; h=.200                        # m
I = b*h**3/12

k0 = 0.04e9                             # N/mm^3 -> N/m^3
E = 10.0e9                              # Pa
k = b * k0                              # N/m^2

β = (k/(4*E*I))**0.25                   # 1/m
print(f"β value = {β} [1/m]")

L = 3.61                                # m
p = -35.0e3                             # N/m   (p)
```

```
β value = 1.1066819197003215 [1/m]
```

For question 1: Clearly, the maximum deflection should be at the center of the distributed force, which is at $x = 0$. Since the observation point is within the span of the distributed load, we need to carry out two integrations, using Eq. (6.30). The code is given as follows:

```
# Define the functions for computing v at point x, caused by p:
# for the part of p on the left-hand-side of x:
vR = lambda x: beamOnFndn_s(x, p, β, k, side='R')[0]

# for the part of p on the right-hand side of x:
vL = lambda x: beamOnFndn_s(x, p, β, k, side='L')[0]

# Compute superimposed solution, when x=0:
vxR, error = si.quad(vR, 0, L/2)  # contribution from vR
vxL, error = si.quad(vL,-L/2, 0)  # contribution from vL

vmax = vxR+vxL
print(f"Maximum deflection is at x=0 (center of p) ={vmax:.4e} [m]")

pmax = vmax * k0
print(f"Maximum interfacial distributed force is at x=0:{pmax:.4e} [Pa]")
```

```
Maximum deflection is at x=0 (center of p) = -9.2414e-03 [m]
Maximum interfacial distributed force is at x=0: -3.6965e+05 [Pa]
```

For question 2: It is not clear where the maximum moment is. One way is to compute it over a wide enough range to locate it. We write the following code for this purpose. In this way, we want to obtain all the solutions in a run, including deflection, rotation, moment, and shear force for $x \geq 0$. The code may also be useful for other related problems.

We first write a general code function to compute these solutions for any given $x > 0$.

```
def beamOnFndn_sp(x, p, L, β, k):
    '''compute the numerical solutions at any given x in a beam on
    foundation of spring k, subject to distributed pressure p∈[-L/2, L/2].
    β = sqrt[4](k/(4*EI))
    return: solutions s = [v, t, M, V], np.array.
            -- deflection, rotation, moment, shear force, at x.
    '''
    s = []
    for n in range(4):
        # functions for solution s caused by p at observation point ξ:
        sR = lambda ξ: beamOnFndn_s(ξ,p,β,k,side='R')[n] # x on right
        sL = lambda ξ: beamOnFndn_s(ξ,p,β,k,side='L')[n]  # x on left
```

```
    if x <= L/2:
        # Compute superimposed solution using Gauss integration:
        sxR = si.quad(sR, 0., L/2+x)[0] # contrib by p on left of x
        sxL = si.quad(sL, x-L/2, 0.)[0] # contrib by p on right of x
        s.append(sxR+sxL)
    else:
        sxR = si.quad(sR,x-L/2,x+L/2)[0] # contrib by p on left of x
        s.append(sxR)
    #print(si.quad(sR, 0., L/2+x)[1])    # check integral accuracy
return np.array(s) #s
```

```
# Take a look of the solution at x = 0:
x = 0
s = beamOnFndn_sp(x, p, L, β, k)
print(f"Solution at x {s}")
```

```
Solution at x [-9.241e-03 0.000e+00 1.765e+03 0.000e+00]
```

The deflection solution is the same as earlier. Other solutions make good sense: rotation and shear force are zero at $x = 0$, and moment has a positive value. Let us use `beamOnFndn_sp()` to compute the solutions for a wide range of $x > 0$. The code is given as follows:

```
# Functions for v caused by p at observation point H: x for load center:
xL = 0;  xR = L
x_data = np.linspace(xL, xR, 200)
s = np.zeros((len(x_data),len(slnR))) # dtype=np.float64)

for i in range(len(x_data)):
    xi = x_data[i]
    s_ = beamOnFndn_sp(xi, p, L, β, k)
    s[i] = np.array(s_)

step = max(int(len(s)/5), 1)
print(f"A glance of some solutions:\n{s[::step]}")
```

```
A glance of some solutions:
[[-3.634e+04 0.000e+00 2.800e+03 0.000e+00]
 [-3.334e+04 8.391e+03 3.021e+03 -2.482e+02]
 [-2.414e+04 1.653e+04 2.089e+03 3.931e+03]
 [-1.146e+04 1.609e+04 -2.314e+03 3.415e+03]
 [-2.663e+03 8.018e+03 -2.678e+03 -1.347e+03]]
```

```
# Functions for v caused by p at observation point H: x for Load center:
xL = 0;   xR = L
x_data = np.linspace(xL, xR, 200)
s = np.zeros((len(x_data),len(slnR))) # dtype=np.float64)

for i in range(len(x_data)):
    xi = x_data[i]
    s_ = beamOnFndn_sp(xi, p, L, β, k)
    s[i] = np.array(s_)

step = max(int(len(s)/5), 1)
print(f"A glance of some solutions:\n{s[::step]}")
```

```
A glance of some solutions:
[[-9.241e-03  0.000e+00  1.765e+03  0.000e+00]
 [-8.514e-03  2.081e-03  2.157e+03 -7.336e+02]
 [-6.123e-03  4.451e-03  1.756e+03  2.998e+03]
 [-2.673e-03  4.298e-03 -1.977e+03  2.643e+03]
 [-4.414e-04  1.859e-03 -2.022e+03 -1.440e+03]]
```

The computation takes less than a second. Let us find the maximum values and their locations for all solutions:

```
smax = s.max(axis=0)
print(f"Maximum values for  [v, t, M, V]: {smax}")
max_id = np.argmax(s,axis=0)
print("Location of the maximum values x:", x_data[max_id])

smin = s.min(axis=0)
print(f"Minimum values for  [v, t, M, V]: {smin}")
min_id = np.argmin(s,axis=0)
print("Location of the minimum values x:", x_data[min_id])
```

```
Maximum values for  [v, t, M, V]: [ 2.561e-04  4.968e-03  2.293e+03  7.737e+03]
Location of the maximum values x: [ 3.610e+00  1.796e+00  1.034e+00  1.796e+00]
Minimum values for  [v, t, M, V]: [-9.241e-03  0.000e+00 -2.364e+03 -1.687e+03]
Location of the minimum values x: [ 0.000e+00  0.000e+00  2.503e+00  3.211e+00]
```

We find a full set of solutions. The maximum moment (in the third column) is $M_{\max} = -2.3636$ kNm at $x = 2.5034$ m. It is very close to the moment $M_{\max} = 2.2931$ kNm at $x = 1.034$.

Using the maximum moment, we can now compute the normal stress. It needs only a single line of code:

```
σmax = smin[2]*(h/2)/I
print(f"Maximum normal stress ={σmax*1e-6} or {-σmax*1e-6} [MPa]")
```

```
Maximum normal stress = -3.545347325951943 or 3.545347325951943 [MPa]
```

For question 3: We write the following code to plot the distribution curves for all solutions. In order to plot them all in one figure, we normalized them using their absolute maximum values. Therefore, the unit for each of the curves is the corresponding maximum value.

```
s_abs_max = abs(s).max(axis=0)
print(f"Abs max values for  [v, t, M, V]: {s_abs_max}")
s_scaled = s/s_abs_max

f_data = [s_scaled[:,0], s_scaled[:,1], s_scaled[:,2], s_scaled[:,3]]

labels=['deflection','rotation','moment','shear force']
f_dataij=[[1, 1, 1, "xl", "yl"]]                    # define figure grids
fij=f_ij[0]

plt.ioff()
#plt.ion()                          # uncomment this to display the plots
fig_s = plt.figure()
gr.plot_fij(fig_s, x_data, f_data, labels, '$x$', '$v$', fij)

plt.scatter(x_data[max_id], smax/s_abs_max, c='r', marker='*', s=20)
plt.scatter(x_data[min_id], smin/s_abs_max, c='k', marker='o', s=10)

plt.savefig('imagesSM/BeamOnFnd_sp.png',dpi=500,bbox_inches='tight')
#plt.show()                         # uncomment this to display the plots
```

```
Abs max values for  [v, t, M, V]: [ 9.241e-03  4.968e-03  2.364e+03  7.737e+03]
```

Figure 6.6 plots the distribution of all the normalized solutions along a thin beam on an elastic foundation: deflection, rotation, moment, and shear force. The discontinuity in the shear force at $x = 1.805$, where the distributed load suddenly drops to zero.

The above code can be modified to compute the solutions for more complex distributed forces $p(x)$. All we need is to define $p(x)$ in the form of the desired function. Since our codes are written in NumPy for numerical computation, $p(x)$ can even be a numerical data set.

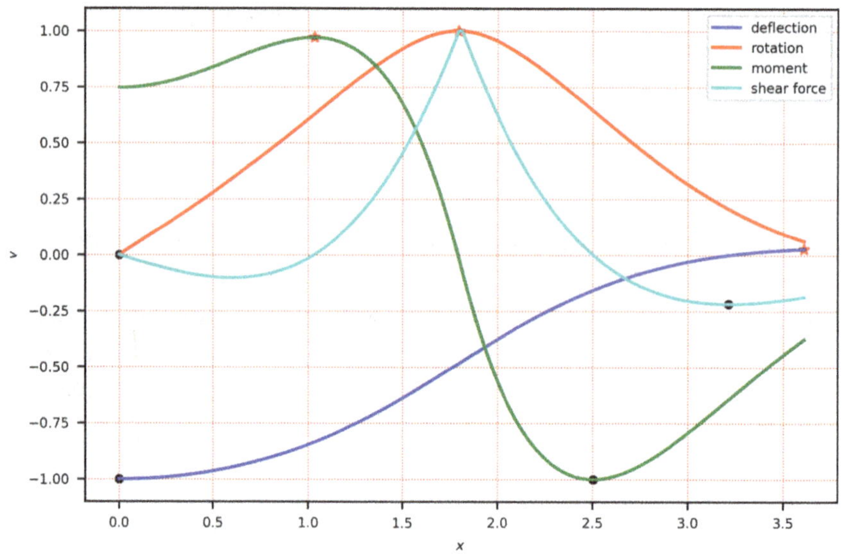

Figure 6.6. Distribution of deflection, rotation, moment, and shear force along a thin beam on an elastic foundation. The beam is subjected to a uniformly distributed load on its top surface over the span of $[-1.805, 1.805]$. $x = 0$ is at the center of the distributed load span. The red asterisks mark the maximum values, and the black dots mark the minimum values for these solutions.

6.2.7 *Semi-infinite beam loaded at an end*

6.2.7.1 *Setting of the problem*

Consider a semi-infinite beam on an elastic foundation. It is subjected to a moment M_0 and point force P_0 at its left end, as shown in Fig. 6.7. We would like to find solutions for deflection, rotation, moment, and shear force in the beam. This time, it is more convenient to choose the origin of the x-axis at the left end of the beam.

6.2.7.2 *Determine the integral constants*

The general solution to this type of problem is given in Eq. (6.16), which contains two constants, c_0 and c_1. The boundary conditions at the left end of the beam are as follows (see also Fig. 6.7):

$$EI_z \frac{d^2 v}{dx^2} = M = -M_0, \quad \text{at } x = 0 \tag{6.32}$$

and

$$-EI_z \frac{d^3 v}{dx^3} = V_y = -P_0, \quad \text{at } x = 0 \tag{6.33}$$

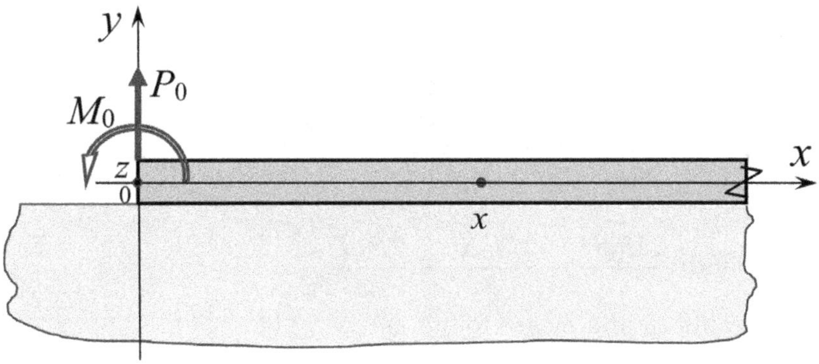

Figure 6.7. Schematic drawing of a thin semi-infinite beam on an elastic foundation, subjected to a moment and point force at its left end.

Let's use Python code to determine the constants c_0 and c_1 based on these two boundary conditions. First, we redefine the function for the general solution of the beam deflection.

```
β, k, EI = sp.symbols('β, k, EI', nonnegative=True)
c0, c1 = symbols('c0, c1')                              # unknown constants
x, M0, P0 = symbols('x, M_0, P_0')
v = sp.Function('v')(x)

# Re-define the deflectoin function:
v  = sp.exp(-β*x)*(c0*sp.sin (β*x) + c1*sp.cos(β*x))
display(Math(f" \\text{{Deflection $v$}}={latex(v)}"))

exp1 = EI*sp.diff(v,(x,2)) + M0    # define the expression for moment BC
display(Math(f" \\text{{Moment BC:}} {latex(exp1)}=0"))

exp2 = EI*sp.diff(v,(x,3)) - P0    # define the expression for force BC
display(Math(f" \\text{{Force BC: }} {latex(exp2)}=0"))
```

Deflection: $v = (c_0 \sin(x\beta) + c_1 \cos(x\beta))\, e^{-x\beta}$

Moment BC: $-2EI\beta^2 (c_0 \cos(x\beta) - c_1 \sin(x\beta))\, e^{-x\beta} + M_0 = 0$

Force BC: $EI\beta^3 (2c_0 \sin(x\beta) + 2c_0 \cos(x\beta) - 2c_1 \sin(x\beta)$

$$+ 2c_1 \cos(x\beta))\, e^{-x\beta} - P_0 = 0$$

The deflection solution has two unknown constants. We define two equations for the boundary conditions (BCs), which are used to determine these constants. The code snippet is as follows:

```
# Use BCs to solve for unknown constants c0 and c1:

sc = sp.solve((exp1, exp2), (c0, c1))
sc0 = sc[c0].simplify().expand().subs({x:0}).subs({EI:k/(4*β**4)})
sc1 = sc[c1].simplify().expand().subs({x:0}).subs({EI:k/(4*β**4)})
display(Math(f" \\text{{Constants found: }}{latex(sc0)}; {latex(sc1)}"))
```

Constants found: $\frac{2M_0\beta^2}{k}; -\frac{2M_0\beta^2}{k} + \frac{2P_0\beta}{k}$

The constants c_0 and c_1 are found as

$$\begin{aligned} c_0 &= \frac{2M_0\beta^2}{k} \\ c_1 &= -\frac{2M_0\beta^2}{k} + \frac{2P_0\beta}{k} \end{aligned} \tag{6.34}$$

6.2.7.3 *Final solution*

Substituting c_0 and c_1 back into the deflection expression v in Eq. (6.16), we obtain:

```
v = v.subs({c0:sc0, c1:sc1})
v
```

$$\left(\frac{2M_0\beta^2 \sin(x\beta)}{k} + \left(-\frac{2M_0\beta^2}{k} + \frac{2P_0\beta}{k}\right)\cos(x\beta)\right)e^{-x\beta}$$

We obtain the deflection solution. It is rearranged as

$$v(x) = \frac{2\beta}{k}e^{-x\beta}\left(M_0\beta\sin\beta x + (-M_0\beta + P_0)\cos\beta x\right)\tag{6.35}$$

```
# Redefine the expression for the deflection in Python:
v = (2*β/k)*sp.exp(-x*β)*(M0*β*sp.sin(x*β) + (-M0*β + P0)*sp.cos(x*β))
```

Once the deflection is obtained, other variables can be found by taking higher-order derivatives of the deflection function. These derivatives can be computed easily using the following code. The rotation of the beam's cross-section is given by the first derivative of the deflection, as shown in Eq. (6.6):

```
θ = diff(v, x).simplify()
display(Math(f" \\text{{Rotation found: }}{latex(θ)}"))
```

Rotation found: $\frac{2\beta^2\left(2M_0\beta\cos(x\beta) - \sqrt{2}P_0\sin\left(x\beta + \frac{\pi}{4}\right)\right)e^{-x\beta}}{k}$

The moment on the cross-section is given by the second-order derivative of the deflection, as shown in Eq. (6.7):

```
Mz  = (EI*diff(v, (x,2))).subs({EI:k/(4*β**4)})
display(Math(f" \\text{{Moment found: }}{latex(Mz)}"))
```

Moment found: $-\dfrac{(M_0\beta\cos(x\beta) + (M_0\beta - P_0)\sin(x\beta))\, e^{-x\beta}}{\beta}$

Clearly, it becomes $-M_0$ at $x = 0$, satisfying the condition in Eq. (6.32). The shear force on the cross-section is given by the third-order derivative of the deflection, as shown in Eq. (6.8):

```
Vy = (-EI*diff(v, (x,3))).subs({EI:k/(4*β**4)}).simplify()
display(Math(f" \\text{{Moment found: }}{latex(Vy)}"))
```

Moment found: $-\left(2M_0\beta\sin(x\beta) + \sqrt{2}P_0\cos\left(x\beta + \frac{\pi}{4}\right)\right) e^{-x\beta}$

Clearly, it is $-P_0$ at $x = 0$, satisfying the condition in Eq. (6.33).

Finally, the interfacial force on the beam is given by the fourth-order derivative of the deflection, using Eq. (6.9):

```
qy = (EI*diff(v, (x,4))).subs({EI:k/(4*β**4)})
qy.simplify()  #expand().factor(P)
```

$-2\beta\,(M_0\beta\sin(x\beta) - (M_0\beta - P_0)\cos(x\beta))\, e^{-x\beta}$

which equals $-kv$, and we recovered Eq. (6.10). Our solution is thus verified as correct.

We have now obtained a full set of solutions for a semi-infinite beam on an elastic foundation, when the left end of the beam is loaded with M_0 and P_0.

6.2.7.4 *Plots of the solutions*

We now use the following code to plot the distribution of all these solutions, by setting all the parameters to unit, so that we can view the profiles of all these four solutions:

```
slns = [v, θ, Mz, Vy]                           # put the solutions in a list

# Set all parameters as the unit for plots:
slns1 = [slns[i].subs({M0:1,P0:1,β:1,k:1}) for i in range(len(slns))]
display(Math(f" \\text{{$v(x)$}}={latex(slns1[0])}"))
display(Math(f" \\text{{$θ(x)$}}={latex(slns1[1])}"))
display(Math(f" \\text{{$M_z(x)$}}={latex(slns1[2])}"))
display(Math(f" \\text{{$V_y(x)$}}={latex(slns1[3])}"))
```

$$v(x) = 2e^{-x}\sin(x)$$

$$\theta(x) = 2\left(-\sqrt{2}\sin\left(x+\frac{\pi}{4}\right)+2\cos(x)\right)e^{-x}$$

$$M_z(x) = -e^{-x}\cos(x)$$

$$V_y(x) = -\left(2\sin(x)+\sqrt{2}\cos\left(x+\frac{\pi}{4}\right)\right)e^{-x}$$

```
title = [["Deflection", "Rotation"], ["Moment", "Shear force"]]

for i in range(len(title)):
    gr.plot2curveS(slns1[2*i:2*(i+1)], xL=0., xR=5., title=title[i])

#fig.tight_layout()
#plt.savefig('imagesSM/beam_cq.png', dpi=500)     # save the plot to file
plt.show()
```

```
Maximum Deflection value=6.448e-01, at x=0.79
Maximum Rotation value=2.000e+00, at x=0.0
Extreme Deflection values=[-2.786e-02],at x=[ 3.930e+00]

Critical Rotation values=[ 1.672e-04],at x=[ 3.930e+00]

Deflection values at boundary =(0.0, -0.0129223618776334O4)
Rotation values at boundary =(1.9999999999999996, 0.016744963420225346)

Maximum Moment value=6.702e-02, at x=2.36
Maximum Shear force value=4.321e-02, at x=3.14
Extreme Moment values=[ 6.702e-02],at x=[ 2.360e+00]

Critical Shear force values=[ 5.081e-04],at x=[ 2.360e+00]

Moment values at boundary =(-1.0, -0.0019113007712959706)
Shear force values at boundary =(-1.0000000000000002, 0.004549880167520731)
```

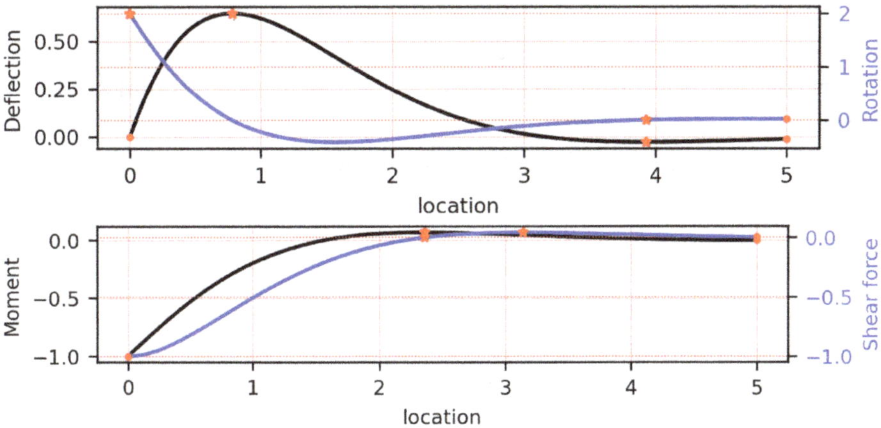

6.2.7.5 *Code for semi-infinite beams on foundation*

We write the following general code for computing numerical solutions in a semi-infinite beam on a foundation with spring constant k, subject to force M_0 and/or P_0 at $x = 0$:

```
def semiBmOnFndn_P(x, M0, P0, β, k):
    '''compute numerical solutions in a semi-infinite beam on foundation
    of spring k, subject to force P0 at x=0. β = sqrt[4](k/(4*EI))
        return: v, t, M, V -- deflection rotation, moment, shear force
    '''
    bx = β*x
    ebx = np.exp(-bx); cbx = np.cos(bx); sbx = np.sin(bx)
    sbx4 = np.sin(bx+np.pi/4); cbx4 = np.cos(bx+np.pi/4)
    v = 2*β/k * ebx*(M0*β*sbx+(-M0*β+P0)*cbx)               # deflection
    t = (2*β**2)/k* ebx*(2*M0*β*cbx-np.sqrt(2)*P0*sbx4)      # rotation
    M = -ebx/β*(M0*β*cbx+(M0*β-P0)*sbx)                        # moment
    V = -ebx*(2*M0*β*sbx+np.sqrt(2)*P0*cbx4)                 # shear force

    return np.array([v, t, M, V]) #(v, t, M, V)
```

6.2.7.6 *Example: A semi-infinite beam, point force at an end*

Consider an example studied in the textbook [2] (Example 10.5). A thin I-beam made of steel with Young's modulus of $E = 200$ GPa. It has a height $h = 102$ mm, base $b = 68$ mm, and is 4 m in length. The second moment of inertia of the beam is given as $I_z = 2.53 \times 10^6$ mm^4. It is perfectly attached to a rubber foundation with $k_0 = 0.350$ N/mm^3. A concentrated load $P_0 = 30.0$ kN and $M_0 = 0$ is applied at the left end of the beam (see Fig. 6.7).

1. Determine the maximum deflection in the beam and the interfacial force between the beam and the foundation.
2. Determine the maximum moment and the maximum normal stress in the beam.
3. Plot the distribution of the deflection, rotation, moment, and shear force.

```
b = .068; h=.102                    # m
I = 2.53e-6                         # m^4

k0 = 0.350e9                        # N/mm^3 -> N/m^3
k = b * k0                          # N/m^2
E = 200.0e9                         # Pa
β = (k/(4*E*I))**0.25               # 1/m
print(f"β value = {β} [1/m]; k value = {k} [N/m^2]")  # 1.852  #23.8e6

L = 4.0                             # m
P0 = -30.0e3                        # N/m  (p)
M0 = 0.
```

```
β value = 1.8517894786230953 [1/m]; k value = 23800000.0 [N/m^2]
```

```
# Compute the solutions:
x = 0
s = semiBmOnFndn_P(x, M0, P0, β, k)
print(f"solution at x = {x}:  {np.array(s)} ")
print(f"The maximum deflection is at x = {x}:  {s[0]:.4e} [m]")
print(f"The maximum interfacial force is at x = {x}: {s[0]*k:.4e} [N/m]")

# Look solution also at:
x = np.pi/(4*β)
s = semiBmOnFndn_P(x, M0, P0, β, k)
print(f"solution at x = {x:.3e}:  {np.array(s)} ")

σ = s[2]*(h/2)/I
print(f"normal stress at x = {x:.3e}:  {(σ*1e-6):.3e} [MPa]")
```

```
solution at x = 0:  [-4.668e-03  8.645e-03 -0.000e+00  3.000e+04]
The maximum deflection is at x = 0:  -4.6684e-03 [m]
The maximum interfacial force is at x = 0: -1.1111e+05 [N/m]
solution at x = 4.241e-01:  [-1.505e-03  5.574e-03 -5.223e+03  1.184e-12]

normal stress at x = 4.241e-01:  -1.053e+02 [MPa]
```

```
# A glance of solutions at some locations:
xL = 0.;   xR = 3.0; dx = 0.001 # domain of the solution to be plotted
x_data = np.arange(xL, xR, dx)

s = np.array(semiBmOnFndn_P(x_data, M0, P0, β, k)).T

step = max(int(len(s)/5), 1)
print(f"A glance of some solutions:\n{s[::step]}")
```

```
A glance of some solutions:
[[-4.668e-03  8.645e-03 -0.000e+00  3.000e+04]
 [-6.819e-04  3.813e-03 -4.780e+03 -4.469e+03]
 [ 3.067e-04  1.771e-04 -1.396e+03 -4.557e+03]
 [ 1.635e-04 -3.615e-04  1.101e+02 -8.469e+02]
 [ 1.452e-05 -1.248e-04  1.835e+02  2.464e+02]]
```

```
# Compute the maxima and minima:
smax = s.max(axis=0)
print(f"Maxima for [v, t, M, V]: {smax}")
max_id = np.argmax(s,axis=0)
print("Locations of Maxima,  x:", x_data[max_id])

smin = s.min(axis=0)
print(f"Minima for [v, t, M, V]: {smin}")
min_id = np.argmin(s,axis=0)
print("Locations of Maxima,  x:", x_data[min_id])
```

```
Maxima for [v, t, M, V]: [ 3.129e-04  8.645e-03  2.257e+02  3.000e+04]
Locations of Maxima,  x: [ 1.272e+00  0.000e+00  2.121e+00  0.000e+00]
Minima for [v, t, M, V]: [-4.668e-03 -3.736e-04 -5.223e+03 -6.236e+03]
Locations of Maxima,  x: [ 0.000e+00  1.697e+00  4.240e-01  8.480e-01]
```

The computation takes milliseconds. The maximum deflection and interfacial force occur at $x = 0$, where the point force P_0 is applied. This is because the deflection is typically largest where the load is applied, and hence the force transmitted to the foundation should also be the greatest. However, the maximum moment (in the third column) is $M_{\max} = -5.223$ kNm at $x = 0.424$ m.

Using the maximum moment, we can now compute the maximum normal stress. It needs only a single line of code:

```
σmax = smin[2]*(h/2)/I
print(f"Maximum normal stress ={σmax*1e-6} or {-σmax*1e-6} [MPa]")
```

```
Maximum normal stress = -105.28589329151455 or 105.28589329151455 [MPa]
```

For question 3: We write the following code to plot the distribution curves for all solutions. In order to plot all of them in one figure, we normalize them using their absolute largest values. Therefore, the unit for each of the curves is the corresponding maximum value.

```
s_abs_max = abs(s).max(axis=0)
print(f"Abs max for  [v, t, M, V]: {s_abs_max}")
s_scaled = s/s_abs_max

f_data = [s_scaled[:,0], s_scaled[:,1], s_scaled[:,2], s_scaled[:,3]]

labels=['deflection','rotation','moment','shear force']
f_ij=[[1, 1, 1, "xl", "yl"]]                    # define figure grids
fij=f_ij[0]

plt.ioff()
#plt.ion()                         # uncomment this to display the plots
fig_s = plt.figure()
gr.plot_fij(fig_s, x_data, f_data, labels, '$x$', '$v$', fij)

plt.scatter(x_data[max_id], smax/s_abs_max, c='r', marker='*', s=20)
plt.scatter(x_data[min_id], smin/s_abs_max, c='k', marker='o', s=10)

plt.savefig('imagesSM/semiBmOnFnd_sP.png',dpi=500,bbox_inches='tight')
#plt.show()                        # uncomment this to display the plots
```

```
Abs max for  [v, t, M, V]: [ 4.668e-03  8.645e-03  5.223e+03  3.000e+04]
```

Figure 6.8 plots the distribution of all solutions along a thin semi-infinite beam on an elastic foundation: deflection, rotation, moment, and shear force.

6.2.8 *Semi-infinite beam under an interior point force*

6.2.8.1 *Strategy to obtain the solutions*

Consider a semi-infinite beam on an elastic foundation. It is subjected to a point force P inside the beam span with a distance d from its left end, as shown in Fig. 6.9. To obtain solutions to this problem, a superposition method can be used, and it works as follows:

1. Use the solution for a semi-infinite beam subjected to a point force P, which is given in Section 6.2.7. This produces internal forces M_z and V_y at the location of $x = -d$, using the solution formula for $x < 0$ given in Eqs. (6.26) and (6.28). This is shown in Fig. 6.9(a). The results can be computed using the code function beamOnFndn_s().

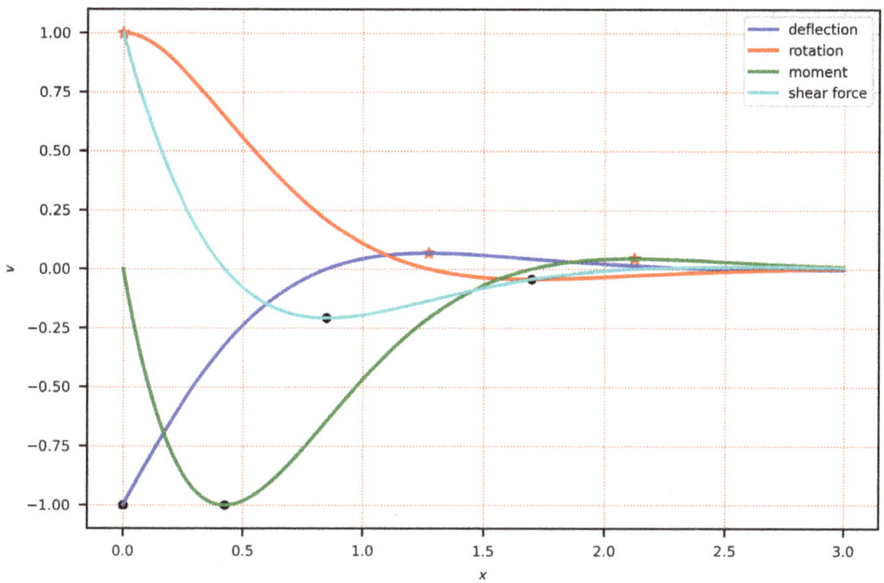

Figure 6.8. Distribution of the normalized solutions of deflection, rotation, moment, and shear force along a thin semi-infinite beam on an elastic foundation. The beam is subjected to a point force at the left end. $x = 0$ is at the left end. The red asterisks mark the maximum values, and the black dots mark the minimum values for these solutions.

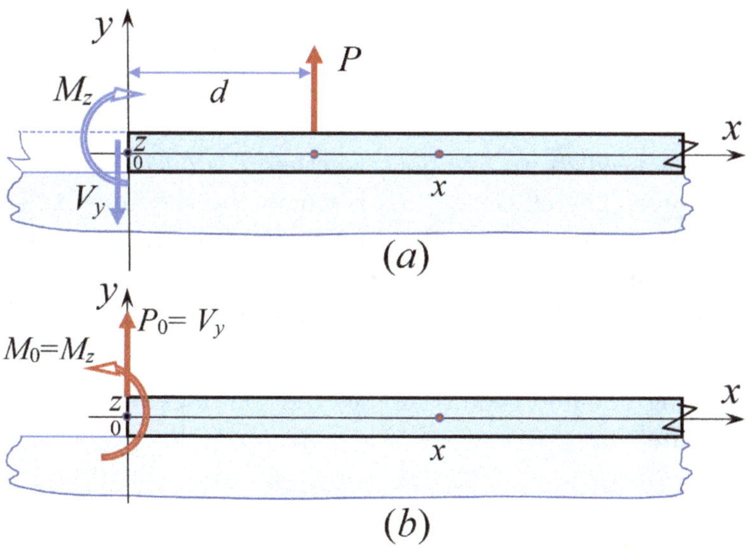

Figure 6.9. A semi-infinite beam on an elastic foundation: (a) subjected to a point force P inside the beam span with a distance d from its left end; (b) subjected to a point force and moment at its left end.

2. Use these internal forces as the external forces and apply them to the semi-infinite beam on an elastic foundation, as shown in Fig. 6.9(b). Then, obtain the solution using formulas Eq. (6.35). The results can be computed using the code function semiBmOnFndn_P().
3. Combine these solutions from steps 1 and 2, leading to the final solution.

6.2.8.2 *Python codes to compute the solutions*

We write the following codes to streamline the application of the formulas derived:

```
def beamOnFndn_sPd(x, d, P, β, k):
    '''Compute the solutions at any given array of x in an infinite beam
    on elastic foundation subjected to point force P. The origin of x
    is set at distance d left to point P. β = sqrt[4](k/(4*EI)).
    Return: s= v, t, M, V -- deflection rotation, moment, shear force
    in an numpy array of (4, len(x)).
    Setting d = 0, it does the same as beamOnFndn_s().
    Less efficient than beamOnFndn_s(), but easier to use.
    '''
    s = np.zeros((4, len(x)))
    for i in range(len(x)):
        if x[i] <= d:
            s[:,i] = beamOnFndn_s(d-x[i],P,β,k, side='L')
        else:
            s[:,i] = beamOnFndn_s(x[i]-d,P,β,k, side='R')
    return s
```

Users do not need to worry about whether x is on the right or left side of the point force. The efficiency loss is caused by the inner `if` statement, which breaks down the vectorization in NumPy. However, since this is a 1D problem, computation time is not a significant issue. The computation time is often measured in milliseconds on the author's old Dell laptop. This code will also be used in other examples.

The following one-line code combines the solutions of the infinite beam and semi-infinite beam:

```
def semiBmOnFndn_sPa(x, d, P, M0, P0, β, k):
    '''Superimposed solution from those of an infinite and a semi-
    infinite-beam. P is d away from the left end of semi-infinite beam.
    Return: s= v, t, M, V -- deflection rotation, moment, shear force
    in an numpy array of (4, len(x))
    '''
    s = beamOnFndn_sPd(x, d, P, β, k) + semiBmOnFndn_P(x, M0, P0, β, k)
    return s
```

6.2.8.3 *Example: A semi-infinite beam, an interior point force*

Consider an example studied in the textbook [2] (Example 10.6). A thin semi-infinite I-beam is subjected to a point force P located at a distance $d = 500\,\text{mm}$ from its left end, as shown in Fig. 6.9(a). All the data are identical to the previous example.

1. Determine the maximum deflection in the beam.
2. Determine the maximum moment and the maximum normal stress in the beam.
3. Plot the distribution of the deflection, rotation, moment, and shear force.

Solution to Questions 1 and 2:

```
# The data are copied from the previous example:
k = 23.8e6                          # N/m^2
β = 1.852                           # 1/m
L = 3.61                            # m
P = -35.0e3                         # N/m   (p)
h = 0.102                           # m
I = 2.53e-6                         # m^4

d = 0.5                             # m    new data
```

```
# Compute the internal forces at d from P on the infinite beam:

x = np.array([0.])
Mz_d = beamOnFndn_sPd(x, d, P, β, k)[2]   # at d on left of P
Vy_d = beamOnFndn_sPd(x, d, P, β, k)[3]
print(f"Internal forces: at d: Mz_d = {Mz_d}, Vy_d = {Vy_d}")
```

```
Internal forces: at d: Mz_d = [ 1.670e+04], Vy_d = [-2.655e+04]
```

```
M0 = Mz_d[0]; P0 = Vy_d[0]  # on the left end, the internal forces are
                            # negative of the external forces.
x = np.array([d])
s = semiBmOnFndn_sPa(x, d, P, M0, P0, β, k).T
print(f"All solution at x = {x}:\n {s} ")
print(f"The deflection is at x = {x}: {s[0,0]:.4e} [m]")
print(f"The interfacial force is at x = {x}: {(s[0,0]*k):.4e} [N/m]\n")
```

```
# Look solution also at:
x = np.array([np.pi/(4*β)])  #x = np.array([.424])
s = semiBmOnFndn_sPa(x, d, P, M0, P0, β, k).T
print(f"All solution at x = {x}:\n {s} ")
print(f"The deflection is at x = {x}: {s[0,0]:.4e} [m]")
print(f"The interfacial force is at x = {x}: {(s[0,0]*k):.4e} [N/m]\n")

σ = s[0,2]*(h/2)/I
print(f"normal stress at x = {x}:  {(σ*1e-6):.4e} [MPa]")
```

```
All solution at x = [ 5.000e-01]:
 [[-1.968e-03  8.490e-03 -9.078e+03 -3.917e+04]]
The deflection is at x = [ 5.000e-01]: -1.9677e-03 [m]
The interfacial force is at x = [ 5.000e-01]: -4.6832e+04 [N/m]

All solution at x = [ 4.241e-01]:
 [[-2.664e-03  1.150e-02 -9.244e+03 -3.989e+04]]
The deflection is at x = [ 4.241e-01]: -2.6645e-03 [m]
The interfacial force is at x = [ 4.241e-01]: -6.3414e+04 [N/m]

normal stress at x = [ 4.241e-01]:  -1.8635e+02 [MPa]
```

These solutions are found at the locations that were used for calculation in the textbook [2] (Example 10.6). Some of the results are different. We shall do a more detailed analysis using the following codes:

```
xL = 0.;  xR = 3.0; dx = 0.001
x_data = np.arange(xL, xR, dx)

s = np.array(semiBmOnFndn_sPa(x_data, d, P, M0, P0, β, k)).T

step = max(int(len(s)/5), 1)
print(f"A glance of some solutions:\n{s[::step]}")
```

```
A glance of some solutions:
[[-8.265e-03  3.566e-02  0.000e+00  7.276e-12]
 [-1.207e-03  6.750e-03 -8.459e+03 -7.912e+03]
 [ 5.431e-04  3.129e-04 -2.470e+03 -8.064e+03]
 [ 2.893e-04 -6.400e-04  1.951e+02 -1.498e+03]
 [ 2.565e-05 -2.208e-04  3.246e+02  4.364e+02]]
```

The computation takes milliseconds. Let us find the maximum values and their locations for all solutions:

```
smax = s.max(axis=0)
print(f"Maxima for [v, t, M, V]: {smax}")
max_id = np.argmax(s,axis=0)
print("Locations of Maxima,  x:", x_data[max_id])

smin = s.min(axis=0)
print(f"\nMinima for [v, t, M, V]: {smin}")
min_id = np.argmin(s,axis=0)
print("Locations of Maxima,  x:", x_data[min_id])
```

```
Maxima for [v, t, M, V]: [ 5.539e-04  3.566e-02  3.995e+02  4.770e+02]
Locations of Maxima,  x: [ 1.272e+00  0.000e+00  2.120e+00  2.544e+00]

Minima for [v, t, M, V]: [-8.265e-03 -6.614e-04 -9.244e+03 -3.989e+04]
Locations of Maxima,  x: [ 0.000e+00  1.696e+00  4.240e-01  4.240e-01]
```

We found a full set of solutions. The maximum deflection (in the first column) is $v_{\max} = -8.265$ mm, at $x = 0$ m. The maximum moment (in the third column) is $M_{\max} = -9.244$ kNm, at $x = 0.424$ m.

Using the maximum moment, we can now compute the normal stress.

```
σmax = smin[2]*(h/2)/I
print(f"Maximum normal stress ={σmax*1e-6} or {-σmax*1e-6} [MPa]")
```

```
Maximum normal stress = -186.3479239530342 or 186.3479239530342 [MPa]
```

Solution to Question 3: We write the following code to plot the distribution curves for all solutions. We normalize it using their absolute largest values. Thus, the unit for each of the curves is its corresponding maximum value.

```
s_abs_max = abs(s).max(axis=0)
print(f"Abs max values for  [v, t, M, V]: {s_abs_max}")
s_scaled = s/s_abs_max

f_data = [s_scaled[:,0], s_scaled[:,1], s_scaled[:,2], s_scaled[:,3]]

labels=['deflection','rotation','moment','shear force']
f_ij=[[1, 1, 1, "xl", "yl"]]                    # define figure grids
fij=f_ij[0]
```

```
#plt.ioff()
plt.ion()
fig_s = plt.figure()
gr.plot_fij(fig_s, x_data, f_data, labels, '$x$', '$v$', fij)

plt.scatter(x_data[max_id], smax/s_abs_max, c='r', marker='*', s=20)
plt.scatter(x_data[min_id], smin/s_abs_max, c='k', marker='o', s=10)

plt.savefig('imagesSM/semiBmOnFnd_sPd.png',dpi=500,bbox_inches='tight')
plt.show()
```

```
Abs max values for  [v, t, M, V]: [ 8.265e-03  3.566e-02  9.244e+03  3.989e+04]
```

Figure 6.10 plots the distribution of all solutions along a thin semi-infinite beam on an elastic foundation: deflection, rotation, moment, and shear force. It is clear that the free boundary conditions at the left end of the beam are satisfied. We also observe a jump in the shear force curve at $x = 0.5$. This is due to the point force P applied there. The point force also causes a slope change in the rotation curve.

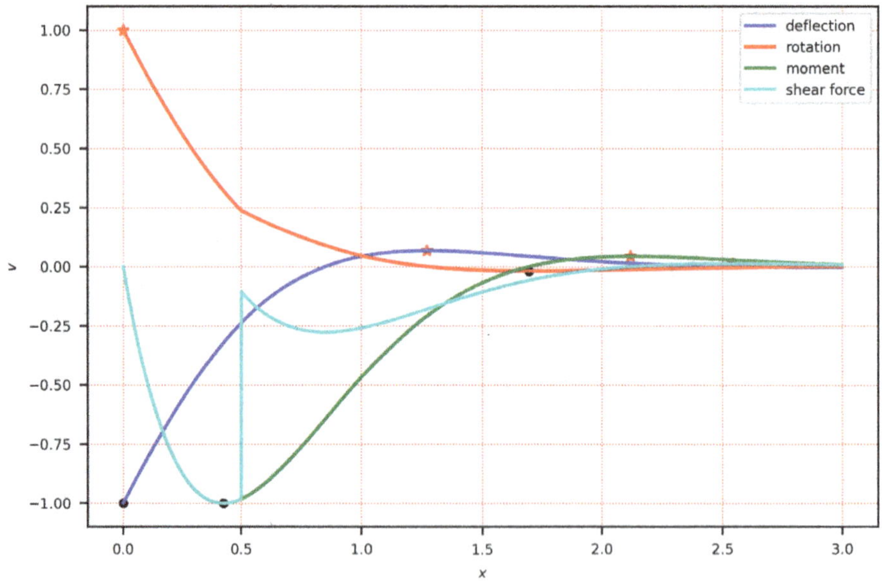

Figure 6.10. Distribution of the normalized solutions of deflection, rotation, moment, and shear force along a thin semi-infinite beam on an elastic foundation. The beam is subjected to a point force P at 500 mm from its left end. $x = 0$ is set at its left end. The red asterisks mark the maximum values, and the black dots mark the minimum values for these solutions. The jump in shear force is due to the point force P.

The above code can be modified to compute the solutions for a distributed force as we did earlier.

6.3 Exercises

1. Consider a beam with finite length L on an elastic foundation. It is subjected to a point force P at the mid-span of the beam. Determine the solution expressions for deflection, rotation, moment, and shear force. [Hints for one possible approach: The general solution given in Eq (6.15) has four constants. A finite-length beam has two boundaries, each with two free-force conditions. This gives a total of four conditions. Since these conditions are all homogeneous, three of these four constants can be written as one constant. Use the force equilibrium conditions, and this constant can then be determined. A good reference is [4].]
2. Consider a semi-infinite beam subjected to two point forces P_1 and P_2, respectively, at d_1 and d_2 from its left end. Using unit data for all variables, plot normalized solutions for deflection, rotation, moment, and shear force. Discuss the results obtained.

6.4 Concluding remarks

This chapter discusses in detail two displacement methods for solving solid mechanics problems of layered half-space or a beam on a foundation defined in Cartesian coordinates. Two typical problems are studied:

- An infinite-layer solid on half-space.
- A beam on a 2D foundation.

Both problems were simple enough to be converted to 1D problems and solved using displacement methods for exact closed-form solutions.

For more complicated layered solids, closed-form solutions can be very challenging. Techniques like the Fourier transformation [5] and general numerical methods [6] need to be used.

Superposition techniques were used to obtain solutions using the fundamental solutions and other available solutions, which are useful for solving complex practical problems.

References

[1] A. P. Boresi and R. J. Schmidt, *Advanced Mechanics of Materials*, Scientific, John Wiley & Sons, New Jersey, 2002.

[2] G.R. Liu, *Mechanics of Materials: Formulations and Solutions with Python*, World Scientific, New Jersey, 2024.

[3] Joseph E. Bowles, *Foundation Analysis and Design, 5th Ed.*, McGraw-Hill, New York, 1996.

[4] M. Hetenyi, *Beams on Elastic Foundation: Theory with Applications in the Fields of Civil and Mechanical Engineering*, University of Michigan, Michigan, 1946.

[5] G.R. Liu and Z.C. Xi, *Elastic waves In Anisotropic Laminates*, RC press, New York, 2001.

[6] G.R. Liu and Siu Sin Quek, *The Finite Element Method: A Practical Course*, Butterworth-Heinemann, London, 2013.

Index

www.ingramcontent.com/pod-product-compliance
Lightning Source LLC
Chambersburg PA
CBHW081220060126
37764CB00006B/326
9789819816422